Sous Vide and Cook–Chill Processing for the Food Industry

Sous Vide and Cook–Chill Processing for the Food Industry

Edited by
SUE GHAZALA
Memorial University of Newfoundland
St John's, NF
Canada

A Chapman & Hall Food Science Book

An Aspen Publication®
Aspen Publishers, Inc.
Gaithersburg, Maryland
1998

Aspen Publishers, Inc., grants permission for photocopying for limited personal
or internal use. This consent does not extend to other kinds of copying, such as
copyright for general distribution, for advertising or promotional purposes, for
creating new collective works, or for resale,
For information, address Aspen Publishers, Inc., Permissions Department,
200 Orchard Ridge Drive, Suite 200, Gaithersburg, Maryland 20878.

Orders: (800) 638-8437
Customer Service: (800) 234-1660

About Aspen Publishers • For more than 35 years, Aspen has been a leading
professional publisher in a variety of disciplines. Aspen's vast information resources
are available in both print and electronic formats. We are committed to providing
the highest quality information available in the most appropriate format for our
customers. Visit Aspen's Internet site for more information resources, directories,
articles, and a searchable version of Aspen's full catalog, including the most recent
publications: **http://www.aspenpub.com**
Aspen Publishers, Inc. • The hallmark of quality in publishing
Member of the worldwide Wolters Kluwer group.

Editorial Resources: Rose Gilliver
Library of Congress Catalog Card Number: 98-70648
ISBN: 0-7514-0433-0

Printed in Great Britain

Contents

8   **The sensory quality, microbiological safety and shelf life of
    packaged foods**   **190**
    IVOR CHURCH

9   **Microbiological safety considerations when using hurdle
    technology with refrigerated processed foods of extended
    durability**   **206**
    LEON G.M. GORRIS and MICHAEL W. PECK

10 Hazards associated with non-proteolytic *Clostridium botulinum* and other spore-formers in extended-life refrigerated foods 234
VIJAY K. JUNEJA

11 Application of combined factors technology in minimally processed foods 274
STELLA MARIS ALZAMORA

12 Hurdle and HACCP concepts in *sous vide* and cook–chill products 294
SUE GHAZALA and ROBERT TRENHOLM

13 Microbiological safety aspects of cook–chill foods 311
CATHERINE J. MOIR and ELIZABETH A. SZABO

Contributors

Stella Maris Alzamora

Departamento de Industrias, Facultad de Ciencias Exactas y Naturales, Universidad de Buenos Aires, Ciudad Universitaria, 1428 Buenos Aires, Argentina

Gail D. Betts

Department of Microbiology, Campden and Chorleywood Food Research Association, Chipping Campden GL55 6LD, UK

Ivor Church

Food Research Group, Leeds Metropolitan University, Leeds, UK

Philip G. Creed

Department of Food and Hospitality Management, The Worshipful Company of Cooks Centre for Culinary Research, Bournemouth University, Fern Barrow, Poole, Dorset BH12 5BB, UK

Judith Evans

Food Refrigeration and Process Engineering Research Centre (FRPERC), University of Bristol, Churchill Building, Langford, Bristol BS18 7DY, UK

Sue Ghazala

Department of Biochemistry, Memorial University of Newfoundland, St. John's, Newfoundland, Canada A1B 3X9

Leon G.M. Gorris

Unit Microbiology & Preservation, Unilever Research Laboratory, Olivier van Noortlaan 120, 3133 AT Vlaardingen, The Netherlands

Tamara Haentjens

Katholieke Universiteit te Leuven, Department of Food and Microbial Technology, Laboratory of Food Technology, Kardinaal Mercierlaan 92, B-3001 Heverlee, Belgium

Marc Henrickx

Katholieke Universiteit te Leuven,
Department of Food and Microbial
Technology, Laboratory of Food
Technology, Kardinaal Mercierlaan 92,
B-3001 Heverlee, Belgium

Vijay K. Juneja

US Department of Agriculture,
Agricultural Research Service, Eastern
Regional Research Center, 600 E. Mermaid
Lane, Wyndmoor, PA 19038, USA

Toon Martens

Alma *Sous vide* Competence Centre,
Catholic University of Leuven, Van
Evenstraat 2C, B-3000 Leuven, Belgium

Catherine J. Moir

Food Science Australia, PO Box 52, North
Ryde, NSW, Australia 2113

Bart Nicolaï

Catholic University of Leuven, Department
of Agro-Engineering and Economics,
Kardinaal Mercierlaan 92, B-3001 Heverlee,
Belgium

Robyn O'Connor-Shaw

1 Boom Court, Birkdale, Queensland,
Australia 4159

Michael W. Peck

The Institute of Food Research, Norwich
Laboratory, Norwich Research Park,
Colney, Norwich NR4 7UA, UK

William Reeve

Department of Food and Hospitality
Management, The Worshipful Company of
Cooks Centre for Culinary Research,
Bournemouth University, Fern Barrow,
Poole, Dorset BH12 5BB, UK

Elizabeth A. Szabo

Food Science Australia, PO Box 52, North
Ryde, NSW, Australia 2113

Robert Trenholm

Center for Aquaculture and Seafood
Development, Marine Institute of Memorial
University, St. John's, Newfoundland,
Canada A1C 5R3

Ann Van Loey

Katholieke Universiteit te Leuven, Department of Food and Microbial Technology, Laboratory of Food Technology, Kardinaal Mercierlaan 92, B-3001 Heverlee, Belgium

Preface

Sous vide and cook–chill foods have been a viable form of alternative processing in European markets for some years providing convenient, top-quality foods to consumers. Large urban centres, close to the manufacturing plants, where large populations are available to consume these products, are the key to the *sous vide*/cook–chill success. The European marketing system has evolved to feasibly allow the environmental controls necessary for the safe production and distribution of these foods. Manufactured foods cannot be subjected to lengthy shipping times and long distances prior to consumption. The North American markets have not yet completely evolved to accept and control these product forms.

Minimally processed refrigerated (MPR) foods will evolve rapidly over the next few years in North America as advances and applications of hurdle technology and packaging technology improve the safety of these foods. Presently, the biggest concern to the food industry, and to the consumers of MPR foods, is their safety aspects. This book addresses this concern through discussions of important topics such as: principles, applications, sensory quality and nutritional aspects, packaging technology, consumer perceptions, handling, computer-integrated manufacturing, time–temperature integrators, critical factors and HACCP, hurdle technology, microbial safety, shelf life expectations, and important marketing practices. MPR foods are discussed in general, and *sous vide*/cook–chill products more specifically. This book is the compilation of the knowledge and experiences of many individual around the world who have been addressing safety as a matter of course in their advancement of the MPR technology.

It is hoped that this book will form the basis for the applications and techniques involved in *sous vide* and cook–chill processing. The information compiled in this book will help to define a starting point for many who may be interested in starting a *sous vide* and/or cook–chill product line and for those interested in starting research into this important product development field. The book is dedicated to the food industry in general, but particularly for those interested and involved in MPR foods. Food technologists will find the book useful both to obtain a good understanding of *sous vide* and cook–chill principles and to generate new product ideas. Research and development scientists and technologists will find this book useful as a starting point for further work and as a source of practical examples of actual food processes presently marketed. Students will find the book very useful in obtaining an excellent understanding of functional MPR food-processing principles and applications.

Steadily positive and productive collaboration with all of the contributing authors during the handling of their manuscripts was invaluable in constructing a well-written, comprehensive text. The experience of compiling this book and sharing knowledge with the contributors has broadened my understanding of this important technology, and I hope that you will share the same experience. Many thanks go to colleagues, assistants and friends both around the world and on my doorstep for providing help, ideas, discussion and unfailing support. I am very grateful for the immense support of the publishers without which this book could not come into being.

Sue Ghazala

1 Consumer perceptions and practice in the handling of chilled foods

JUDITH EVANS

In recent years the numbers of cases of food poisoning in most European countries has risen dramatically (WHO, 1992). In the UK for example, food-poisoning notifications have quadrupled during the 1980s and 1990s (*source*: Public Health Laboratory Service) (Figure 1.1). Increasingly food poisoning incidents have been found to be due to mishandling of food in the home, with insufficient refrigeration or cooling being the most frequent factor causing disease (WHO, 1992). Out of the 1562 cases of food poisoning reported between 1986 and 1988, 970 were from family outbreaks (*sources*: Public Health Laboratory Service and Communicable Disease Surveillance Centre).

Such large increases in reported cases are thought to be due to several factors. The level of reporting of food-poisoning incidents, although still under-reported, has increased due in part to enhanced publicity concerning food-poisoning incidents and hygienic awareness amongst the public. However, changes in technology in the production of food and in lifestyle may also have contributed to greater numbers of food-poisoning cases. The removal of preservatives and additives from certain foods and the trend for convenience meals, more snacking between meals and less shopping has meant that food products are more likely to support bacterial growth and tend to be stored for lengthy periods in the home.

The past decade has seen a considerable increase in legislation defining maximum temperatures during the production, distribution and retailing of chilled food. However, as soon as the food is purchased by the consumer, it is outside of any of these legislative requirements.

After a chilled product is removed from a retail display cabinet it is outside a refrigerated environment while it is carried around the store and then transported home for further storage. In the home it may be left in ambient conditions or stored in the refrigerator until required. There are few published data on consumers' attitudes to chilled food and their handling procedures in the home. The majority of the data quoted in this chapter have been obtained from a survey of 252 households which was funded by the Ministry of Agriculture Fisheries and Food (MAFF) in the UK (Evans *et al.*, 1991). As part of the survey, participants were asked questions to assess their attitude to food poisoning, shopping habits and the

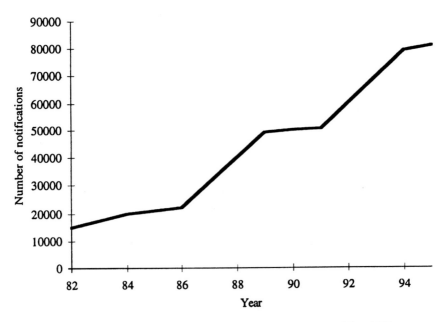

Figure 1.1 Food poisoning notifications in England and Wales 1982 to 1995.

length of time they stored chilled foods in the home. Monitoring was then carried out to determine the facts of these issues. These data were augmented with experimental data from laboratory studies on the performance of refrigerators and temperature changes during transportation to the home.

1.1 Consumer attitudes to food poisoning

In the survey consumers were initially asked about their concern regarding food poisoning and their answers were assessed on a 5-point scale ranging from very concerned to not at all concerned. Answers were restricted to concern about food from shops and did not include concern about food poisoning due to restaurant or fast food type meals or food. The greatest number of participants (56.7%) were either only slightly concerned or not at all concerned about food poisoning. However, 31.7% of participants were concerned or very concerned about food poisoning (Figure 1.2).

When asked to name foods which they considered might constitute a food-poisoning risk, most of the respondents (73%) considered poultry to be a problem. Raw poultry was considered to be a greater risk than cooked poultry. Meat was also considered likely to cause food poisoning with 66.7% of participants mentioning either raw or cooked meat as a potential

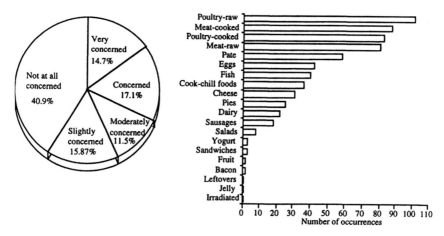

Figure 1.2 Concern about food poisoning and foods considered a risk.

problem. Foods such as eggs, yoghurt and cook–chill meals, on which media attention had recently focused, were not mentioned as much as might have been expected. These foods were only mentioned by a maximum of 16.7% of participants, perhaps demonstrating the transient nature of food scares (Figure 1.2).

1.2 Shopping habits and transport from retail store to the home

The frequency of shopping governs the length of time chilled food is stored in the home. Most consumers, 99.2% of the survey population, shopped on at least one day a week and few (16.3%) less than twice a week for chilled food. The greatest number (33.7%) shopped for food on 3–4 days per week, closely followed by 26.2% who shopped 5–7 days per week and 23.8% who shopped on two days (Figure 1.3).

Generally shopping for food was divided into trips for large quantities (defined as greater than one bag) and small quantities (less than one bag). Large quantities of food were bought by 97.7% of the householders interviewed; 73.8% of these shopped once a week and 23.0% visited the shops twice or more per week. A small proportion (0.8%) shopped less than once a week, most usually monthly or fortnightly. The majority of households (84.5%) shopped for small quantities of chilled food on a variable basis, as required.

Most participants in the survey carried out their main shopping between 1 and 5 miles from their homes and few householders travelled more than 5 miles to shop (Figure 1.4). Most people (85.3%) used a car to transport

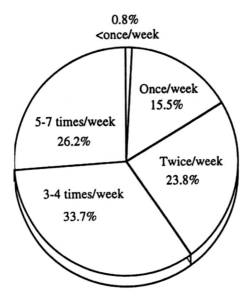

Figure 1.3 Number of shopping trips per week.

their main shopping home. A small percentage used a taxi, bus, bicycle or travelled on foot (Figure 1.4).

Small quantities of food were generally bought close to the home, reflecting the availability of shops in the towns surveyed. Most householders (87.6%) who bought small quantities of food transported it home by either foot or car (Figure 1.4).

Unprotected chilled food will warm up during transportation. Survey results showed that consumers took on average 43 minutes to bring meat, fish or fat items home from the shops and place them in a refrigerator. The

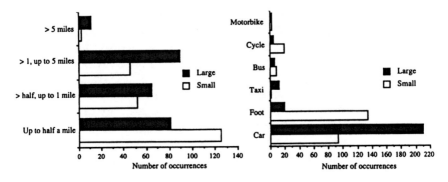

Figure 1.4 Distance from shops and mode of transport used in shopping.

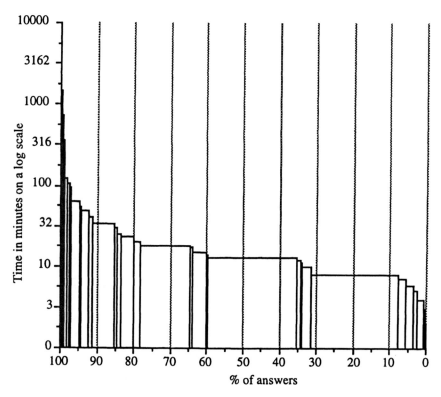

Figure 1.5 Time required to bring food home from the shops and place in refrigerator.

greatest number of items were transported home and placed in a refrigerator within 13 minutes. Although most people bought food home well within 60 minutes, there were a number of items which took far longer to be bought home (up to 2 days) and placed in a refrigerator (Figure 1.5).

Although insulated bags and boxes are widely sold, only a small percentage (12.7%) used them to transport some of their food home. The vast majority (87.3%) of people did not use any means of protecting food from temperature gains during transportation (Figure 1.6).

Increases in product temperatures during transportation can be considerable. In investigations, the temperatures of 19 different types of chilled product were monitored during a simulated journey from the supermarket to home (Evans, 1994). One sample of each product was placed in a precooled insulated box containing eutectic ice packs and the second left loose in the boot/trunk of the car. The car was then driven home and the product removed and placed in a domestic refrigerator after a total journey time of one hour. The external ambient temperature during the journeys ranged from 23 to 27°C. Initial product temperatures measured when the food

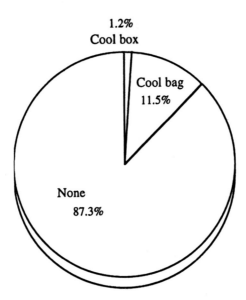

Figure 1.6 Use of cool bags/boxes.

reached the car ranged from 4 to over 20°C. Some product temperatures in samples placed in the boot rose to approaching 40°C during the one-hour car journey while most of the samples placed in the insulated box cooled during the car journey, except for a few at the top of the box which remained at their initial temperature (Figure 1.7).

Thinly sliced products, such as smoked salmon trout, showed the highest temperature changes during transport (Figure 1.8), whereas temperature gains in thicker products, such as chicken, were smaller (Figure 1.9). After being placed in the domestic refrigerator products required approximately 5 hours before the temperature at the surface was reduced below 7°C.

To produce some idea of the likely bacterial growth on food during transportation and storage, the temperature data were entered into a mathematical model. The model assumed that the lag phase of growth had expired and that all foods had an initial contamination of 200 bacteria per square cm. Predicted bacterial growth was always higher on foods transported in ambient conditions than those transported in the cold box (Figure 1.10). Although the cold box prevented growth of *Salmonella* on the chicken it was unable to totally prevent growth of all bacteria on the chilled products. However, the cold box did ensure that bacterial growth was minimal and was substantially less than if transported in ambient conditions.

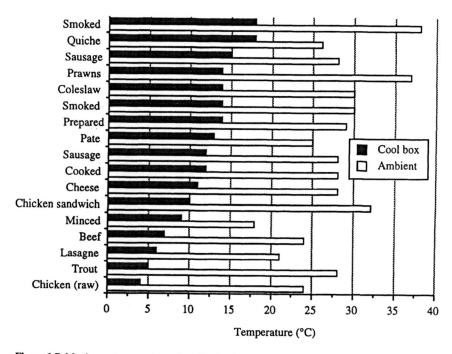

Figure 1.7 Maximum temperatures found in food transported home in cold box and in back of a car during a journey of one hour. (Reprinted with kind permission from James and Evans (1992a).)

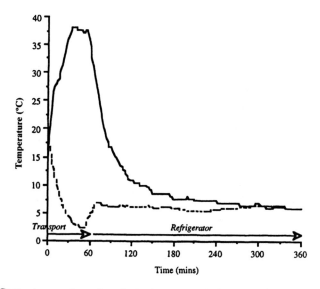

Figure 1.8 Centre temperature of smoked salmon trout during transportation and storage in domestic refrigerator. (Reprinted with kind permission from James and Evans (1992a).)

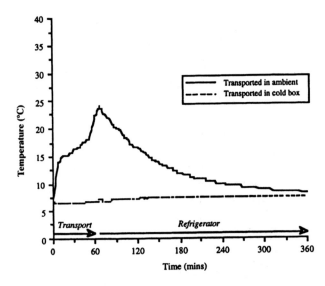

Figure 1.9 Centre temperature of chicken during transportation and storage in domestic refrigerator. (Reprinted with kind permission from James and Evans (1992a).)

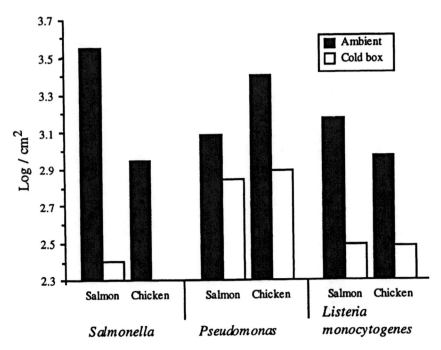

Figure 1.10 Predicted growth of *Salmonella*, *Pseudomonas* and *Listeria monocytogenes* on food transported home in ambient conditions and in a cold box (initial contamination 200 bacteria per cm^2; 2.3 log cm^{-2}).

1.3 Food storage in the home

The length of time consumers store chilled foods after purchase will affect the safety of the purchases. In the survey, consumers thought that the majority of foods (prepared meals, pizza/quiche, cold pies, pâté, sausages, raw poultry, cooked poultry, cooked meat and raw meat) would store well for 2 days. However, a number of people considered that these foods could be stored for greater than 7 days, and sometimes as long as 30 days. Most participants thought that bacon, eggs, cottage cheese and salad could be stored for up to a week, although a few people considered that storage of up to 30 days was acceptable. Fresh fish was generally considered to store least well, with most participants stating that they would only store fish for 1 day or less.

The range in anticipated storage life for different food types varied considerably. Opinions on the storage lives of individual foods ranged from 0.5 to 7 days (range 6.5 days) for cold pies and sausages to between 0.5 and 30 days (range 29.5 days) for pâté, bacon and cottage cheese. The minimum storage life for all foods except eggs (minimum storage life of 2 days) was either a quarter or half a day. A small number of householders thought that they could store chilled foods for periods of up to 30 days. Bacon, cottage cheese, eggs and pâté were all thought to be acceptable after this period by a small proportion of participants. Cooked meat and poultry were also thought to store for up to 21 days by a few householders (Figure 1.11).

It was interesting to note that although poultry and meat were considered a likely cause of food poisoning, participants did not necessarily consider that these foods had short storage lives. It is therefore possible that people did not associate storage time as being related to any food-poisoning problem.

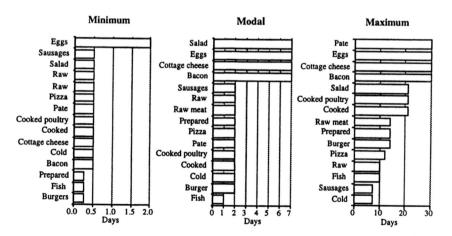

Figure 1.11 Minimum, modal and maximum storage lives stated by householders.

Table 1.1 Mean actual and perceived storage life

Food	Actual mean storage life (days)	Perceived mean storage life (days)
Bacon	8.2	6.6
Cooked meat	5.4	4.5
Cooked poultry	3.9	2.4
Cottage cheese	6.9	4.9
Fish	3.7	2.0
Pâté	10.3	4.1
Pies	3.2	3.3
Raw meat	3.9	2.4
Raw poultry	3.3	2.5
Sausages	5.6	4.1

Consumers do not always store foods in the manner they intend. When the food stored in consumers' refrigerators was examined, actual storage times were generally greater than storage times stated in the questionnaire. Almost 67% of the food was kept for longer periods (Figure 1.12). Actual storage times were greater than the stated storage time for all meat, fish and fat items except pies, which were thought to have an acceptable storage life of 3.3 days and were stored for 3.2 days (Table 1.1).

To provide an estimate of the amount of food consumed after its shelf life had expired, the remaining shelf lives of fats, fish and meats were noted

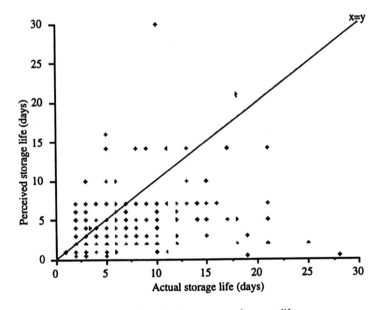

Figure 1.12 Perceived versus actual storage life.

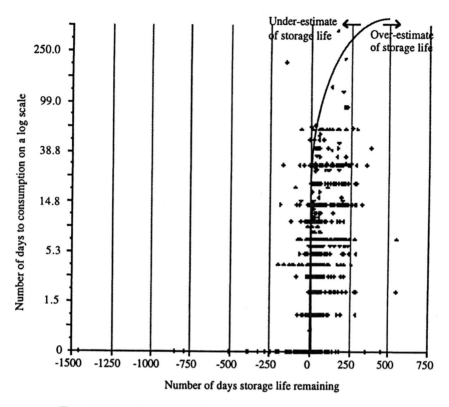

Figure 1.13 Remaining storage life of food in householders' refrigerators.

from the food packaging and compared to the estimated consumption time stated by the householder. The majority of items were intended to be eaten before or on the day the shelf life expired, with more than 50% of items having at least 17 days remaining shelf life at the time of consumption (all data to the right of the line in Figure 1.13).

A number of products (17.1%) were already over their shelf life at the time of examination (all data to the left of the line in Figure 1.13). One product (from the fat category) was 4 years over its shelf life but the householder stated that it would not be consumed. Another product (again from the fat category) had an estimated consumption date 1 year ahead (197 days over its shelf life).

1.4 Temperatures in domestic food storage

The refrigerator is a common household device and very few households do not own a refrigerator or fridge-freezer for storage of chilled foods.

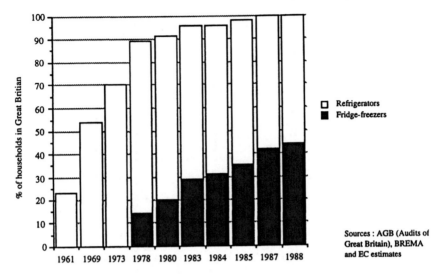

Figure 1.14 Refrigerator and fridge-freezer ownership in Great Britain, 1988 (AGB, BREMA and EC).

Fridge-freezers have become increasingly popular in the last 20 years and now provide almost 50% of the market (Figure 1.14). These figures were almost replicated in the survey where only three of the households did not

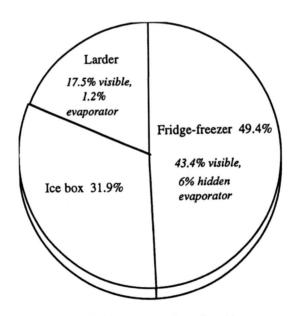

Figure 1.15 Refrigerator types investigated in survey.

own a working refrigerator and 49.4% owned a fridge-freezer. Almost 32% owned an ice-box type refrigerator and the remainder had larder refrigerators (Figure 1.15).

The temperature at which a refrigerator operates is critical for the safe storage of chilled food. Recommendations concerning the microbiological safety of foods advise that maximum temperatures in domestic refrigerators should not exceed 5°C (Richmond, 1991). Consumers in the survey were therefore asked to state the temperature at which they tried to operate their refrigerator. Nearly all participants were unable to name actual temperatures and gave answers based on the method they used to set the temperature dial (Figure 1.16). A large number of people (32.8%) set their refrigerators according to the weather, setting the refrigerator to a lower temperature (higher setting) in the summer. It was interesting to note that although 38 participants had a thermometer in their refrigerator, only 30 actually used the information to set their refrigerator temperature.

To evaluate temperatures within each refrigerator a miniature data logger with three air and two product sensors was placed into the refrigerator to monitor temperatures every 8 seconds and to record mean temperatures

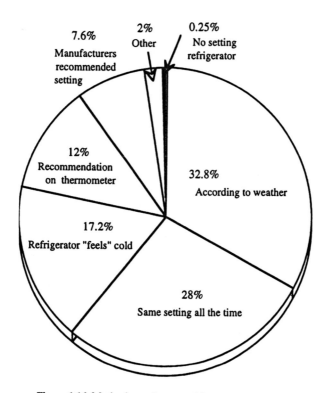

Figure 1.16 Methods used to set refrigerator temperature.

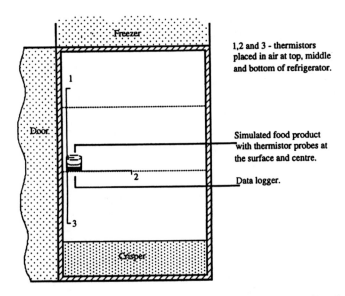

Figure 1.17 Position of miniature data logger and sensors within refrigerator. (Reprinted with kind permission from James and Evans (1992a).)

every 5 minutes for a period in excess of 7 days. Air temperature sensors were positioned in the top, middle and bottom sections of the refrigerator and a simulated food product (87 mm diameter by 28 mm high disc of 'Tylose', a food substitute, in a Petri dish) placed on the middle shelf. Sensors were placed in the geometric centre and centrally on the surface of the Tylose disc (Figure 1.17).

Results showed that the mean temperature over 7 days (evaluated from top, middle and bottom sensors) ranged from –1°C to +11°C. The overall mean air temperature for all the refrigerators in the survey was 6°C, with 70% of refrigerators operating at average temperatures above 5°C (Figure 1.18). An investigation carried out in Northern Ireland found similar results with 71% of refrigerators having a mean internal temperature above 5°C (Flynn *et al.*, 1992).

An analysis of percentage time spent between certain temperatures, calculated for all refrigerators, showed that greatest proportion of time (80.3%) was spent between 3 and 8.9°C. Only small amounts of time were spent above 9°C (Figure 1.19). However, only 4 refrigerators (1.6%) in the whole survey operated below 5°C during all the monitoring period and 33.3% of refrigerators spent all their time above 5°C.

A further analysis showed that in 69.9% of refrigerators the warmest place was in the top and in 45.1% the coolest place was in the middle (Table 1.2). However, the top of the refrigerator was not always the warmest and the bottom not always the coldest place (Table 1.3, Figure 1.20).

The mean temperature range within a refrigerator was found to vary

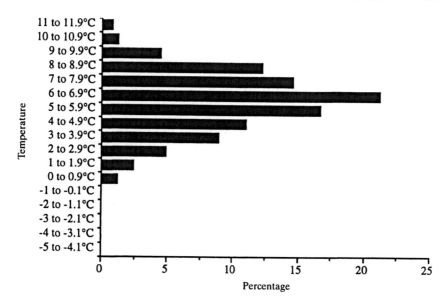

Figure 1.18 Overall mean temperatures for all refrigerators in survey.

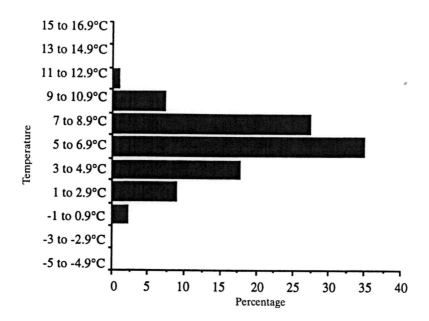

Figure 1.19 Frequency distribution of temperatures in all refrigerators.

Table 1.2 Position of highest temperature within refrigerators investigated

Position	% of refrigerators	
	Highest mean temperature	Lowest mean temperature
Top	69.9	20.3
Middle	8.1	45.1
Bottom	22.0	34.6

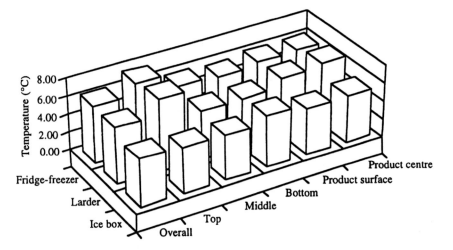

Figure 1.20 Mean temperatures within refrigerator types investigated.

Table 1.3 Positions of lowest and highest mean temperatures in refrigerators investigated

Refrigerator type	% of lowest mean temperatures in:			% of highest mean temperatures in:		
	Top	Middle	Bottom	Top	Middle	Bottom
Ice box	48.1	41.6	10.4	28.6	11.7	59.7
Fridge-freezer	10.6	45.5	43.9	84.6	8.9	6.5
Larder	0.0	50.0	50.0	100.0	0.0	0.0

between refrigerator types. Ice-box refrigerators had the smallest range (average 1.8°C), whereas the range in temperature in fridge-freezers and larder refrigerators was nearly twice as great (average of 3.4°C in fridge-freezers and 3.7°C in larder refrigerators) (Table 1.4, Figure 1.21). A survey carried out in China found higher ranges in temperature within domestic refrigerators with only 2.3% of the refrigerators surveyed operating with a temperature range of less than 6°C (Shixiong and Jing, 1990).

Table 1.4 Temperature range in refrigerator types investigated

Range in temperature (°C)	Ice box	Fridge-freezer	Larder
Minimum temp. range	0.2	0.1	0.5
Maximum temp. range	7.0	12.0	9.0
Mean temp. range	1.8	3.4	3.7

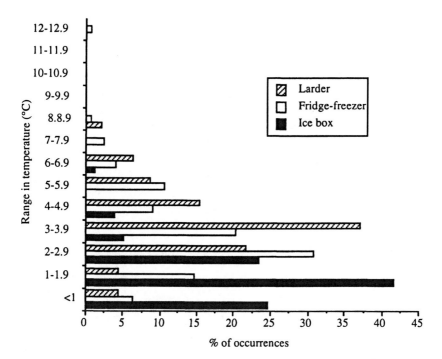

Figure 1.21 Range in mean temperature distribution within refrigerator types investigated.

1.5 Door openings in domestic refrigerators

Most domestic refrigerators are tested by manufacturers with the doors closed. In practice refrigerator doors are opened and food removed or placed into the storage area. The effect of door openings upon temperature performance under normal household use was investigated in a survey. Temperatures, number of door openings and duration of door openings in 60 household refrigerators were measured over a period of between 3 and 7 days. From this information the number of openings, total time opened and average refrigerator door opening time per day were calculated and the effect of door openings upon air and food temperature evaluated.

Households were found to vary widely in their refrigerator use with 65% of households opening their refrigerator less than 30 times per day and 70% opening the door for less than 4 minutes per day. On average refrigerator doors were open for 7.3 seconds (range 1 to 31 seconds) with a mean of 39 door openings (range 1 to 240) and a total door opening time of 3.1 minutes per day (range 0.2 to 11.5 minutes). There was no apparent relationship between the number of times the refrigerator door was opened and the duration the door was open during a day.

Overall, there was a poor correlation between either the number of door openings or the total length of time the door was open per day and the average air temperature at any position or the average food temperatures. Some degree of correlation was found when the analysis was extended to consider the type of refrigerator (box plate, larder or fridge-freezer). A relationship was found between the total number of door openings per day and the maximum temperature ($R^2 = 0.84$) measured in larder refrigerators. A slightly poorer relationship ($R^2 = 0.63$) was also found between the number of door openings and the temperature range within larder refrigerators. Both maximum and range in temperature were increased by greater total daily door-opening time, indicating that larder refrigerators may be more sensitive to door opening time than other refrigerator types.

1.6 Performance testing of domestic refrigerators

After purchase, chilled food can spend between a few hours and many weeks in a domestic refrigerator. However, few data seem to have been published on the temperature performance of domestic refrigerators. Current standards for domestic refrigerators contain some temperature tests that are carried o'.. under controlled conditions on empty, closed refrigerators. In domestic use refrigerator doors are opened, refrigerators are not usually empty but range from nearly empty to crammed full and often food at ambient temperature, or above, is placed in them.

Some data have been published from experiments carried out on examples of three types of refrigerator (James *et al.*, 1989; James and Evans, 1992b). These were a 6 ft^3 dual compressor fridge-freezer (no. 1), a 6 ft^3 single compressor fridge-freezer (no. 2) and a 4 ft^3 free standing domestic refrigerator with an ice box compartment (no. 3).

1.6.1 Performance of empty appliances

When set to the manufacturer's recommended setting, temperatures in the ice-box refrigerator (no. 3) were uniform and low with a minimum of –1.4°C on the bottom shelf and a maximum of 5.9°C in the door, average temperatures were between approximately 0.5 and 1.5°C on the shelves and just above 3°C in the door, with a cycle of less than 0.5°C. There was a much larger temperature range in the two fridge-freezers: 1.7 to 14.3 in no. 1 and −6.7 to 10.7 in no. 2. Average temperatures were far less uniform in the chilled food compartment of the fridge-freezers. In fridge-freezer no. 1 the average temperature of the top shelf was up to 5°C higher than that measured on the middle shelf, which was the coolest area in the appliance. Highest average temperatures of approximately 7.5 and 10°C were measured on the top shelves of the fridge-freezers. In fridge-freezer no. 2 the average temperature on the bottom shelf reached −2°C at the minimum point in the temperature cycle.

1.6.2 Performance of loaded appliances

Loading 12 packs (dimensions 100 × 150 × 25 mm) of Tylose (a simulated food) that had been pre-cooled to 5°C into the ice-box refrigerator reduced the mean temperatures by between 1.2 and 2.0°C (Table 1.5). The temperature change caused by loading was similar in magnitude in fridge-freezer no. 2, where the mean temperature of the top shelf rose by 0.7°C and the mean at other positions dropped by between 0.5 and 1.1°C. It was also noted that the length of the refrigeration cycle increased from approximately 0.75 h to 1.0 h. In fridge-freezer no. 1 the magnitude of the temperature cycle was substantially reduced. The magnitude and position of the maximum temperature was also influenced by loading from a value of 14.3°C located on the top shelf to a reduced value of 9.8°C location on the bottom shelf.

Table 1.5 Maximum, minimum and mean temperatures on shelves and in door of refrigerators

Position		Fridge-freezer no. 1		Fridge-freezer no. 2		Ice box	
		Empty	Loaded	Empty	Loaded	Empty	Loaded
Top shelf	Max.	14.3	6.0	10.7	11.1	2.1	1.2
	Min.	6.6	2.4	4.7	5.2	0.7	−1.2
	Mean	10.2	3.8	7.3	8.0	1.5	0.3
Middle shelf	Max.	8.0	6.9	5.4	4.9	2.2	0.4
	Min.	1.7	4.3	0.9	−0.1	−1.0	−2.6
	Mean	6.3	5.5	3.6	2.9	1.4	−0.6
Bottom shelf	Max.	8.0	9.8	5.0	3.7	1.6	4.0
	Min.	2.4	5.7	−6.7	−5.8	−1.4	−3.0
	Mean	6.7	8.1	2.1	1.0	0.7	−0.6
Door	Max.	8.0	8.4	6.5	6.7	5.9	3.3
	Min.	5.3	0.8	2.2	0.5	0.9	−0.4
	Mean	6.9	3.8	4.2	3.7	3.2	2.0

1.6.3 Effect of loading with 'warm' (20°C) food products

Food is often loaded warm into refrigerators after purchase from retail stores. Loading a small amount of warm (20°C) food, two joints (approximately 17.5 × 7.6 × 3.6 cm, 195 ± 10 g) and two drumsticks (approximately 12 × 6 × 3 cm, 120 ± 10 g) of simulated chicken (Tylose) and two 1-litre cartons of orange juice showed up the poor cooling performance of domestic refrigerators.

Over 2 hours was required in the ice-box refrigerator to reduce the surface temperature of the drumsticks and portions to 7°C compared with over 5 hours in the fridge-freezer (Table 1.6). In all comparisons, packs of orange juice placed in the door compartments took substantially longer (>50%) to cool than those on the top shelf. Drumsticks in the domestic refrigerator always cooled faster than the larger portions. However, in the fridge-freezer, portions on the middle shelf cooled faster than drumsticks positioned on the top shelf.

1.6.4 Effect of door openings

In normal operation, refrigerator doors are opened and left open for different periods while food is loaded and unloaded. In the ice-box refrigerator a single door opening of either 3 or 6 minutes was compared with three or six 1-minute openings over a 1-hour period.

Immediately after a 3-minute door opening the average air temperatures ranged between 5.5 and 16°C compared with a range of 0 to 2°C before the opening (Figure 1.22). Within approximately 1 hour of the door being closed the average temperatures had been reduced to within 1°C of their normal value. With three 1-minute door openings the average temperatures tended to progressively increase with each subsequent opening and the degree of temperature recovery reduced (Figure 1.23).

After a 6-minute door opening the average air temperature rose to between 7.5 and 16.5°C (Figure 1.24). The air temperature slowly recovered over the next 2 hours.

Table 1.6 Time taken (hours) to cool products from 20 to 7°C. (Reprinted with kind permission from James and Evans (1992b)

	Ice box	Fridge-freezer
Orange in door	11.6	15.9
Orange top shelf	5.7	9.0
Surface drumstick	2.2	5.3–8.6
Deep drumstick	2.5	5.4–8.6
Surface portion	2.2	5.1–7.5
Deep portion	3.4	5.9–8.0

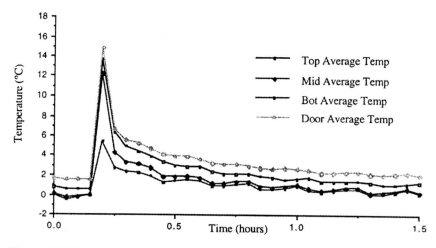

Figure 1.22 Three-minute door opening. (Reprinted with kind permission from James and Evans (1992b).)

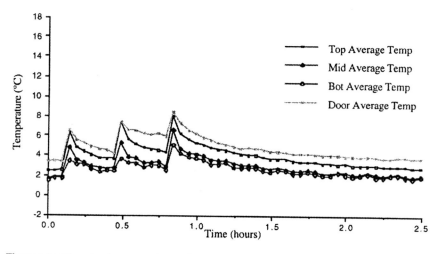

Figure 1.23 Three 1-minute openings at 20-minute intervals. (Reprinted with kind permission from James and Evans (1992b).)

After each of six successive 1-minute door openings the air temperature was warmer and the temperature after 9 minutes of recovery higher (Figure 1.25)

1.7 Conclusions

This chapter considers the two aspects of the chill chain that immediately precede the final preparation and consumption of the chilled food, the

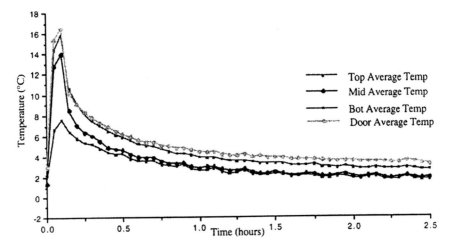

Figure 1.24 Single 6-minute door opening. (Reprinted with kind permission from James and Evans (1992b).)

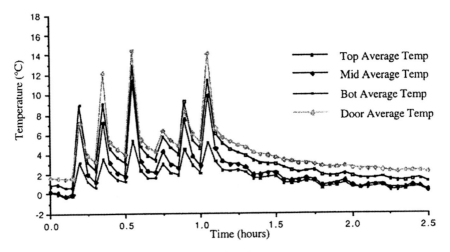

Figure 1.25 Six 1-minute openings at 10-minute intervals. (Reprinted with kind permission from James and Evans (1992b).)

transportation of the food from retail to the home and domestic handling and storage.

It is clear from the data presented that the temperature of chilled foods can rise to unacceptably high values if transported, without insulation, in a car boot. These data were obtained in June 1989, a very sunny period, but higher ambient temperatures are not uncommon in mid-summer. The predictions carried out show that substantial increases in bacterial numbers can occur during transportation and subsequent re-cooling. It is not difficult to think of even worse situations where chilled products reside in the open backs of estate cars for many hours on hot summer days. However, a combination of increased consumer education and the use of insulated/pre-cooled containers would solve this particular problem.

The basic design of the domestic refrigerator has not changed in the last 50 years although their use and, lately, the type, complexity and microbiological sensitivity of the foods stored in them has markedly changed. Designers have responded to market demands for more compact appliances and more features, i.e. chilled drink and ice dispensers, but temperature control is only advertised as a sales point on more expensive multi-compartment refrigerators.

Consumers now purchase and store a wide range of ready meals and other chilled products and they have demanded, and obtained, substantial reductions – and in some cases the total elimination – of preservatives and additives in these products. New chilled products are therefore inherently more bacterially sensitive and require closer temperature control than their predecessors. If predictions that eating habits will change from the current pattern of set meals to all-day grazing, then the consequence is likely to be a demand for and purchase of more pre-prepared chilled foods and more visits to domestic refrigerators.

These results indicate that current refrigerators are unlikely to be able to maintain foods at the temperatures desirable for chilled products when subjected to more frequent door openings and the addition of 'warm' food. The appliances differed substantially in their ability to respond to door openings and loading with warm (20°C) food and both these operations are crucial to domestic operation. Since current and proposed standards only test empty closed appliances, these differences would not be apparent to the consumer who would only know that they were tested to the relevant standard. These limited experiments indicate the problems in providing consumers with general recommendations and simple temperature test procedures. Recommendations such as 'by keeping the top of a fridge at 5°C, the bottom should be at 0°C' (Anon., 1990) may be applicable to the majority of appliances but not pertinent to specific appliances.

The consumer study of 252 households indicated that higher temperatures than desirable are to be found in current domestic refrigerators and that there is a need to educate consumers as to the need for lower temperatures. This need should subsequently create a demand for domestic refrigerators that will maintain low temperatures under normal operating conditions. The data presented on the temperature performance of refrigerators in the laboratory indicate the need for different international standards that relate to food safety, quality and consumer usage of domestic refrigeration.

Acknowledgements

The author would like to thank the Ministry of Agriculture Fisheries and Food for providing funding for much of the above work and Mr S. James, Mr I. Stanton and Mr S. Russell for their valued assistance during the investigations.

References

Anon. (1990) Fridge-freezers. *Which?*, May, 286. The Association for Consumer Research. CDR Review, 21 June 1996. *PHLS*, Vol. 6 , Review No. 7.

Evans, J. A. (1994) The cool option, in *Proc. 14th International Home Economics and Consumer Studies Research Conference: People , Populations and Products*, July.

Evans, J. A., Stanton, J. I., Russell, S. L. and James, S. J. (1991) *Consumer Handling of Chilled Foods: A Survey of Time and Temperature Conditions*, MAFF, PB 0682.

Flynn, O. M. J., Blair, I. and McDowell, D. (1992) The efficiency and consumer operation of domestic refrigerators. *Int. J. Refrig.*, **15**(5).

James, S. J., Evans, J. A. and Stanton, J. I. (1989) The performance of domestic refrigerators, in *Proc. 11th Int. Home Economics Conference*. Sept.

James, S. J. and Evans, J. A. (1992a) Consumer handling of chilled foods: Temperature performance. *Int. J. Refrig.*, **15**(5), 299–306.

James, S. J. and Evans, J. A. (1992b) The temperature performance of domestic refrigerators. *Int. J. Refrig.*, **15**(5), 313–319.

Richmond, M. (1991) *The Microbiological Safety of Food*, Part II. Report of the Committee on the Microbiological Safety of Food, HMSO, London.

Shixiong, B. and Jing, X. (1990) Testing of home refrigerators and measures to improve their performance. IIR Commission B2, C2, D1, D2/3, Dresden, Paper 96.

WHO (1992) *Surveillance Programme for Control of Foodborne Infections and Intoxications in Europe*, Fifth report, 1985–1989.

2 Principles and applications of *sous vide* processed foods

PHILIP G. CREED and WILLIAM REEVE

2.1 Introduction

Changing lifestyles over the last 20 years have provoked much activity in the food-manufacturing and foodservice industries in developing new technologies for processing and packaging food in order to take over many of the tasks of preparing and cooking food which had usually taken place in the home or the restaurant.

In the home, meal preparation was traditionally the duty of the housewife caring for a husband and children, but as career opportunities and expectations have increased for women, the time available for meal preparation has decreased. The percentage of working women over 16 has risen from 49% in 1984 to 53% in 1994, while for men the figures have decreased from 76% in 1984 to 73% in 1994 (CSO, 1996). The concept of the family meal has also been breaking down with family members eating at different times due to different working patterns and personal preferences. Family or household size has also changed, making the once typical family group of two adults and one or two children less common in the United Kingdom, falling from 30% of households in 1961 to 20% in 1994/1995 (CSO, 1996). In many cases, there is much less confidence in preparing raw foods to make a meal as traditional cookery skills once passed on at home from one generation to the next, and reinforced by school activities, have declined. These factors have resulted in the market for convenience foods such as chilled ready meals increasing in the United Kingdom, with the market in France also increasing (Table 2.1).

In the restaurant and institutional foodservice, food has usually been prepared, cooked and served in fairly rapid sequence although the practice of preparing food in advance has long been used for banquets. However, the pressure on organisations to become more efficient by minimising the costs associated with labour, equipment and energy has prompted the introduction of new systems for providing food on a large scale. These are based on centralised production units supplying satellite kitchens for staff feeding or chains of commercial restaurants. The starting material has also changed from traditionally using mostly raw materials such as meat, fish and vegetables to buying in portioned cuts of meat, fish and poultry and peeled,

Table 2.1 The value of the markets for chilled ready meals in the United Kingdom and France (Mitchell, 1997)

	United Kingdom		France	
	Value (£m)	% change	Value (FFr)	% change
1990	–	–	513	–
1991	205	–	562	+9.6
1992	238	+16.1	599	+6.6
1993	267	+12.2	626	+4.5
1994	288	+7.9	689	+10.1
1995	310	+7.6	785	+13.9
1996	338	+9.0	–	–

diced or sliced vegetables ready prepared for cooking. The next logical step in this development is then buying in ready prepared dishes which only need heating before service – the concept of the 'kitchenless' restaurant.

Both at home and in the restaurant, whether at work or for pleasure, demand has increased for a wider range of dishes, perhaps reflecting the meals enjoyed while travelling abroad. With increasing prosperity for the majority of the Western world, a higher disposable income has encouraged a willingness to pay a premium for foods which are considered to have higher 'quality'. The *sous vide* method of food processing and preservation is one way in which these demands may possibly be satisfied and this chapter aims to discuss some of the ways in which this has been done, drawing attention to the strengths and limitations in applying this method.

2.2 The role of the time buffer

The concept of a time buffer has long been an essential part of the chain of food provision for consumers buying food from manufacturers through retailers. The preservation systems of dehydration, canning and freezing have all been familiar methods for providing convenience foods. As consumers have come to prefer chilled foods regarding them perhaps as closer to 'home cooking' than other forms of preservation, the market volume and choice has increased as mentioned previously.

Manufacturers of ready prepared chilled foods aim to provide safe and wholesome products at a given price level. They will also use forms of packaging, processing methods and tightly controlled distribution systems which will enable the consumer to enjoy the full range of organoleptic qualities intended to be present in the reheated product by their product development technologists. Manufacturers and retailers will, of course, take great care to ensure that their part of the time buffer, the cold chain, is secure up to the point of purchase by the consumer. After that, only consumer education and clear instructions will help the remaining part of the time buffer fulfil its potential in maintaining the shelf life of the chilled product. Studies on how consumers treat products after purchase, during transport to the home and in the refrigerator at home do not inspire great confidence in their ability to maintain product quality (James and Evans, 1990).

For the foodservice industry, the time buffer has been a recent concept compared to the food-manufacturing industry. In this case, maintaining the cold chain for ready prepared chilled meals can be more easily ensured where in-house production for consumption is on premises under the organisation's own control (hotels, restaurant chains, transport foodservice, hospitals, schools, etc.) or where products are bought from the chilled food manufacturers for a similar end use. The traditional cook–serve foodservice system means that once inputs of raw food materials enter the system, they have to go through the entire process or else there will be high levels of wastage due to problems in predicting what will be required and the short shelf life once cooked. In addition, skilled staff can only work when food is required; expensive equipment is only used for a small part of the working day; energy use is inefficient and staff have to work unsocial hours.

Introducing a time buffer between the cooking and service stages, means that many of the problems mentioned above should be capable of solution (Figure 2.1). In theory:

- Staff can work normal hours without the peaks of activity associated with meal times.
- Staff can be organised to work more efficiently with predictable and planned production.
- Less equipment can be used for longer periods to provide the same output of meals.
- Food preparation and cooking can be centralised to avoid duplication of equipment and skilled labour in a larger number of smaller kitchens.
- Energy can be used more efficiently with fewer warming-up periods.
- Economies of scale can cut food costs.
- Meals can be reheated as and when required, leading to lower food wastage levels.
- Less skilled staff can be employed to reheat and serve food in satellite kitchens after the time buffer.

However, all these advantages depend on finding a reliable method of preserving and distributing the cooked food during the time buffer which will have a minimal effect on its microbiological, sensory and nutritional qualities. The acceptability of the reheated product to the final consumer should be the criterion for judging the success of a preservation system. Safety problems can arise from the potential for growth of pathogenic and spoilage microorganisms; sensory problems can be caused by colour changes, separation of food components, drying out, rancidity, development of off-flavours

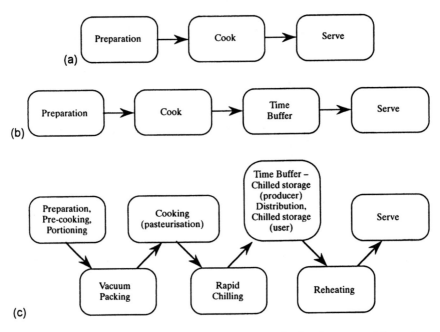

Figure 2.1 Systems diagram for (a) a traditional cook–serve foodservice system, (b) a modern foodservice system incorporating a time buffer and (c) a *sous vide* system.

and nutritional levels can fall during the time buffer if food is not carefully handled. Many of these factors will be discussed in other chapters but for *sous vide*, the main subject of this chapter, an adequate pasteurisation heat treatment followed by rapid chilling and chilled storage, combined with completely enclosed packaging to prevent contamination, will address safety problems; the packaging enclosing the food will also physically retain components of food flavour, nutrients and moisture and at the same time exclude oxygen and its influence in the formation of off-flavours and the relatively low cooking temperature itself will lead to lower breakdown rates for some nutrients and flavour components.

2.2.1 Creating the time buffer

Some of the first systematic efforts to incorporate a time buffer for food-service use were the cook-freeze systems at the Leeds Hospital for Women and for the City of Leeds Schools Meals Service in the late 1960s (Millross *et al.*, 1973). Freezing is a reliable method of preservation as it includes the 'barrier' of the latent heat plateau as shown on any time–temperature graph during the thawing of frozen food. The time required for the food temperature to rise through the latent heat range during the phase change from ice to water in the food provides some sort of safety factor against temperature abuse, although any partially thawed food at the surface would cause potential food safety and appearance problems.

Since then, the emphasis has moved to using chilled foods. In the eyes of the consumer, these have the aura of being 'fresh' and 'natural', whatever those words mean. Compared to frozen foods, chilled foods are inherently more susceptible to temperature abuse because there is only a slight 'barrier' to temperature change. This depends on the thermal conductivity of the food and its packaging compared to frozen products where the phase change slows down the rise in internal food temperature. However, these disadvantages for chilled foods regarding temperature abuse would become advantages when reheating is required at the point of consumption; chilled foods can be reheated more quickly and uniformly than frozen prepared foods (Bangay, 1996).

2.2.2 Incorporating the time buffer of chilling into foodservice systems

One of the first systems to use chilling and chilled storage as the time buffer was introduced in a Swedish hospital: the Nacka system in the 1960s was an early attempt at cook–chill. The procedure consisted of heating food to 80°C, filling it into plastic bags which were then sealed using a vacuum, placed in boiling water for 3 minutes to pasteurise the food and then cooled (Bjorkman and Delphin, 1966). It was therefore essentially a two-stage cooking process. In the United States, the AGS system (a forerunner of *sous vide* also intended for hospital feeding) consisted of assembling raw or part-cooked food, vacuum packing it before pasteurisation in water baths, followed by cooling in iced water and refrigerated storage. Reheating the pouches in a water bath or plated meals in a microwave ensured hot food for the patients, who found that the food was 'hot, smells good and is tasty' (McGuckian, 1969). Another more recent variant is CapKold; a bulk system of cooking in computer-controlled kettles, packing into casings, clipping to provide the seal and then rapid chilling by tumbling the packs in cold water followed by chilled storage (Daniels, 1988). The *sous vide* method (Figure 2.1(c)) made possible by the development of heat-resistant plastic packaging in the USA, was developed in France in the 1970s and, as will be explained, has continued with variable success in other parts of the world.

However, the most widely used system has been cook–chill, where foods are cooked conventionally and then portioned and cooled before chilled storage and distribution for reheating in the satellite kitchen. In the United Kingdom, the system started in the 1980s and was strongly promoted by manufacturers of large-scale cooking and refrigeration equipment working with the electricity industry who stressed its energy efficiency (Electricity Council, 1982). Similarly for the *sous vide* method, manufacturers of specialised equipment for vacuum packing, cooking, cooling, chilled storage, reheating and packaging have co-operated in putting complete systems together in order to encourage their adoption. The package usually included training (Anon., 1989a).

The question that must be asked when introducing any foodservice system, is whether a specific combination of equipment is the optimal choice in providing the menu range for the consumers' needs in that particular foodservice operation. It may have particular constraints, thus losing the anticipated flexibility of operation and choice of menu items.

2.2.3 Other minimal processing technologies for the time buffer

The examples above show that refrigeration by freezing or chilling has been widely used to maintain quality during the time buffer, but more recently, other minimal processing technologies have been used or suggested in order to provide additional safeguards. Techniques including modified atmosphere packaging (for fresh foods), irradiation or combinations of techniques in conjunction with the cook–chill and *sous vide* methods could provide the 'hurdles' to microbial growth and lead to a lower risk of food safety problems (Leistner and Gorris, 1994). Hydrostatic pressure treatment may have an application for low market volume, high-quality expensive foods but products have yet to be launched on the European market (Mertens, 1995).

2.3 Production of *sous vide* foods

The production of *sous vide* foods may be divided into two categories: firstly, in-house, craft-based producers for small- to medium-scale production supplying, for example, hotels and restaurants and, secondly, food manufacturing for large-scale industrial production of products to be sold to the foodservice industry or to the consumer through retail sale. However, this categorisation is somewhat blurred as some large-scale producers may be craft-based and medium-scale producers more manufacturing-based. More extensive details on the quality management aspects and equipment requirements for *sous vide* processing are discussed in other chapters of this book.

2.3.1 In-house, small-scale production

On a small scale, it is likely that the in-house producer using a restaurant or hotel kitchen will already have much of the necessary equipment such as a combination steam and forced convection cooker for heat treatment and reheating, bains-marie for reheating and chilled storage rooms but will need to invest in vacuum-packing equipment, air blast chillers or iced water immersion chillers for rapid cooling of *sous vide* packs and temperature monitoring and recording equipment. Specialised equipment for reheating is becoming available for incorporation into foodservice outlets. These are

based on using pouches surrounded by hot water at a controlled temperature into which chilled *sous vide* packs can be placed and removed when the
built-in timers provide a signal (Baird, 1990; Lehr, 1990).

Alongside these needs are the higher standards required by regulatory
authorities for the physical surroundings and buildings, staff training and
appropriate quality management systems, all of which must be managed
and monitored by competent people. This quantum leap in standards is not
a question to be underestimated for a small- to medium-scale operation. It
is, however, an opportunity for companies that has been encouraged by the
European Union's funding of research projects into the *sous vide* method
(Martens, 1996).

2.3.2 *Food manufacturing, large-scale production*

In contrast to the small- to medium-scale producers, production of *sous vide*
foods on a large scale by food manufacturers should not impose such a great
change since investing a large amount of capital in equipment, training and
quality management systems is not so unusual. Industrial scale machinery
is currently available for the preparation of raw materials by dicing, slicing
and grating, and for pre-preparation steps such as blanching or for browning meats before vacuum packaging.

For vacuum packing, the move from manual single or multi-station
machines using pouches for small- to medium-scale producers requires
investment in highly automated continuous packing systems where the base
of the pack is formed on-line followed by filling and vacuum sealing under
a top layer of packaging such as the Darfresh system (Pré, 1992).

The heat treatment step for pasteurising *sous vide* products on a large
scale can be accomplished by a variety of equipment which often incorporate the rapid-chilling process. Several methods are used, most making use
of computers for precise temperature control and controlling the sequence
and timing of operations. Some examples have been reviewed in detail by
Pinot (1988), Dréano (1988), Sicot (1990), Urch (1991), Martens (1995) and
Peyron (1996):

- The Lagarde system pasteurises packs stacked on trolleys using a mixture
  of water vapour and compressed air followed by cooling with streams of
  water in the same vessel.
- The Armor Inox Thermix system uses immersion in hot water for pasteurisation which is replaced by iced water for rapid chilling. The circu-
  lation system aims to minimise energy consumption.
- The Barriquand Steriflow system uses cascading water to heat and then
  chill *sous vide* packs in an enclosed vessel. The water circulates at a high
  rate through a closed loop to minimise the risk of contamination and uses
  a heat exchanger for efficient heat transfer.

2.4 The application of *sous vide* foods to foodservice problems

Articles published highlighting the practical use of the *sous vide* method cover a very wide range of sectors of the food and foodservice industries. Table 2.2 summarises world activity in the areas of production, application, education and research related to *sous vide* processed foods. The following sections illustrate examples of some of these applications.

2.4.1 Industrial manufacturing

A major use of the *sous vide* method can be adding value: food processors can extend their operations and market from providing portioned raw materials for other food manufacturers, foodservice outlets or retailers to manufacturing 'gourmet' dishes themselves. The implications of this are a higher investment in the necessary equipment, training and management.

(a) Meat processing. Vacuum packing has long been a useful packaging method for large joints of fresh and cured meats. It has also been used with low-temperature water baths to produce cooked joints for use in foodservice or for retail sale of sliced cooked meats (Van der Leest, 1985). Wiesel (1987) described the economics of the continuous production of sausages pasteurised in vacuum packs. These applications could be considered to use a similar mode of operation to the *sous vide* method. The use of vacuum-packing as part of the *sous vide* system has been suggested for eliminating problems of oxidation found with the conventional preparation of many meat-based prepared meals (Anon., 1987) and adding value to meat by selling it as prepared meals of higher quality at a higher profit (Beauchemin, 1990; Pröller, 1990).

(b) Fruit and vegetable processing. Fruits and vegetables prepared using the *sous vide* method are considered to be an improvement over frozen or sterilised versions as long as processing temperatures are not too low (Varoquaux and Nguyen-The, 1994) but a modified process is claimed to improve sensory qualities: sliced or diced vegetables are put under a vacuum of 10 mb to remove oxygen, followed by injection of steam at 950 mb which heats the surface temperature to 98°C until the required pasteurisation value is attained, followed by using controlled vacuum cooling to 40–60 mb then 5 mb to chill the vegetables to 1–2°C. The process has also been studied for producing pre-cooked lentils. Temperatures below 90°C for less than 2 hours maintained firmness and wholeness (Varoquaux *et al.*, 1995).

(c) Seafood processing. Urch (1991) noted the use of Barriquand Steriflow equipment for producing consumer packs of *sous vide* processed salmon with pasta for retail sale as part of a low-calorie range of products in France.

Vacuum packing salmon slices so that the end user can then cook them 'in the bag' is one idea suggested by packaging manufacturers (Varney-Burch, 1991). Hackney *et al.* (1991) suggest *sous vide* can extend the shelf life of crab meat. Ghazala (1994) considered that the *sous vide* method offered the fish industry many opportunities for providing chilled convenience products, providing that stringent working practices and pasteurisation levels are in place. Bergslien (1996) is of a similar opinion and points out some specific problems of an unacceptable white layer of precipitated protein on salmon during *sous vide* processing which can be reduced using a sodium carbonate dip pre-treatment.

(d) Supermarket retailing. In contrast to the United Kingdom and to some extent the United States, the retail sale of *sous vide* products in France and Belgium has been well tried and tested. Martens (1995) attributes this to the greater emphasis on food safety in the UK and the USA as opposed to the requirement for foods with better sensory qualities in France and Belgium. In the USA, the FDA, the regulatory authority, considered that *sous vide* products made by foodservice operators were not acceptable for retail sale because of the lack of stringent food safety controls found in large-scale food manufacture (Schwarz, 1988). Riell (1988) briefly describes plans for 'boutiques' selling *sous vide* products made by a manufacturer: Rice (1991) mentions market testing of GourmetFresh products in grocery outlets and Millstein (1990) discusses the problems of reliable delivery of Culinary Brands' *sous vide* products to supermarkets in the USA. More recently, optimism has been expressed that supermarket sales of manufactured *sous vide* dishes will rise once the selling price can provide a profit (Jones, 1996).

Research in Belgium and France has shown that the typical consumers for retail *sous vide* products would be single people, dual career families, the younger age group and city dwellers. In 1991 there were 87 *sous vide* manufacturers selling to this market through the retail and foodservice industries (Martens, 1996). In a Belgian consumer survey, a range of complete and semi-complete *sous vide* dishes from eight producers led to the emergence of three 'Best Buys': beef bourgignon with potatoes, chicken with mushrooms and diced chicken breast with potatoes (Anon., 1993a). These facts illustrate the general acceptance in France and Belgium of *sous vide* products at retail level. So far, there do not seem to have been any food safety problems associated with consumers using this type of product.

2.4.2 Hotels and restaurants

The *sous vide* method was developed for restaurant use by a French chef, Georges Pralus, making use of a packaging material and system patented in USA (Ready, 1971). Its support by the French culinary establishment ensured that the claimed high levels of sensory quality for the *sous vide*

Table 2.2 Summary of world activity in the production, application, education and research on *sous vide* processed foods from 1984–1997. (Data from Creed, 1996; Martens, 1995.)

	Production		Application				Education	Research
	Manufacture	In-house	Hotels & restaurants	Institutional	Retail	Transport		
Africa								
Senegal								
South Africa		X	X				X	
America								
Argentina	X							
Canada	X	X	X		X	X		X
Mexico							X	
United States	X	X	X		X		X	X
Asia								
Japan		X	X				X	X
Singapore		X	X					
Taiwan								X
Australasia								
Australia	X	X	X		X		X	X
New Zealand		X	X					

Europe							
Austria							X
Belgium	X	X	X		X	X	X
Czech Republic						X	X
Denmark	X	X	X	X	X	X	X
Finland							X
France	X	X	X	X	X	X	X
Germany	X						X
Greece							
Iceland							X
Ireland	X	X		X	X	X	X
Italy						X	X
Luxembourg							
Netherlands	X	X	X	X		X	X
Norway	X	X		X		X	X
Portugal							X
Spain	X	X	X		X		X
Sweden	X	X			X		X
Switzerland	X	X	X	X	X	X	X
United Kingdom	X	X		X	X	X	X

method would appeal to the owners and staff of up-market restaurants and hotels around the world as an innovative marketing idea following on from other less technologically based fashions such as *'nouvelle cuisine'*, *'cuisine minceur'*, etc.

(a) Restaurant applications. Many restaurants using *sous vide* products do not advertise the fact as most customers would probably feel cheated if they realised that a real chef was not present in the kitchen (Bristol, 1989; Gledhill, 1991). However, some manufacturers have set up centralised production kitchens to supply restaurant chains, examples being Le Petit Cuisinier in France supplying the Flunch restaurant chain (Goussault, 1992) and Grace Culinary systems supplying American Café (Swientek, 1989). The benefits experienced can be financial, for example, reducing the costs of food and labour. The Petroleum Club in New Orleans cut food costs by 2% because of reduced wastage and lower weight loss during cooking (Bertagnoli, 1987). Another US operator, Gerard's of Burlington, Vermont, claimed that kitchenless restaurants supplied from a central *sous vide* production unit cut investment costs by two-thirds (Scarpa, 1988) while enabling one member of staff to serve 125 dinners in 20 minutes (Bertagnoli, 1987). In the UK, *sous vide* dishes have been bought in from a supplier across the English Channel for use in a chain of restaurants, the advantages being no problems with 'mercurial chefs', good portion control, low wastage, a wide menu range and speedy service (Glyn, 1993).

With fashions in restaurant styles changing so frequently, the flexibility of a wide range of prepared meals, the tight portion control and low food wastage provided by the *sous vide* method offer the restaurant operator significant tangible advantages while maintaining those more intangible advantages of quality and customer satisfaction.

(b) Hotel room service and banquets. At the Gatwick Hilton, UK, being able to prepare a batch of meals while the raw materials are most fresh on delivery, meant less wastage (Stacey, 1985); while at the Brussels Hilton, *sous vide* dishes provided easier and more flexible room service at all hours (Bertagnoli, 1987). Marriott Hotels decided to make use of *sous vide* products from commercial manufacturers to overcome a shortage in skilled labour, insisting on freezing to ensure food safety with time–temperature indicators to detect temperature abuse. They also initiated training with live demonstrations to overcome staff resistance and ensure familiarity (Schecter, 1990).

Another important hotel activity, banqueting – often only possible by using a make-shift cook–chill system (rechauffé) – has been greatly improved and made less risky by the use of *sous vide* processing. Advantages include: easier preparation for banquets at the Scharzenberg Palace, Vienna (Raffael, 1984); more banqueting business at the Brussels Hilton due to improved food quality; providing the right number of meals instead of guessing numbers with the consequent risk of high wastage (Coomes, 1994); easier organisation of banquet work and outside events at the Petroleum Club, New Orleans, and lower costs by not using expensive chefs for weekend work (Scarpa, 1988; Bertagnoli, 1987).

A wide-ranging case study on using the *sous vide* method by Johns *et al.* (1992) discussed the financial and labour benefits to be gained for room service, banqueting and restaurants in Holiday Inns in the UK. These included better use of capital and control of high-revenue areas, easier staff recruitment, more consistent assurance of food safety and quality, better career development for chefs and higher productivity.

2.4.3 Transport applications

Lack of space on aircraft and, to a lesser extent, trains means that meals must either be kept hot (a method now out of use) or be capable of reheating for service on-board. Marine transport will have more space for food-service but, especially on long cruises, will need to be self-sufficient in food supplies and so can make use of the same advantages of the *sous vide* method as a land-based restaurant.

(a) In-flight foodservice Competition between airlines can often depend on the standard of meals offered, especially for first-class travellers. The *sous vide* method can offer the requirements of high quality, the flexibility of a wide menu range and a convenient method of reheating and has been taken up with enthusiasm by the airlines but with strict attention to eliminating any risk of food poisoning (Anon., 1988c; Gostelow, 1989).

Using *sous vide* products from Culinary Brands in the USA has been seen as one way to persuade passengers into first class with higher quality prepared dishes (Anon., 1989b). Gehrig (1990) reported on its use for supplying airlines (Iberia and Swissair) at Malaga Airport, Spain from a central production unit at Madrid, stating that it met all requirements 'in terms of quality, flexibility, profitability and hygiene'. Others mention the use of the *sous vide* method by American Airlines (Coomes, 1994), Air France (Anon., 1988c) and British Airways (Raffael, 1985a).

(b) Railway foodservice Like the airlines, railways are always under pressure to provide higher quality meals. In France, SNCF, the national railway, has used *sous vide* meals on its high-speed TGV trains (Raffael, 1984; De Liagre, 1985; Chauvel, 1992). In Canada, VIA Rail used the system (Kalinowski, 1988; Bristol, 1989) and Irish Rail has used products manufactured by Ovac Star in Dublin to 'broaden the variety of dishes available' (Bacon, 1989). Gostelow (1990) mentions the use of *sous vide* meals by Swiss Railways with their policy of discarding unused pouches after two journeys to ensure no risks in food safety.

(c) Marine foodservice   An unusual application for *sous vide* foods has been for a yacht crossing the Atlantic with 30 days' supplies stored in the hull and kept cool by the sea water outside (MacNeil, 1987). A more conventional use has been on ferries operating between France and Corsica (Moisy, 1990) and ferries across the English Channel using products from Thomas Morel in the UK (Thomas Morel Ltd, 1994).

2.4.4 Institutional applications

The image of *sous vide* foods providing up-market restaurant dishes seems somewhat incongruous for use with institutional foodservice applications where costs are under tight control. However, when foodservice is contracted to third parties, buying in prepared meals such as *sous vide* products can be a cost-effective option depending on the amount of financial subsidy.

(a) Military foodservice   Especially when external contact has to be limited, providing a wide range of high-quality food for the military is very important for morale. *Sous vide* products would be easy to use in the field as well as tasting better than the usual military rations (Baird, 1990). In France, the air force has updated its foodservice and set up a 'pilot' in-house production unit to provide *sous vide* products for half of its output at first, hoping to increase this to 90% later (Schamberger, 1991).

(b) Industrial foodservice   In France, at the Peugeot-Talbot car factory at Poissy, the emphasis on productivity has been extended to include staff foodservice and the use of the *sous vide* method to achieve this. The advantages given were less repetitive preparation, reduced costs through improved productivity and better organoleptic, nutritional and hygienic qualities (Eustache, 1988; Anon., 1991; Defais and Elman, 1989). The central kitchen of the European Commission, Brussels, also makes satisfactory use of the *sous vide* process as part of its staff foodservice, although the recipes and quality control of raw materials need further development (Mouligneau, 1996).

(c) Hospital foodservice   For hospital patients, meals can often become the highlight of the day and assume an importance much greater than in normal life. The *sous vide* method has possibilities in satisfying patients' needs for high-quality foods as well as some of the medical and operational needs of the hospital itself. Sessions (1987) has pointed out some of the advantages of the *sous vide* method for hospitals; as a complement to cook–chill systems, providing special diet dishes and also as an extra source of income by catering for banquets, a common function in large teaching hospitals where research conferences are held. Choain and Noel (1989) also suggest a wide-ranging list of specific benefits for its use in hospitals. These

include easier supply of special dietary requirements such as low calorie, low fat or low salt meals, more flexibility by using meal assembly, higher quality of foods for patients with a stronger, more 'natural' taste and better portion control.

A French health clinic at Deauville has used a neighbouring hotel to supply it with *sous vide* foods where dietary constraints can be accommodated as well as providing meals which will encourage a healthier attitude towards eating (Raffael, 1985b). At the Carolus and Liduina Hospitals at Den Bosch and Boxtel, Netherlands, where *sous vide* products are produced in house, the emphasis is on providing a high quality of service to the customers (patients and staff) (Verbraken, 1993).

(d) School and college meal foodservice   In France, a centralised production for school meals at Nice includes using the *sous vide* method as well as other methods. Of the production of 17 000 to 18 000 meals per day, 60% are cooked with automated cooking equipment using the *sous vide* method, for up to 170 different dishes (Anon., 1988a; Ward, 1988). Another unit at Lyon designed to replace 56 school kitchens, supplies 200 school dining rooms in the area using traditionally and *sous vide* cooked foods with dieticians consulting children on their opinion of the food (Lepage, 1990).

The Catholic University of Leuven, Belgium, has a centralised production unit run by Alma University Restaurants to provide around 8000 meals per day for staff and students. They use the *sous vide* method extensively, with computers controlling not only the cooking process but the scheduling of food production and recipe formulation. Some claimed advantages are: enhanced flavour; lower weight loss; 20 to 30% for traditional versus 5% for *sous vide* foods (Martens, 1993; Wolthuis, 1993). Its use at the State University at Ghent, Belgium, has led to the advantages for the student customers of improved colour, taste, aromas, texture and nutrient retention in meals and, for the restaurateur, lower cooking losses (for meat, 8% for *sous vide* compared with 20% for conventional cooking), easier organisation of work but the necessity for strict hygiene training for all staff (Van Oyen, 1993).

2.5 Examples of the advantages and disadvantages of *sous vide* foods

These can be summarised from the previous sections:

2.5.1 Advantages

(a) Cost benefits

- Adding value to basic raw materials such as meat and fish by producing chilled prepared meals.

- Cutting food costs by reducing materials for enhancing flavours.
- Tighter portion control.
- Lower weight through retention of moisture in packaging.
- Lower wastage due to quicker response to consumer demand.
- Extension of shelf life leading to lower wastage during storage.
- Economies of scale for variable food costs using a centralised production unit to supply several outlets.
- Lower capital costs by centralising location and use of equipment for production.
- Option to eliminate central production unit by buying in meals from a manufacturer.

(b) Labour cost benefits

- Using less-skilled or fewer skilled staff.
- Easier staff recruitment.
- More rapid service of food to the consumer.
- Easier room service and banqueting preparation.
- Higher productivity in terms of food produced per man-hour.
- Easy incorporation into a meal assembly system.

(c) Quality benefits for food and service

- Reducing problems due to oxygen by vacuum packing.
- Reducing risk of contamination through use of vacuum packing.
- Claimed superior sensory qualities.
- Claimed superior nutritional qualities relevant to hospital patients.
- Enabling production of pre-cooked foods otherwise needing long processing times (e.g. dried vegetables).
- Higher quality of service due to less time on repetitive food preparation work.
- Higher perceived quality of food due to more time for presentation of food.
- Better career development for chefs.

(d) Marketing advantages

- Strong image of '*haute cuisine*' restaurant dishes with endorsement by respected chefs.
- Flexibility offered by a wide menu range.
- Innovative concept for the use of technology.
- Wide range of sizes from individual to multi-portion for foodservice.

2.5.2 Disadvantages

(a) Extra costs

- Cost associated with equipment and systems required for more stringent quality management systems (discussed in other chapters) to ensure food safety through monitoring raw materials, heat treatment storage, distribution, etc.
- Higher capital costs of equipment for preparation, vacuum packing, pasteurising, chilling and storage.
- Higher costs for packaging.
- Higher costs for using devices for detecting temperature abuse.

(b) Labour costs

- Costs of training for staff at all levels as part of the more stringent quality management systems.
- Developing suitable recipes.
- Overcoming staff resistance relating to deskilling.
- Costs of staff to liaise with suppliers and customers as part of the quality management system.

(c) Quality costs

- Increased risk of food poisoning if *sous vide* packs are subjected to temperature abuse.

(d) Marketing costs

- Ensuring reliable delivery of chilled foods over a wide geographical area.
- Overcoming fears of the consumer on food safety through education.
- Overcoming confusion between pasteurised *sous vide* products and sterilised 'boil-in-the-bag' products.
- Setting price of *sous vide* foods to ensure return on capital invested in equipment, training, etc.

2.6 Success and failure in the use of *sous vide* foods

The reservations regarding food safety discussed in other chapters have probably been among the main reasons that the *sous vide* method has not been the resounding commercial success anticipated by many in the United Kingdom and North America. Several manufacturing companies such as Home Rouxl in the UK, Culinary Brands in the USA and many in Canada have failed in the last few years.

In order to determine some of the factors which have influenced success and failure in the production and use of *sous vide* processed foods, it is worth considering how some manufacturers went out of business or how others took a different approach and have been more successful. These factors are often those which are less amenable to scientific research; factors determined by human behaviour, optimism, enthusiasm and resistance to change.

2.6.1 Australia

The *sous vide* method was under discussion in 1989 regarding its advantages and disadvantages, equipment and training requirements (Loughran, 1989). Interest in the *sous vide* method was sufficient for a conference of 100 delegates (Sumner, 1990) as well as training courses. Production companies such as Pasta Master, My Personal Chef and SuVide Australia were in operation aiming mainly at the foodservice sector (McHenry, 1990). Experience of the method in France, the United States and the United Kingdom was used as a framework for starting the development of a code of practice for manufacturers (Dunn, 1990) and for export controls by AQIS (Australian Quarantine and Inspection Service) based on HACCP methods (Abhayaratna, 1990). By 1992, the AQIS Code of Practice had reached draft form (Pickard, 1992). HACCP was also recommended for use by Australian producers with specific details for processing times and temperatures (Warne, 1990). The need for specialised training for manufacturers, users, educators and consumers was also emphasised by Dunn (1990).

In 1992, SuVide Australia, based in Victoria, was expecting to produce 5000 meals per day concentrating on foodservice users after trials at retail level had concluded that the marketing investment was too high. Overall the *sous vide* method had been 'very slow to become accepted for the mass market' (Pickard, 1992) although food companies were expecting its use to increase.

Pacific Dunlop thought that it could avoid the usual pitfalls by buying in expertise from the French *sous vide* company, Fleury Michon, by using its recipes, technology, distribution systems and chefs for its 'Chef to You' range. However, the A\$10 million factory at Echucha, Victoria, closed in 1996 after only one year of operation as a result of: supply problems, due to being in the wrong location; high costs, due to lack of automation; and poor sales, due to consumer perception that the product quality did not justify the selling price (Shoebridge, 1996).

More recent views (Caffin *et al.*, 1996) suggest that the problems in maintaining a reliable distribution and retailing, consumer scepticism and scaling up to industrial-scale production have held back the market in Australia. The increasing acceptance of quality management systems and recognising the benefits may change this.

2.6.2 Belgium

Alma University Restaurants at the Catholic University of Leuven have been pioneers in applying the *sous vide* method based on their scientific expertise to a real foodservice environment, student and staff feeding. Their approach has been to use a high degree of computer control to ensure consistent processing of foods and to use sophisticated computer modelling to predict the shelf life of particular *sous vide* products linked in to predicting heat transfer and quality characteristics. They also act as consultants to *sous vide* producers, having set up a competence centre to help communicate information more easily (Martens, 1996).

There are many examples of companies in Belgium using the *sous vide* method: Seffelaar and Looyen, a foodservice company, use a computer controlled system of tanks with hot or cold water for manufacturing *sous vide* products. Other *sous vide* operations are Het Bilt (Friese) and VACO (Geel) (Wijnia, 1992). Hot Cuisine (Gent) offer a range of *sous vide* components to allow the chef to assemble and garnish complete dishes as required (Hot Cuisine, 1996). *Sous vide* products from Ancora Cuisine (Grobbendonk) cover a range of needs from retail to foodservice including assembly to individual packs of meals for 7 days (Ancora Cuisine, 1996). The use of *sous vide* foods in Belgium can be gauged, for example, from the wide range of *sous vide* products available in many sizes, from individual 'snack' to multi-portion sizes from the Deliva company of Genk. This range comprises 19 soups, 30 snacks, 10 stews, 19 meat dishes, 19 meat dishes with sauce, 10 fish dishes, 3 fish dishes with sauce, 11 starch-based dishes, 35 sauces, 30 vegetables and 6 vegetables with sauce. Of the 38 meat-based dishes, 15 are beef, 11 veal, 5 lamb and 7 pork; of the poultry-based dishes, 8 are chicken, 11 turkey and 1 guinea fowl; and of the fish-based dishes, 2 are labelled fish, 5 are salmon, 5 cod, 2 halibut, 3 pollack, 3 dolphinfish (also known as dorade or mahi-mahi) and 2 scampi (Deliva, 1996).

2.6.3 Canada

In Canada, especially Quebec, the French influence meant that the pioneering work of Georges Pralus was quickly taken up so that by 1989 several companies had been formed. Cuisifrance and Charcuterie La Tour Eiffel produced *sous vide* products as part of their output, and others such as Letel, Cuisifrance, Jean Goubin Canada, Gastronomie Sous-Vide and Le Banquetier concentrated on chilled *sous vide* dishes with one company, New York Corp., specialising in frozen *sous vide* entrées. These companies decided to market products towards institutional use, as in France, but were only working at 30% capacity. Indicators for success were thought to be approval from the regulatory authority (Agriculture and Agri-Food Canada), using highly qualified personnel and a realistic advertising budget

(Bristol, 1989; Sacharow, 1988; Bangay, 1996). However, the technology of *sous vide* was being heavily used by some companies for marketing when foodservice operators only wanted 'a good product for the right price' (Bangay, 1996).

By 1991 only four of the ten Canadian *sous vide* companies started since 1986 were still in business (Barrett, 1991a). The European market for chilled foods was studied, noting especially the retailer-driven initiatives, distribution methods, consumer needs and culture influencing the chilled meals market. This led to the conclusion that only a strong partnership between manufacturer, distributor and retailer would bring success (Barrett, 1991b). Other reasons given for the failure of *sous vide* products in the market were due to hotels and restaurants often equating *sous vide* with 'boil-in-the-bag' sterilised foods of mediocre sensory quality (Campbell, 1993). Many companies had turned to freezing to overcome the potential food safety risks of temperature abuse in the cold chain, but all the companies mentioned above are now out of business. Only Sous Vide Canada based at Barrie, Ontario, is operational in Canada producing chilled and frozen *sous vide* products and offering training courses to foodservice operators (Bangay, 1996).

However, the *sous vide* method is being planned for use at a hotel restaurant, at the Inn at Manitou, Ontario, to satisfy health-conscious Canadians (Steen, 1995).

2.6.4 France

In France, the strong influence of the Ministry of Agriculture ensured that the *sous vide* method was covered by the regulations concerning the hygiene of premises for food production, vacuum packing of food and chilled foods. Premises for production of *sous vide* food are licensed and regulations enforced and monitored by the veterinary services from their regional laboratories (Hrdina-Dubsky, 1989; Baird, 1990; Majewski, 1990). The Ministry also gave details on the relevant temperatures to be monitored to emphasise the degree of control necessary during the preparation, cooking, cooling and storage of *sous vide* food (Beaufort and Guiliani, 1988) and the relationship between pasteurisation values and shelf-life dating methods (Beaufort and Rosset, 1989). They also publicised the importance of food hygiene in the restaurant sector and the requirement for *sous vide* food producers to obtain the *'marque de salubrité'* from the veterinary service, showing approval for their procedures (Ministère de l'Agriculture, 1990).

The 'unqualified success' of *sous vide* products in France with the food and foodservice industry has been contrasted with the situation in the United Kingdom of 'a crashing lack of interest' (Raffael, 1990). Various types of industrially prepared foods (including *sous vide*) are used in a wide range of outlets, allowing restaurateurs to assemble various meal components and so personalise the dishes available (Lacaberats, 1989; Anon., 1989c, 1991; Méhu, 1989; Tjomb, 1990; Cordier 1992). Prominent chefs such as Joël Robuchon, Michel Troisgros and Michel Oliver have also been influential in developing recipes (Anon., 1988b). As in North America, the choice of using chilled or frozen preservation for *sous vide* products has been discussed (Auliac, 1988) as well as the concept of the 'kitchenless restaurant' (Defais and Elman, 1989: Anon., 1990b)

Many producers have studied the market for *sous vide* products, finding that 35% was bought by single people and 33% by those without children: 65% of the volume sold was in individual portions and 35% as multi-portion products (Hrdina-Dubsky, 1989). These figures are comparable to those for Belgian consumers (Martens, 1996).

The overall figures for production of manufactured chilled foods show large increases. Falconnet and Litman (1996) report a rise in production of chilled cooked prepared meals by industrial manufacturers from 19 015 tonnes in 1990 to 28 629 tonnes in 1994, a 51% increase.

2.6.5 Irish Republic

Apart from a Dublin-based manufacturer, Ovac Star, supplying *sous vide* products to Irish Rail for use on its services from Dublin to Cork and Dublin to Tralee (Bacon, 1989), activity seems to be limited to some hotel use and plans for setting up production companies.

2.6.6 Japan

Georges Pralus had opened a school for *sous vide* training, prompting interest in the *sous vide* method (Pralus, 1993). Research has followed into studying the sensory, chemical, microbiological and nutritional characteristics of *sous vide* versions of Japanese dishes based on rice, vegetables, chicken and pork as well as Western-style restaurant dishes using beef, lamb, chicken, guinea fowl, salmon, scallop and mixed vegetables (Miyazawa *et al.*, 1994; Goto *et al.*, 1995; Yoshimura *et al.*, 1995). The style of Japanese food with its attention to flavours and textures combined with fresh products is seen to fit in well with the *sous vide* process.

2.6.7 Norway

The fishing industry is very important in Norway, especially for supplying canneries, but as traditional fishing has declined, farming salmon has increased to 20 000 tonnes per year (Bergslien, 1996). The *sous vide* process has been used to add value to farmed salmon by a US operator, Vie de France, as a frozen product which is claimed to have a much fresher taste (Coomes, 1994). There are also two producers of chilled and one of frozen *sous vide* products, supplying the Norwegian foodservice industry. In hotels and restaurants, there are estimated to be between 10 and 20 in-house users (Bergslien, 1997).

2.6.8 Spain

As well as its use for in-flight catering mentioned earlier, a central production unit (Anfitrios) has been set up for a trade show in Madrid. This unit supplies 31 outlets on the site, a variety of restaurants, cafés, etc. The *sous vide* system is used alongside other methods and helps to even out production for busy days. Retailing of *sous vide* products has been unsuccessful in Spain due to low sales volume but restaurants, schools, colleges and hotels are still being supplied (Hernandez, 1996). Another company has a small factory to produce *sous vide* cooked chicken breast intended for use in fast-food restaurants, etc. (Domenech-Pol and Calonge-Fornells, 1996).

2.6.9 United Kingdom

In 1986, Base One started manufacturing *sous vide* products for a few years supplying restaurants around London (Whitehall, 1987). At the same time, Albert Roux, a well-known restaurateur, opened a *sous vide* restaurant in London, Rouxl Britannia, with the intention to 'concentrate on friendly service assisted by the complete deskilling of the food preparation in the restaurant' (Hyam, 1986). This restaurant was supplied by a small production unit in London, later superseded by the larger Home Rouxl factory which hoped to supply 'a dozen or more' restaurants with a wide range of *sous vide* dishes (Anon., 1986; Arbose, 1987; Pring, 1986). Since 1983, Albert Roux had already been running a *sous vide* factory in south-west France, supplying SNCF and British Airways (Raffael, 1985b). In 1991, the Home Rouxl company was said to have 95% of the UK *sous vide* market with a range of 900 'classical French dishes' (Tutunjian, 1991). There were high expectations of market growth due to the 50% stake put into Home Rouxl by Scott's Hotels (Anon., 1990a). An extensive case study on the strategies of the operation was published by Johns *et al.* (1992). It was estimated that despite increased food costs, the net profit could rise from 10% to 29% by eliminating waste and halving the proportion of labour costs. Another estimate showed that a chef could spend half the time on presentation and marketing and a quarter on preparation and cooking with a *sous vide* operation instead of the reverse for a conventional operation.

The company finally decided to close in 1993. The main reasons for this were given as the unwillingness of the market to accept the *sous vide* concept and the losses requiring substantial funding in the poor economic climate then prevailing (Anon., 1993b). Other factors were consumer concern heightened by the hostile attitudes of the popular and technical press towards *sous vide* products (Edwards, 1989; Lacey, 1989).

The only current British manufacturer, Thomas Morel, using experience gained at Home Rouxl, started up in 1989 (Anon., 1993b). They have supplied a range of branded pubs, hotel and restaurant chains and ferry companies with chilled or frozen *sous vide* products (Thomas Morel Ltd, 1994). By concentrating on braised and stewed products which can stand more intense pasteurisation treatments and still maintain a high level of sensory quality and so offer a good margin of safety, they seem to have found a way to minimise fears on food safety. Four of their lamb dishes have also won prizes from meat-marketing organisations for product innovation. Their range includes 12 sauces and 28 entrées of which 9 are based on poultry, 3 on beef, 1 on venison, 1 on sausage and beer, 5 on vegetables, 4 on lamb, 3 on pork and 2 on seafood with 3 complete meals which include vegetables. These are made to order and distributed by a specialist food distributor (Larderfresh, 1997). This range is somewhat smaller than the range from Belgium mentioned earlier, perhaps reflecting the different states of the market in the two countries.

2.6.10 United States

The first commercial organisation to experiment with *sous vide* is said to have been the Holiday Inn near Greenville, South Carolina, working with the inventor of the AGS system, Ambrose T. McGuckian in 1970 (Farkas, 1988). The enterprise did not get commercial backing and, at the time, distributed transport was set up for frozen or unrefrigerated products, not chilled foods. In 1984, Vacutech of Vermont was an early producer of *sous vide* products supplying Gerard's, a restaurant in a large Radisson Hotel (Bangay, 1996). On the West Coast, after two years developing *sous vide* products and running abuse tests in collaboration with the University of California, Davis, Culinary Brands with backing from Nestlé, opened a large production plant in 1989 to much publicity (Lawrence, 1989; Hrdina-Dubsky, 1989; Otto, 1989). They advocated the concept of meal assembly to create menus from modular components to give their clients flexibility. At first, their business supplied the West Coast using time–temperature indicators on each pouch to detect any possible temperature abuse during the distribution chain (Parsons, 1989). They tried supplying several retailers on the East Coast but despite good reaction to the quality of the products, the costs of chilled distribution from their plant in California to the East Coast and ensuring continuity of supply proved to be too great and the experiment was ended (Millstein, 1990). To allay consumer and client fears on food safety, they moved towards freezing the *sous vide* products to supply Marriott hotels over a wider area (Anon., 1990c) and Nestlé's foodservice division, Stouffer's, regretting that they had not started out like this.

To increase sales, they also cut wholesale prices of portions to the $2 to $3.50 range compared with the initial start-up prices in 1989 of $3.50 to $6 per serving. Eventually the company split up in 1992 due to 'bad luck and bungled marketing' but with optimism that the process would ultimately be successful (Birmingham, 1992).

On the East Coast, Grace Culinary Services, a subsidiary of W.R. Grace, the packaging manufacturers, began production in 1989 using CapKold and *sous vide* processes. They produced 15 000 pounds of food per day covering a wide range of entrées, sauces, soups, pasta, salads and bakery products, using cook–chill, cook–freeze and *sous vide* processes, supplying customers in restaurants, hotels and supermarkets as well as their own chain of American Café restaurants (Lingle, 1991; Lyman, 1991; Hudson, 1993). They have more recently moved to focusing only on the foodservice market (Bangay, 1996).

GourmetFresh of Florida also set up operation in 1989 and were hoping to produce 6000 units per hour with 70% for retail and 30% to foodservice (Hauck, 1992). They used preformed trays for single portions and pouches for multi-portions and emphasised their commitment to safety with USDA on-site inspection of products before and after cooking combined with the use of time–temperature indicators on packs to detect any handling abuse during distribution (Rice, 1991).

In 1990, Vie de France opened a *sous vide* factory next to their bakery in Virginia with a peak capacity of 70 000 meals per day, relying on their experience of *sous vide* production in France since 1982. Initially, their relatively small range of 20 dishes was forecast to be priced at between $1.25 and $5.50 per meal (Lehr, 1990). Like GourmetFresh, they have placed great importance on quality management during production and distribution, training and use of computers for control with most of the output frozen using liquid nitrogen (Przybyla, 1990). Consumers are said to be unable to detect any difference between chilled and frozen products (Birmingham, 1992). Knowing that the reheating stage is crucial to ensure that high quality is retained for the consumer, Vie de France have designed a reheating unit for foodservice users for heating up to 200 meals per hour (Baird, 1990; Lehr, 1990).

In 1994, it was estimated that the *sous vide* market was $30 million annually with Vie de France taking two-thirds with the remainder divided between Grace Culinary (now Townsend Culinary) and Gerard's of Vermont (Coomes, 1994).

2.6.11 Reasons for success and failure

To summarise, the reasons contributing to failure were:

- Low market acceptance due to reservations of users and retailers on food safety.

- An emphasis on marketing the technology and not its benefits.
- Incorrect price/quality ratio giving poor profit margins.
- Lack of profitability due to high start-up investment costs linked to expected short pay-back periods.
- Professional resistance to pre-cooked foods.
- Costs and reliability of distribution over a wide area.
- Consumer concerns on food safety and lack of product knowledge.

Some reasons contributing to success were:

- A realistic attitude towards likely sales volume.
- Supplying foods meeting customer requirements whether *sous vide* processed or not.
- Emphasising advantages of portion control, reduction of labour costs and increased convenience as opposed to the '*haute cuisine*' image.
- Selling the product at a realistic price.

Light and Walker (1990) have investigated many aspects of implementing cook–chill systems, based on evidence from a survey. These covered the management factors affecting the successful introduction of a cook–chill system, the financial aspects associated with the introduction of cook–chill units, how products and menus were developed as well as the technical aspects. Their findings would also be applicable to the *sous vide* method.

2.7 Conclusions

The *sous vide* method has been shown to have many advantages for the foodservice operator because of the many gains in flexibility for meal provision and reducing costs but at the cost of a much higher standard of quality management to overcome the undoubted microbiological risks. However, the implementation of a quality management system is now almost a necessity for any organisation in the food industry.

Many companies have made much of their connections with well-known chefs or were run by people with a very high level of culinary and gastronomic expertise but in the end, the question arises as to whether consumers really care that much about the technology behind how a food is produced. Certainly, many restaurateurs have been unwilling to admit they were using the products (Bacon, 1990) but as long as the meal fulfils or exceeds consumers' expectations based on their perception of the balance between quality and cost, consumers would probably be satisfied.

Particularly for the companies starting up from scratch, their enthusiasm may have influenced their business approach and expectations of what premium could be put on the selling price in order to finance the necessary capital costs. These would include the equipment for vacuum-packing, low-temperature cooking, rapid cooling, reliable refrigerated storage, transport and reheating with all the monitoring and recording equipment necessary to fulfil 'due diligence' requirements for quality management systems required by regulatory authorities and often by the large retailers. In addition, the recruitment and training of personnel had to be that much more careful and thorough than usual to ensure the high level of hygiene and management.

For marketing, companies have relied very much on using the well-known chefs to develop and endorse a wide range of recipe dishes, hoping to bring to the mass market the gastronomic excellence normally only to be found at a top-class restaurant at a much higher price. This method worked well in France but not in the United Kingdom (Raffael, 1990). However, if the best chefs are given the best ingredients, they are bound to produce high-quality food no matter what system of preparation they are using.

In the hotels where *sous vide* has been used for room service, breakfasts and banquets, and in transport foodservice, consumer expectations take into account that there is unlikely to be a chef producing their meals to order, but in the restaurant, they expect a chef, so the 'kitchenless' restaurant is probably not yet acceptable. *Sous vide* has focused on improving the sensory quality of foods, but work on determining the relative importance of those factors influencing consumer attitudes towards the 'eating out' experience has indicated that the actual sensory quality of the meal can often be only a small part of that totality of factors that also go to make up the overall experience that is remembered. These other factors are often social – the company of others at the meal, the attitude of the waiting staff, the ambience and decoration (Bell *et al.*, 1994; Reeve *et al.*, 1994; Pierson *et al.*, 1995). These sensory aspects are discussed in another chapter of this book.

In contrast to the experience in the UK and North America, foodservice operators in France seem to have taken a much more pragmatic view towards providing consumer satisfaction. They have moved towards the meal assembly concept (Anon., 1989c, 1991; Cordier, 1992) although Nestlé are offering their COMPASS meal assembly system in Canada (Campbell, 1993). Instead of trying to force all foods through the *sous vide* system, French foodservice operators have combined what is best for each type of food chosen from the wide range of chilled, frozen and sterilised prepared foods available from many companies at various levels of preparation and convenience. In addition, many new types of minimally processed food may come onto the market to augment this range. Chefs can then use their expertise in presentation to assemble dishes according to the consumers' needs at many different market levels from restaurant chains to institutional and staff feeding. This offers the flexibility often lost by a 'system' which is technology-led rather than related to consumer needs. The availability of a relatively small range of base dishes of protein and starch foods with a range of sauces and garnishes, can provide a huge range of dishes for the consumer while still allowing good management of quality and tight control over wastage. Alongside providing high-quality dishes where the cost to the consumer is not so important, the future for the successful use of *sous vide* processed foods may lie in meal assembly.

Acknowledgement

The authors wish to acknowledge with thanks, the financial support of the Worshipful Company of Cooks, London.

References

Abhayaratna, N. (1990) *Sous vide*: Regulations for export, in Sumner, J., *op. cit.*, pp. 58–62.

Ancora Cuisine (1996) Publicity material, Grobbendonk, Belgium.

Anon. (1986) Sweet talk at Chef's Conference. *Chef*, April, 8–9.

Anon. (1987) Sealed with a hiss. *Meat Processing*, July, 66.

Anon. (1988a) Cuisine Centrales – Les Solutions Sogeres. *Néo Magazine*, (186), May, 61–63.

Anon. (1988b) Technologie nouvelle pour nouvelles mentalités. *Néo Restauration*, (183), February, 61, 64, 66.

Anon. (1988c) *Sous vide*: the next step? *I.F.C.A. (In-Flight Catering Association) Quarterly Review*, 3 (3), July, 11–12.

Anon. (1989a) *Sous vide*. *Catering*, November, 130.

Anon. (1989b) *Sous vide*: A new look at meal in a pouch. *Airline, Ship & Catering Onboard Services*, 21 (4), April, 42–43.

Anon. (1989c) Plats cuisinés – Un service à la carte. *Néo Magazine*, (199), June, 91–95.

Anon. (1990a) *Sous-vide*? Be serious . . . *Caterer & Hotelkeeper*, 13 September, 9.

Anon. (1990b) Restauration sans Cuisson – Une Offre Complete. *Néo Restauration*, (213), 8 June, 46–47.

Anon. (1990c) Supplier to Marriott opts for frozen, citing safety concerns over chilled. *Frozen Food Age*, 38 (10), 4, 62.

Anon. (1991) Plats Cuisinés – L'essor des produits-services. *Néo Restauration*, (237), 18 October, 50–55.

Anon. (1993a) Test van luchtledig bereide gerechten. *Test-Aankoop Magazine*, (352), 14–20.

Anon. (1993b) Packing it in. *Caterer & Hotelkeeper*, 30 September, 48–50, 52.

Arbose, J. (1987) À Belle Époque for fast foods? *International Management*, July/August, 38–39.

Auliac, A. (1988) Le conditionnement des plats cuisinés *sous vide*: un choix stratégique. *Industries Alimentaires et Agricoles*, 105 (4), 271–273.

Bacon, F. (1989) *Sous-vide* on track. *Caterer & Hotelkeeper*, 2 November, 14.

Bacon, F. (1990) Working to Rouxl. *Caterer & Hotelkeeper*, 22 February, 45–47.

Baird, B. (1990) *Sous vide*: What's all the excitement about? *Food Technology*, 44 (11), 92, 94, 96.

Bangay, L. (1996) The state of *sous vide* in North America, in *Proceedings of Second European Symposium on Sous Vide*, 10–12 April 1996, Alma University Restaurants/FAIR, University of Leuven, Belgium, pp. 239–249.

Barrett, H. (1991a) *Sous Vide* chilled foods in Canada, in *Proceedings of CAP '91 – 6th International Conference on Controlled/Modified Atmosphere/ Vacuum Packaging*, 9–11 January, The San Diego Hilton, San Diego, California, USA, 1991, Schotland Business Research Inc., Princeton, USA, pp. 217–228.

Barrett, H. (1991b) Transferring the European success to North America: consumer marketing insights, in *Proceedings of Pack Alimentaire '91 – 5th Annual Food & Beverage Packaging Expo and Conference*, 23–25 April, The Rivergate, New Orleans, LA., USA, 1991, Innovative Expositions Inc., Princeton, USA, Session C-2, 8 pp.

Beauchemin, M. (1990) *Sous-vide* technology. *Q.C.R & D Research Bulletin*, **17** (1), 1–11.

Beaufort, A. and Guiliani, L. (1988) Allongement du delai de consommation de plats cuisiné à l'avance. *Industries Alimentaires et Agricoles*, **105** (4), 245–248.

Beaufort, A. and Rosset, R. (1989) Durée de vie des plats cuisinés *sous vide* réfrigérés – Adaptation de la réglementation française. *Industries Alimentaires et Agricoles*, **106** (6), 475–477.

Bell, R., Meiselman, H.L., Pierson, B.J. and Reeve, W.G. (1994) The effects of adding an Italian theme to a restaurant on the perceived ethnicity, acceptability, and selection of foods. *Appetite*, **22**, 11–24.

Bergslien, H. (1996) *Sous vide* treatment of salmon (*Salmon salar*), in *Proceedings of Second European Symposium on Sous Vide*, 10–12 April 1996, Alma University Restaurants/FAIR, University of Leuven, Belgium, pp. 281–291.

Bergslien, H. (1997) Personal communication, Norconserv, Stavanger, Norway, 9 May.

Bertagnoli, L. (1987) Pouches gain ground in foodservice. *Restaurants & Institutions*, **97** (10), 178–180, 184.

Birmingham, J. (1992) Whatever happened to *sous vide*? *Restaurant Business*, **91** (6), 64, 65, 68, 72.

Bjorkman, A. and Delphin, K.A. (1966) Sweden's Nacka hospital food system centralizes food preparation and distribution. *Cornell Hotel & Restaurant Administration Quarterly*, **7** (3), 84–87.

Bristol, P. (1989) *Sous vide* – Gourmet meals in a vacuum package. *Food in Canada*, **49** (6), 15–18.

CSO (Central Statistical Office) (1996) *Social Trends*, HMSO, London.

Caffin, N.A., Morrison, P.A. and Atkinson, L. (1996) An Australian perspective on *sous vide* processing, in *Proceedings of Second European Symposium on Sous Vide*, 10–12 April 1996, Alma University Restaurants/FAIR, University of Leuven, Belgium, pp. 309–310.

Campbell, L. (1993) Swingin' with *Sous-Vide*. *Foodservice and Hospitality*. **26** (2), 88, 92, 96.

Chauvel, A. (1992) Servair sur tous les fronts. *Néo Restauration*, (243), 21 February, 28–30.

Choain, F. and Noel, P. (1989) *Le Sous-Vide en Cuisine*. – *Des fondements scientifiques aux applications pratiques*, Editions Jacques Lanore: Malalkoff, France, 207 pp.

Coomes, S. (1994) *Sous Vide* – Closing the Gap Between Truth and Reality. *The Consultant*, Winter, 27–29, 58.

Cordier, F. (1992) Dossier Plats Cuisinés. *Néo Restauration*, (249), 27 May, 53–55, 58–60, 63–64.

Creed, P.G. (1996) *The Sous Vide Method – An Annotated Bibliography*. Worshipful Company of Cooks Centre for Culinary Research, Bournemouth University, Poole, UK, 156 pp.

Daniels, D. (1988) *Sous Vide* & CapKold . . . Two approaches to cooking 'Under Vacuum'. *The Consultant*, **21** (2), Spring, 26–28.

De Liagre, C. (1985) Train fare. *Home and Garden*, **157**, November, 202–205, 271–272.

Defais, C. and Elman, F. (1989) La Restauration sans Cuisson. *Néo Magazine*, (203), 15 October, 53–66.

Deliva (1996) Price list, April 1996, Genk, Belgium, 16 pp.

Domenech-Pol, N. and Calonge-Fornells, R. (1996) Simple and practical HACCP implementation for the *sous vide* chilled high quality product 'Natural chicken breast with skin', in *Proceedings of Second European Symposium on Sous Vide*, 10–12 April 1996, Alma University Restaurants/FAIR, University of Leuven, Belgium, pp. 337–338.

Dréano, C. (1988) Cuisson de plats cuisinés *sous vide*. Procédé de cuisson-refroidissement Thermix. *Industries Alimentaires et Agricoles*, **105** (4), 277–279.

Dunn L. (1990) *Sous vide*: Regulatory, storage and training, in Sumner, J., *op. cit.*, pp. 51–57.

Edwards, D. (1989) *Sous vide* is potentially lethal. *Booker Food*, (16), May, 43.

Electricity Council (1982) *Planning for Cook-Chill*, Electricity Council, London.

Eustache, D. (1988) Après la 4e gamme: Les plats cuisinés frais *sous vide*: une realité au quotidien. *Industries Alimentaires et Agricoles (IAA)*, **105** (4), 267–268.

Falconnet, F. and Litman, S. (1996) Le marché des produits *sous vide*, in *Proceedings of Second European Symposium on Sous Vide*, 10–12 April 1996, Alma University Restaurants/FAIR, University of Leuven, Belgium, pp. 231–238.

Farkas, D. (1988) Fresher under pressure. *Restaurant Hospitality*, **72** (7), 126–128.

Gehrig, B.J. (1990) Prinzip und Einsatzmöglichkeiten des Sous-Vide-Verfahrens. *Mitteilungen aus dem Gebiete der Lebensmitteluntersuchung und Hygiene*, **81** (6), 593–601.

Ghazala, S. (1994) New packaging technology for seafood preservation: shelf life extension and pathogen control, in *Fisheries Processing: Biotechnological Applications* (ed. A.M. Martin), Chapman & Hall, London, pp. 82–110.

Gledhill, B. (1991) Filling a vacuum. *Caterer & Hotelkeeper*, 4 July, 69, 70, 72.

Glyn, I.R.H. (1993) *Sous vide* far and wide. Letter to *Caterer & Hotelkeeper*, 14 October, 18.

Gostelow, M. (1989) Onboard Service . . .The Swiss Way. *Airline, Ship & Catering Onboard Services*, 21 (5), May/June, 12.

Gostelow, M. (1990) Swiss Railway Catering Adopts a New Look. *Airline, Ship & Catering Onboard Services*, 22 (2), February, 40.

Goto, M., Hashimoto, K. and Yamada, K. (1995) Difference of ascorbic acid content among some vegetables, texture in chicken and pork, and sensory evaluation score in some dishes between vacuum and ordinary cooking. *Nippon Shokuhin Kagaku Kogaku Kaishi*, 42 (1), 50–54.

Goussault, B. (1992) Le défi cuisson enjeux technologies et marketing . . . en viandes cuites et plats cuisines frais. Paper presented at *Les Journées d'Études du CETEVIC*, Paris, 24 and 25 March 1992, 8 pp.

Hackney, C.R., Rippen, T.E. and Ward, D.R. (1991) Principles of pasteurisation and minimally processed seafoods, in *Microbiology of Marine Food Products* (eds D.R. Ward and C.R. Hackney), Van Nostrand Reinhold, New York, pp. 355–371.

Hauck, K. (1992) Fresh by any other name. *Prepared Foods*, 161 (11), 43.

Hernandez, C. (1996) *Sous vide* practice at Anfitrios, in *Proceedings of Second European Symposium on Sous Vide*, 10–12 April 1996, Alma University Restaurants/FAIR, University of Leuven, Belgium, pp. 251–256.

Hot Cuisine (1996) Culinary Assembly brochure, Gent, Belgium.

Hrdina-Dubsky, D.L. (1989) *Sous vide* finds its niche. *Food Engineering International*, 14 (7), 40–42, 44, 46, 48.

Hudson, B.T. (1993) Industrial Cuisine. *Cornell Hotel & Restaurant Administration Quarterly*, December, 73–79.

Hyam, J. (1986). Rouxl Britannia opens. *Caterer & Hotelkeeper*, 24 July, 7.

James, S.J. and Evans, J.A. (1990) Consumer handing of chilled foods, in *Process Engineering in the Food Industry – 2, Convenience Foods and Quality Assurance* (eds R.W. Field and J.A. Howell), Elsevier Applied Science, London, pp. 98–109.

Johns, N., Wheeler, K. and Cowe, P. (1992) Productivity angles on *sous vide* production (Home Rouxl and Scott's Hotels Ltd), in *Managing projects in hospitality organisations* (eds R. Teare, D. Adams and S. Messenger), Cassell, London, pp. 146–168.

Jones, S. (1996) The return of *sous vide*. *Progressive Grocer*, (75), May, 187.

Kalinowski, T. (1988) Ready, set, serve. *Canadian Hotel & Restaurant*, 66 (6), 24–26.

Lacaberats, R. (1989) Du plat cuisiné cuit *sous vide* aux entrées et plats préparés industriels. *Viandes et Produits Carnes*, 10 (4), 149–153.

Lacey, R. (1989) *Safe Shopping, Safe Cooking, Safe Eating*, Penguin Books, London, pp. 41–42.

Larderfresh (1997) Price list, 26th January, Warley, West Midlands, UK, pp. 14–15.

Lawrence, B. (1989) *Sous vide* speeds up. *Food Business*, 19 June, 28–30.

Lehr, H. (1990) Vie de France: creating a new market for *sous vide*. *Food Engineering International*, 15 (5), 45–48.

Leistner, L. and Gorris, L.G.M. (eds) (1994) *Food Preservation by Combined Processes – Final Report of FLAIR Concerted Action No. 7, Subgroup B*, European Commission, DG XII (EUR 15776 EN). 100 pp.

Lepage, V. (1990) Cuisine Centrale de Lyon – Des Petits Plats en Salle Blanche. *Néo Magazine*, (206), January, 44, 53.

Light, N. and Walker, A. (1990) *Cook–Chill Catering: Technology and Management*, Elsevier Applied Science, London.

Lingle, R. (1991) New generation foods? Just say Grace. *Prepared Foods*, 160 (7), 58, 59, 61, 63, 64.

Loughran, F. (1989) *Sous vide – Panacea or Gimmick?* The Centre for Hospitality Research and Service, Department of Hotel and Tourism Management, Faculty of Business, Footscray Institute of Technology, Melbourne, Australia, 25 pp.

Lyman, W. (1991) What's it going to be: flexible, rigid, microwaveable, dual ovenable or what?, in *Foodplas VIII – 91: Plastics in food packaging*, Plastics Institute of America, 5–7 March, Orlando, Florida, Technomic Publishing: Lancaster, PA, USA, pp. 161–174.

MacNeil, K. (1987) French Cuisine Sets to Sea in Pouches. *New York Times,* Wednesday, 29 April.

Majewski, C. (1990) *Sous vide* – New Technology Catering. *Environmental Health,* **98** (4), 100–102.

Martens, T. (1993) JIT, CIM and '*sous vide*', in *Proceedings of First European Symposium – 'Sous Vide' Cooking,* 25–26 March 1993, Alma University Restaurants/FLAIR, University of Leuven, Belgium, 10 pp.

Martens, T. (1995) Current Status of *Sous Vide* in Europe, in *Principles of Modified-Atmosphere and Sous Vide Product Packaging* (eds J.M. Farber and K.L. Dodds), Technomic Publishing, Lancaster, USA, pp. 37–68.

Martens, T. (1996) *Sous vide*: an opportunity for small and medium sized companies, in *Proceedings of Second European Symposium on Sous Vide,* 10–12 April 1996, Alma University Restaurants/FAIR, University of Leuven, Belgium, pp. 259–270.

McGuckian, A.T. (1969) The A.G.S. Food System – Chilled pasteurised food. *Cornell Hotel & Restaurant Administration Quarterly,* May, 87–92, 99.

McHenry, M. (1990) *Sous vide*: Overview, in Sumner, J., *op. cit.,* pp. 1–7.

Mehu, J. (1989) *Sous vide*: premier bilan. *Revue Industries Agro-Alimentaires (RIA),* (417), 12–13.

Mertens, B. (1995) Hydrostatic pressure treatment of food: equipment and processing, in *New Methods of Food Preservation* (ed. G.W. Gould), Blackie Academic & Professional, London, pp. 135–158.

Millross, J., Speht, A., Holdsworth, K. and Glew, G. (1973) *The Utilization of the Cook-Freeze Catering System for School Meals,* University of Leeds, Leeds, UK.

Millstein, M. (1990) Culinary Brands drops bicoastal delivery. *Supermarket News,* **40** (26), 25 June, 36, 38.

Ministère de l'Agriculture (1990) *Une compétance et une experiénce au service de la restauration hors foyer,* Services Vétérinaires de la DDAF de la Gironde, Bordeaux. 4 pp.

Mitchell, A. (1997) The market for chilled prepared meals, *Seminar on Chilled Prepared Meals,* March 1997, Leatherhead Food Research Association, Leatherhead, Surrey, UK.

Miyazawa, F., Eto, K., Kanai, M., Kashima, M., Sakai, H., Koike, Y. and Tani, T. (1994) Microbial contaminants in foods prepared by vacuum-packed pouch cooking (*sous-vide*). *Journal of the Hygienic Society of Japan,* **35** (5), 530–537.

Moisy, S. (1990) La SNCM choisit le *sous-vide*. *Néo Restauration,* (216), 21 September, 47.

Mouligneau, M. cited by Gorris, L.G.M. (1996) 2nd European Symposium on *Sous Vide. Trends in Food Science and Technology,* **7** (9), 303–306.

Otto, A. (1989) *Sous-vide* or not *sous-vide*? *Prepared Foods,* **158** (6), 38–39.

Parsons, H. (1989) Seven the Hard Way. *Meat Processing,* **28** (10), 16, 21, 22, 24, 46.

Peyron, A. (1996) Comparaison entre plusieurs équipements de chauffage, in *Proceedings of Second European Symposium on Sous vide,* 10–12 April 1996, Alma University Restaurants/FAIR, University of Leuven, Belgium, pp. 29–42.

Pickard, D. (1992) *Sous-vide* technique taking rapid hold. *Food Industry* (Australia), January, 17–18, 20.

Pierson, B.J., Reeve, W.G. and Creed, P.G. (1995) 'The Quality Experience' in the food service industry. *Food Quality and Preference,* **6** (3), 209–212.

Pinot, J.-P. (1988) Cuisson *sous vide* par le systeme Barriquand. *Industries Alimentaires et Agricoles,* **105** (4), 305–307.

Pralus, G. (1993) '*Sous vide*' cooking, in *Proceedings of First European Symposium – 'Sous Vide' Cooking,* 25–26 March 1993, Alma University Restaurants/FLAIR, University of Leuven, Belgium, pp. 142–150.

Pré, G. (1992) Trends in processing and packaging technologies, in *Proceedings of 'Packaging of Food',* London, March, Euro Food Pack Group, PIRA (Packaging Industry Research Association), Leatherhead, Surrey, UK, 12 pp.

Pring, A. (1986) *Sous-vide*: Boon or threat? *Caterer & Hotelkeeper,* 24 July, 19.

Pröller, T. (1990) Frisch-Gar-System für die Fleischerei: *Sous vide* – Frische, die überzeugt. *Fleischerei,* **41** (5), 346–348.

Przybyla, A.E. (1990) New *sous-vide* plant will have 70,000 meal per day capacity. *Food Engineering,* **62** (8), 103–104.

Raffael, M. (1984) Revolution in the kitchen. *Caterer & Hotelkeeper,* 16 August, 34, 35.

Raffael, M. (1985a) Cooking in a vacuum – perhaps the best thing since sliced brioche. *Guardian*, London, 27 September, 23.

Raffael, M. (1985b) How Normandy Hotel serves a health centre. *Caterer & Hotelkeeper*, 7 November, 73, 75.

Raffael, M. (1990) *Sous-vide* – a British vacuum. *Caterer & Hotelkeeper*, 15 November, 22.

Ready, C.A. (1971) Method of preparing and preserving ready-to-eat foods. US Patent 3,607,312, 4 pp.

Reeve, W.G., Creed, P.G. and Pierson, B.J. (1994) The restaurant – a laboratory for assessing meal acceptability. *International Journal of Contemporary Hospitality Management*, 6 (4), iv–vii.

Rice, J. (1991) *Sous vide* – new breed retail products. *Food Processing (USA)*, 52 (5), 39–40.

Riell, H. (1988) Lateline – Phoenix rise to occasion with a new *sous vide* chain. *Restaurant Business*, 87 (3), 10 February, 302.

Sacharow, S. (1988) Flexible plans. *Packaging Week*, 4 (14), 17.

Scarpa, J. (1988) *Sous vide* – Fresh Technology. *Restaurant Business*, 87 (4), 136, 138, 140.

Schamberger, J.-Ch. (1991) Décollage de la cuisson *sous vide*. *Néo Restauration*, (226), 19 April, 12.

Schecter, M. (1990). A Revolution in a Pouch? *Food Management*, 25 (10), 88, 92, 96.

Schwarz, T. (1988) *Sous vide* guidelines. *Restaurant Business*, 87 (6), 10 April, 94, 96, 97.

Sessions, P. (1987) Sensible *sous vide*. *Hospital Caterer*, May/June, 14.

Shoebridge, N. (1996) The convenience food that got left in the cold. *Business Review Weekly* (Australia), 22 April, 108, 110.

Sicot, D. (1990) Cuire au Dixième de Degré Pres. *Néo Restauration*, (206), 16 March, 112, 115, 117, 118.

Stacey, C. (1985) Hilton Pioneers. *Caterer & Hotelkeeper*, 7 November, 67, 70.

Steen, P. (1995) Northern harvest. *Foodservice and Hospitality*, 28 (6), 32–34.

Sumner, J. (ed.) (1990) *Proceedings of a conference – Sous vide in Australia*, Clunies Ross House, Melbourne, 14 November 1990, M & S Food Consultants, Albert Park, Victoria, Australia.

Swientek, R.J. (1989) *Sous vide* – refrigerated upscale food in a vacuum pouch. *Food Processing* (USA), 50 (6), 34, 36, 38–40.

Thomas Morel Ltd. (1994) *Newsletter*, No. 1, Redditch, UK, p. 4.

Tjomb, P. (1990) Cuisson *sous vide* soft ou hard. *Revue Industries Agro-Alimentaires (RIA)*, (441), 30–31.

Tutunjian, J. (1991) Hotelympia '92. *Canadian Hotel & Restaurant*, 69 (12), 26–27.

Urch, M. (1991) Processing developments on show in France. *Seafood Processing & Packaging International*, 5 (3), 26–27.

Van der Leest, A.J. (1985) Pasteurised meat products in co-extruded bags, in *Proceedings of International Symposium – Trends in Modern Meat Technology*, 30 October–1 November 1984, Wageningen, Netherlands. Pudoc, Wageningen, pp. 93–95.

Van Oyen, P. (1993) Wat is vacuumkoken? *De Officiële Horeca*, 4, 7–9.

Varney-Burch, A. (1991) Macropak reflects 'green' concerns. *Seafood Processing & Packaging International*, 5 (3), 18–19.

Varoquaux, P., Offant, P. and Varoquaux, F. (1995) Firmness, seed wholeness and water uptake during the cooking of lentils *(Lens culinaris cv. anicia)* for 'sous vide' and catering preparations. *International Journal of Food Science & Technology*, 30 (1), 215–220.

Varoquaux, P.J.A. and Nguyen-The, C. (1994) Vacuum processing: a new concept for precooked fruit and vegetables. *Food Science & Technology Today*, 8 (1), 42–49.

Verbraken, E. (1993) De hele voedingslijn in handen van de keuken. *Grootkeuken*, (7/8), 17–19.

Ward, P. (1988) Foreign Flair. *Caterer & Hotelkeeper*, 15 December, 53.

Warne, D. (1990) *Sous vide*: The HACCP approach, in Sumner, J., *op. cit.*, pp. 17–32.

Whitehall, B. (1987) *Sous-vide* report – all systems go! *Caterer & Hotelkeeper*, 23 July, 48, 49, 51.

Wiesel, V. (1987) Einsatz kontinuierlich arbeitender Koch-, Kühl- und Pasteurisieranlagen für Wurst- Kaliberware und Würstchen-Packungen. *Fleischwirtschaft*, 67 (4), 386, 388–390, 452.

Wijnia, J. (1992) Produktveiligheid staat centraal bij *sous vide* koken. *Food Management*, (2), 23, 24, 27.

Wolthuis, A. (1993) Achtduizend *sous vide* maaltijden per dag. *Grootkeuken*, (6), 19–22.
Yoshimura, M., Shono, Y. and Yamauchi, N. (1995) Influence of vacuum packaging and processing temperature on the quality change of Mitsuba. *Nippon Shokuhin Kogaku Kaishi*, **42** (8), 588–593.

3 Sensory and nutritional aspects of *sous vide* processed foods

PHILIP G. CREED

3.1 Introduction

The *sous vide* method for preparing chilled meals came to prominence due to the efforts of respected chefs. After the initial enthusiasm about the products' sensory and nutritional qualities came the realisation that under poorly managed conditions, the potential risk of food poisoning could be significant enough to warrant studies into the microbiological safety aspects of the process, the subject of other chapters in this book. This aspect has taken over research in this area to a large extent with about 70 publications in scientific journals up to 1995 compared with 30 on the sensory and nutritional aspects of *sous vide* processed foods (Creed, 1996). Regulators have brought in ever more stringent combinations of times and temperatures to be as near certain as possible that food poisoning could not occur. Many of these treatments would have a deleterious effect on those sensory aspects which first brought it to the food industry's attention. This chapter attempts to assess whether the sensory and nutritional qualities of *sous vide* processed foods are as claimed and how far the effects of ensuring a 'safe' heat treatment act against these 'beneficial' qualities. Indeed, this point of conflict for *sous vide* processing has been seen by Martens (1995) as the difference in attitude between those who believe food safety is paramount (United Kingdom, United States of America, Canada and Australia) and those who believe that enjoying food is more important (France and Belgium).

3.2 Background to the *sous vide* process

The discovery of the *sous vide* method for food preparation in the mid-1970s is generally credited to Georges Pralus, a chef from Briennon in France. However, a patent by Ready a few years earlier in 1971 assigned to W.R. Grace, the large American packaging company, contained the basic concepts of the process: vacuum packing raw food materials in laminated plastic packaging material able to withstand the temperatures of the hot water used for cooking, followed by cooling and storage. Other systems used before this time contained some elements of the *sous vide* method, for example: Nacka (Bjorkman and Delphin, 1966) developed in Sweden and AGS (McGuckian, 1969) in the United States. Later systems used for producing ready-prepared potatoes (Poulsen, 1978), cooked meats (Buck *et al.*, 1979) or large-scale production of meals such as CapKold (Daniels, 1988) also contained some similar elements.

Despite this, Pralus must be given the credit for applying the concept to practical problems using his culinary expertise to overcome areas of recipe development which would have been beyond the reach of a food technologist working in a food-manufacturing environment. As the *sous vide* method started in this way, its first protagonists were the French culinary establishment. This association with *haute cuisine* has continued ever since and perhaps is one reason why the *sous vide* method is looked at differently to other systems-based approaches for food preparation such as cook–chill and cook–freeze which do not have the same 'up-market' reputation.

Pralus himself adopted an emotional attitude towards the process, referring to it as '*Une Histoire d'Amour*' (a love story) in the title of the book on his development work and recipes (Pralus, 1985). He emphasised the requirement for the highest quality raw materials, strict hygiene standards and correct procedures and temperatures for cooking, cooling, storage and reheating. He condensed these ideas into 'ten commandments' in his search to produce the perfect meal as well as all the other benefits of an interrupted foodservice system. His training schools have also been responsible for spreading knowledge of the *sous vide* method to chefs in many countries.

In the 20 years since Pralus's first work, the *sous vide* process has often been acclaimed for the excellent sensory qualities of the dishes produced but also held up as a process which is bringing too much technology into food production with consequently an increased risk of the growth of pathogenic bacteria. The first example of the latter was in 1988 when the FDA in the USA issued warnings about its use and the potential dangers to public health. They used legislation, originally brought in to discourage canning at home with the associated risk of botulism, to prevent the production of *sous vide* products for retail sale. The exception were manufacturers using experienced personnel in a professionally designed and managed environment, i.e. a factory (Schwarz, 1988). The emphasis on *sous vide* then turned much more from trying to produce food with excellent sensory characteristics to deciding what combination of time and temperature was necessary in the core of the product during the cooking or pasteurisation stage for the process to produce 'safe' food. These figures were usually based on pure microbiological data for various types of bacterium and included many safety factors (Table 3.1). At one time, 2 minutes at 70°C was considered adequate (DoH, 1989): later it was suggested that 90°C for 4.5 minutes was

Table 3.1 Some heat treatments recommended or suggested for *sous vide* products

Treatment (Specified centre temperature and time)	Intention	Source	Target organism
70°C for 40 minutes 70°C for 100 minutes 70°C for 1000 minutes	6 days chilled shelf life 21 days chilled shelf life 42 days chilled shelf life	Ministère de l'Agriculture (1974, 1988 – French regulations)	*Enterococcus faecalis*
70°C for 2 minutes	5 days chilled shelf life	DoH (1989)	*Listeria monocytogenes*
80°C for 26 minutes or 90°C for 4.5 minutes, etc.	up to 8 days chilled shelf life	SVAC (1991)	*Clostridium botulinum* type E
90°C for 10 minutes	>10 days chilled shelf life	ACMSF (1992)	*Clostridium botulinum*
70°C for 2 minutes	short shelf life and reliable storage temperature	Gould (1996) (ECFF Botulinum Working Party)	*Listeria monocytogenes*
90°C for 10 minutes	longer shelf life		*Clostridium botulinum*

safer (SVAC, 1991): and later 90°C for 10 minutes (ACMSF, 1992). French figures using a different target bacterium have often been less severe (Goussault, 1993). However, sensory quality was often rarely mentioned and how it was affected by these various heat treatments. So the main factor which brought *sous vide* to prominence seemed almost secondary in the research efforts to find a method to guarantee product safety.

The *sous vide* method has received a large amount of coverage in the technical press with applications ranging from commercial and institutional caterers, in-flight catering and schools, to retailers and food processors. Most coverage has concentrated on the risks of food poisoning and the steps needed to ensure the safety of the process but a significant proportion has shown an almost evangelical attitude in claiming the improved sensory and nutritional qualities of *sous vide* foods compared to those produced by conventional means (Table 3.2).

3.3 Claims for the *sous vide* process

3.3.1 Description and definition of the process

The term '*sous vide*', originally from French, meaning simply 'under vacuum', has come to mean a process where the food is vacuum packed and cooked in that state at low temperature. It has also had a large number of other names such as pouch cooking, vacuum cooking, etc. which have also

Table 3.2 Some comments on the sensory and nutritional qualities of *sous vide* products

Comments	Source
'... no flavour is lost into the surrounding water or steam ...'	Anon. (1987a)
'... retains all of its natural flavour, along with more of its nutritional qualities ...'	Anon. (1987b)
'... the food retains all its flavours and fresh taste ...'	Bacon (1990)
'... all the nutrients, flavor, texture and aroma of the food are locked in ...'	Baird (1990)
'... there is no opportunity for loss of volatile flavour notes ...'	Bauler (1990)
'... raw or lightly cooked food product retains almost all its color, flavor and nutrients ...'	Bertagnoli (1987)
'... every delicate morsel of flavor is retained in the food ...'	Campbell (1993)
'... amplifies the food's flavor; if the product is fresh, it tastes wonderful. If it's not fresh, it tastes and smells twice as bad as if you cooked it conventionally ...'	Coomes (1994)
'... the only possible problem can be one of too much flavor. You've got to know how to make certain things lose their flavor ...'	De Liagre (1985)
'... enhances flavor and aroma ...'	Ivany (1988)
'... the integrity and taste of the food is generally considered superb ...'	Kalinowski (1988)
'... the best lamb they'd ever tasted ...'	Levine and Rossant (1987)
'... tasted like real food ...'	Levy (1986)
'... the process intensifies flavour ...'	Manser (1988)
'... food that tastes like it was freshly made ...'	Petit (1990)
'... the flavours can't escape ... more taste and smell ... the texture of the food is constant ...'	Pring (1986)
'... seals in flavour, juices and nutrients ...'	Raffael (1984)
'... does not harm the color, texture or flavour of food ...'	Scarpa (1988)
'... its flavour is highly praised in haute cuisine circles ...'	Sellers (1990)
'... retain flavouring ... reduce the loss of vitamins and nutrients in the cooking process ...'	Sessions (1987)
'... the quality of the final products are often far superior to foods prepared in the traditional manner ...'	Sornay (1990)

sometimes referred to foods which have been cooked conventionally then vacuum-packed or foods which have simply been vacuum-packed. A widely accepted definition has been one put forward by SVAC (*Sous Vide* Advisory Committee) a group formed in 1989 in the United Kingdom:

> *sous vide* (also known as *Cuisine en Papillote Sous Vide*) is an interrupted catering system in which raw or par-cooked food is sealed into a vacuumised laminated plastic pouch or container, heat treated by controlled cooking, rapidly cooled and then reheated for service after a period of chilled storage. (SVAC, 1991)

Many sources including chapters in this book, provide more detail on the various advantages and disadvantages of this system from many different aspects (Baird, 1990; Sheard and Church, 1992; Schellekens and Martens, 1992a, b).

3.3.2 *Claims for excellence*

The comments shown in Table 3.2 provide ample anecdotal evidence that the *sous vide* method produces food of a high sensory quality with the added supposition that its nutritional qualities must be equally high. However, there is, as yet, little evidence for these opinions which can be supported by objective and scientific experiment.

The views in Table 3.2 can be summarised as: the *sous vide* method produces a food with a better flavour, colour, texture and nutrient retention than conventionally cooked foods. It should, however, be remembered that this evidence was put forward mostly on the basis of professional judgement by those who have already made a commitment to the *sous vide* method in time, resources and capital.

3.4 Sensory aspects of *sous vide* foods

The views on *sous vide* foods in Table 3.2 often involve the emotional and psychological responses linked to many factors associated with the eating environment. In contrast, the scientific study of the sensory properties of foods by sensory analysts usually excludes these types of response to foods through the tight control of the environment and experimental design.

3.4.1 *Techniques for assessing the sensory quality of food*

Sensory analysis is now regarded as an objective science due to improved training programmes for sensory assessors, a wider range of sensory tests and statistical techniques for data analysis, and the introduction of sophisticated computer software packages, particularly in graphics for improved interpretation and presentation of the results. The sensory analyst can construct models or images of products, to relate the results of laboratory tests to acceptability ratings from consumer studies and to predict the effect of changes in a product's sensory profile on consumer acceptability. The most frequently used methods available are shown in Table 3.3.

Many sources have provided outlines of how and when the various tests should be used as well as updates in the field of sensory analysis (Amerine *et al.*, 1965; Piggott, 1988; Thomson, 1988). The objectives of these tests can be summarised as:

- Using difference tests to locate and identify differences between products.
- Applying descriptive analysis to describe products quantitatively in terms of their sensory attributes and displaying the results graphically in two- and three-dimensional models.
- Using preference/acceptability tests to identify and quantify consumer acceptability for products.
- Relating results from descriptive analysis and acceptability tests to predict consumer responses to changes in products by identifying those sensory attributes contributing to the direction and magnitude of the response.
- Applying time-intensity measurements to investigate the appearance, duration, disappearance and linger of sensory attributes thereby assessing their relative contribution to the eating experience.

Difference tests would be appropriate to determine if a dish cooked traditionally could be distinguished from the same dish cooked using the *sous vide* method. They would not, however, enable the investigator to know what those differences were or their magnitude on an absolute scale. Techniques for this could involve quantitative descriptive analysis (QDA) defined by Powers (1988) as:

> developing a list of descriptive terms, screening would-be assessors for possible membership of a panel, training judges, using sufficient replication so that the performance of the assessors, the effectiveness of descriptive terms, product differences and possible interaction effects may be isolated and evaluated by statistical analysis, and expressing the results graphically as well as numerically.

The QDA technique has been used for *sous vide* bolognaise (Armstrong,

Table 3.3 Procedures available to the sensory analyst

Difference/preference tests	Descriptive tests
Triangle	Conventional Profiling
Paired Comparison	Free choice Profiling
Duo–Trio	Differential Profiling
Ranking	Time-Intensity Measurements
Grouping	

1996), *sous vide* chicken ballotine, vegetable rice and dauphinoise potatoes (Church, 1990) as well as more homogeneous foods such as beer (Mielgaard *et al.*, 1979), wine (Vedel *et al.*, 1972), whisky (Shortreed *et al.*, 1979), strawberries (Shamaila *et al.*, 1992), apple juice (Dürr, 1979) and cider and perry (Williams, 1975).

3.4.2 Reviews of the sensory quality of chilled foods

Several reviews have covered chilled foods but only a few specifically on the *sous vide* method. Leadbetter (1989) concentrated on the microbiological hazards but also provided a brief history of *sous vide* and its related systems; Nacka (Bjorkman and Delphin, 1966), AGS (McGuckian, 1969) and CapKold (Daniels, 1988), packaging methods, equipment and the benefits of the system.

Church and Parsons (1993) considered that in comparison with conventional cook–chill methods, the *sous vide* method offered shelf life extension and eating quality benefits, but presented an increased microbiological risk. Their review covered vacuum packaging, prime cooking/pasteurisation, chilling, chilled storage and reheating. They concluded that, although there is some theoretical foundation for the claims of excellence, they have not yet been substantiated by research.

Creed (1995) reviewed research on nutritional and sensory properties of *sous vide* foods. He also concluded that few consistent data are available to provide quantitative scientific evidence for this method's gastronomic appeal and that, similarly, fewer data are available to support the supposed superiority in retention of vitamins.

Other reviews on the use of pre-cooked chilled foods in catering have also covered food produced by the *sous vide* method: Robson and Collison (1989) reviewed sensory aspects; Glew (1990) covered sensory, nutritional and microbiological aspects, somewhat mistakenly outlining the method as being in use for the last 30 years and a variant of the now obsolete Nacka system; Mason *et al.* (1990) reviewed the sensory aspects of chilled foods, concluding that vacuum-packing may decrease juiciness and that the potential for enhancing the eating quality of particular products was real but exaggerated by the popular and catering press. They also highlighted the problems of comparing sensory studies due to the lack of standardisation in product preparation, assessors, experimental design and test design; a comprehensive chapter on the *sous vide* method was also included in an extensive book on the technology and management of cook–chill catering (Light and Walker, 1990). Ghazala (1994), reviewing the microbiological hazards affecting fish and how this related to the different preservation techniques, included a section on *sous vide* processing, providing a short history and outline of the process, noting the main temperature requirements and its advantages and disadvantages.

3.4.3 *Sensory quality of specific* sous vide *foods*

In the following sections, there is some overlap of categories in an attempt to compare work on similar food groups together.

(a) Meat-based products   Lefort *et al.* (1993) studied the quality of pigs of four genetic groups differing in stress resistance to provide data on yield of carcass and primal cuts, meat colour, preparation and processing losses, sensory properties and suitability for manufacture of meat products (roast pork, *sous vide* pork, sausages, cooked ham and bacon). For *sous vide* cooked pork, Large Whites gave the best results for colour, flavour and tenderness using a simple sensory test with weight losses for *sous vide* cooked pork approximately half those for pork roasted in the oven.

Hansen *et al.* (1995) studied the sensory, technological and microbiological changes of *sous vide* processed roast beef for two heat treatments (59 and 62°C), two storage times (1–2 and 17 days) before processing, using spices with different microbial loads and two storage temperatures (2 and 10°C). Shear force values (Warner-Bratzler) were significantly lower for 62°C than 59°C treatments. Warmed-over flavour measured by TBARS (thiobarbituric acid reactive substances, a test for lipid oxidation) did not increase during storage but did increase for sliced roast beef stored for 20 days. Sensory analysis of off-flavours showed only minor changes.

Bertelsen and Juncher (1996) focused on the oxidative stability and sensory quality of some *sous vide* processed products. In pre-cooked meat products, warmed-over flavour developed within a few hours as a result of oxidation of the phospholipids in the products. In *sous vide* meat products, however, the oxidative processes were inhibited as long as the packages were unbroken, otherwise the *sous vide* cooked meat developed warmed-over flavour more rapidly than conventionally cooked beef resulting in a poorer sensory quality. Various possibilities to overcome this problem were discussed.

Training a panel of assessors for *sous vide* bolognaise using QDA (Quantitative Descriptive Analysis) was reported by Armstrong (1996). After screening using a set of discrimination tests, assessors went through a six-part training course. This consisted of orientation to the process, producing lists of descriptors for a product, agreeing on a set of descriptors, using reference samples with extreme attributes, practice scoring on in-house (*sous vide*) and commercial bolognaise and training on a computerised sensory system. Statistical analysis of the results showed that this method was accurate and meaningful.

Marinating beef before *sous vide* processing was found to have no effect on tenderness by Thorsell (1996). Only small changes in scores for the tenderness, taste, overall impression and sour odour of various cuts of beef were found during storage at 3 or 7°C for up to 6 weeks.

(b) Poultry-based products   Nazaire (1987) posed questions about how the *sous vide* process has been mythologised regarding optimal times and temperatures to use. He described a research programme to determine optimum conditions for chicken which would give the best sensory quality for consumers and yet provide the necessary microbiological security. It was concluded that 20 minutes at 95°C was optimal from the range of conditions tested (20 and 180 minutes at 65, 80 or 95°C). This result contradicted the conditions usually recommended for this product, based on subjective judgements, of 30 minutes at 80°C.

Work by Light *et al.* (1988) based on earlier work by Schafheitle *et al.* (1986) showed that chicken à la king was acceptable up to 14 days but courgettes provençale only up to 7 days. A consumer panel was used to discriminate between samples freshly made and those stored for 7, 14 and 21 days. A trained panel assessed pungency, crispness, strength of flavour as appropriate for the two dishes, giving mean quality scores for different storage times.

Later work by Schafheitle and Light (1989a, b) showed that chicken ballotine remained acceptable for up to 21 days stored at 1–3°C and that the stored product was different from the freshly produced dish using triangle tests but little difference was found using descriptive analysis. However, the times that the food was held at 80°C (the chosen pasteurisation temperature) varied from 8 to 23 minutes, having already taken from 15 to 39 minutes to attain this temperature. This variation in heating conditions and the possible variation in raw material could have been the reason for the mean scores for appearance, odour, juiciness, flavour, chicken and vegetable texture showing no significant difference during chilled storage from 0 to 21 days. The mean scores were, however, all at the positive end of the scale used. The conclusion that combination ovens do not provide even heat transfer is similar to that from Sheard and Church (1992) on variation in oven performance.

Choain and Noel (1989) compared traditionally cooked and *sous vide* processed dishes such as rare roast beef and blanquette of veal. Using a taste panel, appearance, aroma, taste and texture were assessed separately on the basis of being pleasant, ordinary or unpleasant/doubtful with a number of descriptive words to tick for each category. They found that the *sous vide* beef was more moist and pale in appearance with a consistently coloured cut surface, had more flavour and a juicy texture compared with the traditional version. For the veal dish, the *sous vide* version had better colour, more flavour and smoother texture with better retention of the texture of vegetables.

Church (1990) has provided brief details of comparative trials where a large proportion of assessors could distinguish *sous vide* from traditionally cooked chicken ballotine, vegetable rice and dauphinoise potatoes. This was confirmed by QDA as differences in 'chicken juiciness, filling moistness, initial flavour and aroma depth'.

Smith and Fullum-Bouchard (1990) compared Chicken Velouté prepared in cook–chill, cook–freeze and *sous vide* systems with a fresh sample at 1, 3 and 6 days storage at 4°C or –14°C and found no significant differences in aroma, appearance, flavour and tenderness.

Unpublished work (Schafheitle and Pierson, 1992) covered multi-component foods (Chicken/prawn, chicken/bacon/pepper, haddock, minced lamb) encased in pastry with sauce, vacuum packed in trays on an automatic machine, pasteurised and stored chilled (0 to 3°C) or frozen (–18°C) for one week. No difference was found between scores for chilled and frozen samples on acceptability of appearance, flavour, texture and aftertaste: all scores except for the lamb being at least 4.5 on a 6 point scale, a level deemed satisfactory.

As part of a study on the use of irradiation on *sous vide* products, chicken breasts were assessed for off-flavour and odours by Shamsuzzaman *et al.* (1992). It was found that the electron-beam treatment had little effect on these two factors up to 55 days of storage at 2°C, whereas untreated packs had spoiled between 30 and 42 days of storage.

Studies on chicken in red wine sauce by Creed *et al.* (1993) compared the freshly prepared product with the *sous vide* version immediately after cooking and after 5 and 12 days of chilled storage. An untrained panel could distinguish the two treatments using the Duo–Trio test ($P < 0.001$) due to the sauce becoming lighter during storage. If freshly made sauce accompanied the chicken, no difference was observed ($P = 0.05$).

Goto *et al.* (1995) compared *sous vide* cooking with normal cooking for a range of dishes. The values for hardness, springiness, gumminess and chewiness (instrumental values) were significantly lower ($P < 0.001$) for *sous vide* cooked chicken breast and pork fillet than ordinary cooking in a gas oven at 200°C.

Further work by Shamsuzzaman *et al.* (1995) on irradiated *sous vide* chicken breast showed few effects on odour, flavour and texture during storage at 8°C compared to frozen controls. Products treated with 2.0 and 3.0 kGy were still acceptable at 42 and 56 days respectively.

The effects of adding sodium lactate, heating rate, internal final cooking temperature and storage time on the colour, flavour and texture of *sous vide* processed brined chicken breasts were evaluated by Turner and Larick (1996). Chicken cooked to 94°C was significantly tougher, less juicy but less soapy/bitter ($P < 0.05$) than chicken cooked to 77°C. Meaty flavour decreased and warmed-over flavour increased during storage for 14 days ($P < 0.05$) and the addition of sodium lactate to the brine resulted in significant increases in meaty and salty flavours ($P < 0.05$).

Chicken breast with sliced potatoes in cream prepared by cook–chill and *sous vide* methods were compared by Church (1996). The heat treatment was 80°C for 10 minutes followed by rapid chilling to 5°C within 60 minutes and storage at less than 5°C for up to 7 days. Assessment by an attribute scaling method showed that for freshly cooked products, *sous vide* significantly increased the flavour intensity of both chicken and potatoes, the juiciness of the chicken and the moistness of the potatoes ($P < 0.05$). These effects decreased with storage time. Hedonic appeal was no different for freshly cooked cook–chill and *sous vide* dishes but *sous vide* products maintained their appeal during storage whereas cook–chill products became discoloured.

The diffusion coefficients of sodium chloride, free amino acids, organic acids and nucleic acids were studied during the *sous vide* cooking of chicken. These coefficients were less than those in conventional cooking, indicating that the release of these substances was slower and so perhaps causing the better taste of *sous vide* cooked chicken (Odake, 1996).

The concept of fuzzy mathematics was used by Xie *et al.* (1996) to analyse the often ambiguous results obtained from panels of assessors. Four different *sous vide* treatments (SVAC, 1991) on three dishes (chicken, carrot and salmon) were assessed in duplicate samples by 12 assessors for colour, juiciness, texture, flavour and overall acceptability on 8-point scales. The procedure helped to determine which treatment was optimal and which sensory attribute had the most influence in determining the overall acceptability. These preliminary results showed varying levels of agreement between the fuzzy model and the taste panel results.

(c) Fish-based products   Choain and Noel (1989) compared traditionally cooked and *sous vide* versions of trout stuffed with cucumber and fish stock. They found that the *sous vide* trout had more aroma and pronounced flavour compared with the traditional version. The *sous vide* fish stock also had more aroma and brought out the taste of the fish used compared to traditionally made fish stock.

Preliminary results of a state-financed study in France, on the effect of the temperature of pasteurisation on shelf life and final temperature on the texture of salmon (fatty fish) and whiting (white fish) were given by Picoche (1991). Enzymatic changes causing off-flavours were avoided by core temperatures during processing above 85°C. Problems of lower sensory quality if higher temperatures were required for a longer shelf life, could perhaps be solved by vacuum packing ready-to-cook fish, freezing then cooking directly as needed.

Gittleson *et al.* (1992) assessed the sensory qualities of commercially produced *sous vide* salmon packed in two types of plastic packaging. Using QDA techniques, colour, flakiness, crumbliness, fish odour and overall acceptability were measured on 9-point scales over 100 days of storage in ice at 0 to 4°C. Overall acceptability was negatively correlated with fish odour and satisfactory up to 12 weeks of storage ($P < 0.01$). Instrumental texture measurements of shear force value (Instron) showed no significant difference between package types possibly due to product variability and maintaining the orientation of the fish muscle between the shear blades.

Sarli *et al.* (1993) investigated the microbiological quality of eviscerated sea bass and gilthead sea bream, vacuum packaged with seasonings, wine and fish stock, cooked at 57, 60, 62 or 65°C, and stored at 0–5°C for up to 40 days. It was concluded that these relatively mild heat treatments gave products with good sensory properties and a long shelf life.

Bergslien (1996) investigated the effect of processing conditions and storage time on the microbiological and sensory qualities of *sous vide* processed salmon with pre-treatment by dipping in salt solutions containing sodium carbonate at 2.5, 5 or 10%. Heat treatments were 65, 70 and 75°C for 10, 12.5 and 20 minutes respectively. Appearance, smell, taste, rancidity and texture were assessed at weekly intervals during storage at 2°C. It was concluded that pre-treatment with 5% sodium carbonate solution and heat treatment at 65°C for 10 minutes gave an acceptable product for up to 3 weeks storage and packaging using an aluminium barrier inhibited rancidity.

Determining the breakdown products of ATP (adenosine triphosphate) has been tested as an indicator of raw fish quality before processing (hot smoking or *sous vide* processing) products such as whitefish (Hattula and Kiesvaara, 1996).

(d) Fruit- and vegetable-based products   Choain and Noel (1989) also compared traditionally cooked and *sous vide* versions of poached pears. They found that the *sous vide* pears had more aroma and flavour and a better texture compared with the traditional version.

Research on vegetables by Picoche (1991) focused on textural changes dependent on the types of polysaccharide found in vegetables and fruit and their breakdown at the higher temperatures (80–90°C) needed for food safety. Studies on apple slices, half onions, and cylinders of carrot and pear using temperatures at the core from 65 to 95°C were used to try to predict the effect on texture. At the lower temperatures of the range studied, enzymatic changes made the product unacceptable so higher temperatures were suggested for a long shelf life.

Petersen (1993) studied the effect of *sous vide* processing, steaming and traditional boiling of broccoli florets on the retention of vitamins and the sensory properties of freshly prepared samples. Sensory evaluation showed that *sous vide* cooked and steamed broccoli florets generally had higher acceptability than boiled samples. *Sous vide* samples also had significantly more flavour, but were more bitter and softer than the boiled samples ($P < 0.05$).

Varoquaux *et al.* (1995) investigated the cooking of lentils, concluding that using temperatures below 90°C for up to 2 hours maintained the required firmness and wholeness necessary for use in foodservice.

Goto *et al.* (1995) compared *sous vide* cooking with normal cooking for a range of dishes. Sensory evaluation (judged by appearance, colour, flavour, taste, softness and overall preference) of takikomigohan (rice cooked with other ingredients), saba-nituke (boiled mackerel with soy sauce) and rolled cabbage showed mainly no significant differences between *sous vide* and ordinary cooking. For nikujaga (boiled potato and beef with soy sauce), chicken boiled in cream and fig compote, the *sous vide* version was most often preferred. The green colour of kiwi fruit sauce was also maintained by *sous vide* cooking.

The colour, aroma, flavour and overall acceptability of mitsuba were investigated by Yoshimura *et al.* (1995). They found that the aroma and flavour of the *sous vide* version, cooked for 1.5 minutes at 100°C, were superior to that cooked conventionally at 100°C for 1 minute. Mitsuba is a kind of Japanese herb that looks like a large chervil and is often used in soup. Its length is about 20 cm with a straight, leafless stalk and three green leaves at the top, hence its name in Japanese: three (mitsu) leaf (ba) (Endo, 1996).

3.4.4 Sensory quality of commercially manufactured sous vide foods

Informal taste panels have been conducted by Levy (1986) and Manser (1988) on *sous vide* products provided by Home Rouxl, the UK producer. Levy commented favourably on the texture of scallop and chicken liver mousses and the flavour of brill, fish mousse and poulet chasseur but unfavourably on vegetable dishes. Manser commented favourably on the intensity of flavour of carrots and lamb and all the sensory attributes of duck breast.

While giving impressions of new *sous vide* plants in the United States, Mayer (1990) included some uncomplimentary comments on various previously frozen then thawed *sous vide* dishes from Vie de France kept warm in a waterbath such as, 'mealy and dry' snapper, mis-shapen pork tenderloin with 'no distinctive taste and dry', 'gray and chewy' steak au poivre and 'gray and tough' rack of lamb. Better results occurred with chicken breast from Grace Culinary Systems, 'moist and tender' but dishes were not hot enough.

In a Belgian consumer survey, a range of complete and semi-complete *sous vide* dishes from eight producers was reviewed, summarising weights, nutritional and energy content, bacteriological and sensory quality and price. Fifteen complete meals and 14 semi-complete dishes (meat, poultry, or fish-based, potatoes and paella) were assessed by professionals, who gave brief comments. Three 'Best Buys' emerged: beef bourgignon with potatoes ('the meat and the sauce tasted good but the potatoes were slightly watery') and chicken with mushrooms ('chicken with unnatural colour but overall pleasant taste') from Tastefin and diced chicken breast with potatoes ('very attractive, mild taste, very tasty potatoes') from Delhaize (Anon., 1993).

3.4.5 *Sensory quality of foods produced by methods similar to* sous vide

Some methods of food preparation have been in use for several years and use the same principle as the *sous vide* method of cooking the vacuum-packaged raw food at low temperatures.

(a) Meat-based products   The cook-in-bag system has been applied mainly to meats. Buck *et al.* (1979) found that vacuum-packed beef roasts cooked to 60°C in a waterbath at 60–61°C were significantly more tender ($P < 0.01$) than matching roasts cooked to the same temperature in a conventional oven at 94°C, scoring on average 6.1 and 4.4 respectively on an 8-point scale (8, extremely tender; 1, extremely tough).

Dinardo *et al.* (1984) studied beef muscles prepared in either a 60°C waterbath or a 94°C conventional oven. Waterbath-cooked muscles were placed in nylon bags and vacuum packed before cooking; some samples were held in the bath for 2 or 4 additional hours after reaching internal end-point temperature. Yields were greatest for waterbath-prepared samples removed from the bath immediately upon reaching internal end-point temperature. Extended cooking times increased collagen solubilisation and decreased yields, overall rareness, panel scores for juiciness and flavour and Warner-Bratzler shear force values for texture.

A study by Jones *et al.* (1987) compared control pork joints wrapped in PVC, stored at –20°C then cooked conventionally with joints vacuum packed, cooked to 70°C in a 100°C waterbath then stored for 14 and 28 days at 4° or –20°C then reheated. At 28 days, the frozen stored joints were less juicy than the control and chilled stored joints ($P < 0.05$); the tenderness of frozen stored joints decreased over the period ($P < 0.05$); the intensity of pork flavour of the chilled stored joints also decreased and again the intensity of off-flavour and Warner-Bratzler shear force values did not differ significantly ($P > 0.05$) over the storage period.

In other work, beef chuck roasts were vacuum packed and then heated to 65°C in a 71°C waterbath and compared with conventionally cooked roasts. No significant differences in cooking loss, tenderness, juiciness, intensity of beef flavour and off-flavour or Warner-Bratzler shear force values were observed ($P > 0.05$) either immediately or after storage for 14 or 28 days at 4°C. The only significant change was in the less intense beef flavour of the vacuum-packed roasts ($P < 0.05$) (Stites *et al.*, 1989).

Cooksey *et al.* (1990) studied beef loin roasts prepared by three procedures that simulated home, foodservice or commercial methods for precooking meat prior to vacuum packaging. Roasts were prepared by electric oven (176.5°C), cook-in package waterbath (70°C water), or smokehouse (ambient temperature) heating. The yield, thiamin, fat, water content, and Instron tenderness of roasts were determined. No differences were observed between the cooking procedures evaluated ($P > 0.05$), suggesting that most heat-processing treatments used for pre-cooked packaged meat result in similar quality products.

Pre-cooked pork chops and roasts from 30 pigs given no supplemental vitamin E (Control) or supplemented with 100 mg vitamin E per kg diet were evaluated for lipid oxidation (TBA values), microbial growth (total plate count), sensory characteristics (tenderness, juiciness, pork-flavour intensity and off-flavour intensity), cooking storage losses and reheating losses by Cannon *et al.* (1995). Chops and roasts were vacuum packed in cook-in-bags, cooked to 60°C internal temperature and stored at 2°C for 0, 7, 14, 28 or 56 days. Lipid oxidation was lower in chops and roasts from pigs on the supplemented diet than the controls. Off-flavour intensity was lower and tenderness scores higher for vitamin supplemented pork roasts than the control but not for the chops. The conclusions were that a vitamin E supplement helped to reduce lipid oxidation and hence warmed-over flavour in pre-cooked vacuum-packed pork, giving a palatable product up to 56 days.

(b) Poultry-based products   Commercially produced turkey breast rolls prepared by vacuum packing and then cooked to an internal temperature of 71°C were studied over 87 days storage at 4°C. Warner-Bratzler shear force values from rolls produced at one of two factories decreased significantly ($P < 0.05$) (Smith and Alvarez, 1988).

Rosinski *et al.* (1989) formed comminuted chicken breast meat into cylinders with various amounts of salt and phosphate types, vacuum packed and then cooked them to an internal temperature of 68°C using water at 60, 65 then 70°C for 30 minutes at each temperature and then cooled them in a waterbath at 0°C for 15 minutes. Shear force values, measured on slices of product and adhesion of packaging material to the meat, increased with salt concentration.

(c) Vegetable-based products   Another process similar to the *sous vide* method has been applied to potatoes: Poulsen (1978) described the production of vacuum-packed cooked potatoes using potatoes which were steam peeled, vacuum packed, pasteurised at 95–98°C for 40 minutes in water, then cooled, packed in cartons and stored at 0–5°C. Sensory tests showed a significant preference for vacuum-packed potatoes when compared with canned potatoes. It was concluded that vacuum-packed cooked potatoes offered a new possibility for ready-to-heat products as long as they were kept refrigerated at 0–4°C with a shelf life limited to 5 weeks.

3.4.6 Summary of sensory work

Some of the findings emerging from the preceding sections can be summarised as:

- There are few significant differences in texture, aroma, flavour and appearance between chilled and frozen *sous vide* foods.
- Heating to a higher end-point temperature increases toughness of meats.
- Flavour is intensified but the effect decreases with time in storage.
- Shelf life is increased by the *sous vide* method.
- Higher processing temperatures are needed for acceptable texture in vegetables.

Apart from these points, it is difficult to provide general conclusions from the considerable amount of research on sensory aspects, some reasons for this being:

- Experiments have been designed to determine sensory changes caused by:
  (i) different storage temperatures
  (ii) different cooking temperatures
  or made comparisons between:
  (i) *sous vide* and conventionally cooked versions of the same food
  (ii) *sous vide* and versions of the same dish produced by other food-service systems.
- Some experiments have used recommended heat treatments which may be inherently too severe.
- Different statistical tests and procedures have been used, making comparison difficult, e.g. using different scales and attributes.
- Many experiments have problems of reproducibility when making comparisons, e.g. repeated production of a freshly prepared product to compare with a stored product.
- Results could often be specific for the piece of equipment used.
- Results are often product specific.

Referring to the last point, the number of products manufactured commercially or for use in-house is vast: Pralus (1985) gave about 130 recipes and commercial producers commonly have 50 different products which vary according to the season. Trying to determine an overall view regarding changes in sensory qualities over time, or to make a comparison with conventionally cooked dishes while also covering all the variations in cooking times and temperatures, cooling times, storage temperatures and storage times, would mean an experimental design incorporating perhaps thousands of combinations. As in microbiology, moving towards the concept of predictive modelling based on product attributes and treatment conditions may eventually be the answer.

3.4.7 *Relevance of comparisons made in sensory studies of* sous vide *foods*

Sensory analysis has not clearly confirmed the perceived sensory aspects shown in Table 3.2, often finding conflicting results. If heat treatments used for sensory studies have been in accordance with those of Table 3.1, any high levels of sensory quality may well have been destroyed. Genigeorgis (1993) thought that the difference between *sous vide* products processed at 60°C for 12 minutes at the centre and those processed for a 6 decimal destruction of non-proteolytic *Clostridium botulinum*, was like 'day and night'. He suggested that milder heat treatments combined with additives such as lactic acid and its salts or other hurdles would be a solution.

Goussault (1993) has discussed how cooking at low temperatures affects the sensory quality of *sous vide* foods. He emphasised the need for strict control of cooking temperature, best achieved through using water or water spray methods, to optimise breakdown of connective tissue proteins and so increase sensory quality.

Tjomb (1990) discussed finding the right balance between increased microbiological risk at low pasteurisation temperatures (below 68°C) and the loss of organoleptic quality at high temperatures (over 80°C). He later discussed the contrast between the attitudes of the chefs advising *sous vide* product manufacturers and the food industry (Tjomb, 1992). For example, Georges Pralus emphasised that *sous vide* is a cooking technique and extending shelf life will be to the detriment of organoleptic quality, while ADRIA (a food research organisation) recommended that a severe heat treatment was necessary, causing a texture deterioration which would not adversely affect products with sauce, such as *sous vide*.

Langley-Danysz (1992) has provided a useful diagram credited to Bernard Barlet (Casino) to illustrate how sensory quality and levels of food safety vary according to the level of heat treatment. In Figure 3.1, sensory quality will reach a peak at point A when food safety is at a medium level: at point B, the situation is reversed. At point C, lies a band of heat treatment or pasteurisation values which will be the best compromise. However, this diagram could be different for each recipe and possibly for each recipe component, so a producer faces a large amount of research and development work to optimise individual dishes.

Assuming that the heat treatment has not been severe, the problem may be that the chefs and scientists have been measuring two different attributes. The chefs take into account their emotional attitude towards flavours, odours and textures which have been startlingly different from the usual products of a time-interrupted foodservice system; the sensory analysts, on the other hand, have by the usual reductionist approach, tried to measure only what they can measure with confidence using the techniques mentioned earlier. What is required is a method of assessing food products with quality attributes which, some say, approaches 'perfection'. This could incorporate some measure of this emotional attitude which can modify the sensory scales usually assumed to be linear. If this emotional factor can be included, it may help to resolve the conflict between the views from science and craft-based judges on the quality of foods with 'perfect' sensory attributes.

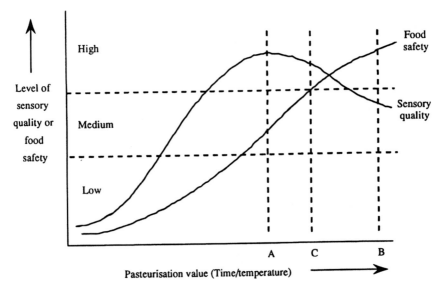

Figure 3.1 Optimising heat treatment for sensory quality and food safety of *sous vide* processed foods (A, high sensory quality/medium safety; B, medium sensory quality/high safety; C, optimum high sensory quality/high safety. (Adapted from Langley-Danysz, 1992.)

3.5 Nutritional aspects of *sous vide* foods

Unlike the sensory attributes of food, nutritional properties are intangible and rely on instrumental analysis to provide information. However, consumers have become more health conscious and retailers and restaurateurs are marketing products to satisfy that need. As detailed in another chapter, the forerunners of the *sous vide* method were designed for hospital use (AGS, Nacka) and, since then, *sous vide* products have been used in hospitals for special diets. Unless consumers were dependent for their nutritional requirements on an institutional foodservice system which used *sous vide* products, such as a school or elderly feeding programme, the fact of whether the *sous vide* method does actually retain more nutrients than conventional methods could be considered somewhat academic.

3.5.1 Techniques for assessing the nutritional quality of food

Quantifying the nutritional content of foods has long been of interest to many areas of human medicine. More recently, due to requirements for nutritional labelling, this may eventually cover cooked and chilled meals such as *sous vide*, produced for sale to third parties.

The vitamins of interest in this case will be those which are liable to decrease during the *sous vide* processing method, i.e. those sensitive to heat treatment: thiamin (vitamin B_1), riboflavin (vitamin B_2) and ascorbic acid (vitamin C). Losses of minerals are not caused by heat but can occur by leaching into cooking water (British Nutrition Foundation, 1987). In the case of the *sous vide* method, the mineral content of the fresh food would probably be maintained. The most important ones, calcium and sodium, are both of interest to the consumer as regards maintaining an adequate supply of the former and limiting the intake of the latter.

The strengths and weaknesses of various methods for determining the vitamin content of foods have been discussed by Lumley (1993), Eitenmiller (1990) and Gregory (1983). These methods include:

- thiochrome for thiamin
- microbiological methods measuring the dose response of microorganisms
- the radioimmune (ELISA) method
- biological methods based on the growth response in rats and chicks
- High Performance Liquid Chromatography (HPLC).

Of these methods, HPLC seems to be the most developed and advancing method although Speirs (1996) is developing a microbiological method specifically for detecting thiamin, the least stable water-soluble vitamin, in *sous vide* foods.

Many HPLC methods have been devised to determine the amount of vitamins and other substances present in food (Saxby, 1978). A very wide range of foods have been analysed using different extraction procedures to determine various combinations of vitamins. Examples of its use have been reviewed by Polesello and Rizzolo (1986) for water-soluble vitamins, and by Van Niekerk (1988).

Most vitamin determinations using HPLC have analysed meal components with few analyses of prepared dishes such as those provided by the *sous vide* method. Hare (1996) has reported on a rapid method using HPLC for the detection of folic acid and its derivates in *sous vide* foods so that changes in the nutritional status of products caused by different heat treatments can be more easily measured. Two other examples using HPLC for analysis have been for white sauce (Nandhasri and Suksangpleng, 1986) and turkey bologna (Tuan *et al.*, 1987). Other determinations of vitamins in meal items, such as meat loaf (Dahl and Matthews, 1980), beef patties and fried fish (Ang *et al.*, 1978), pot roast and gravy, beans and frankfurters (Ang *et al.*, 1975), beef stew, chicken à la king, shrimp Newburg and peas in cream sauce (Kahn and Livingston, 1970), cod au gratin (Jonsson and Danielsson, 1981), Italian spaghetti (Khan *et al.*, 1982), beef stew (Nicholanco and Matthews, 1978) and pork casserole in retort pouches (Uribe-Saucedo and Ryley, 1982), generally made use of chemical methods recommended by AOAC (1984).

3.5.2 *Reviews of the nutritional quality of chilled foods*

Few nutritional data are available for chilled foods prepared by the *sous vide* method but a large amount of information is available on changes occurring during processing (Bender, 1987) and foods prepared in cook–chill systems of which the *sous vide* method is a variant. Bognar (1980) studied the vitamin A, B_1, B_2 content of chilled meals prepared conventionally or by cook–chill, cook–freeze or sterilisation methods and the effects of chilling, storage and reheating conditions. Hunt (1984) reviewed nutrient losses in cook–chill and cook–freeze systems, concluding that vitamin C losses in cook–chill were not large compared with conventional systems. Later work on cook–chill systems found changes in vitamin content but not in the protein, fat, carbohydrates or mineral content of chilled and pasteurised chilled meals after 10 and 28 days storage at 2°C followed by reheating. Data are available for many types of products at various times from raw to the point of service, noting that vitamins C and B_1 are the most labile in food processing and the easiest to quantify. Vitamin C is often used as an indicator of vitamin retention because, if its content decreases, other vitamins may have also been destroyed (Bognar *et al.*, 1990). Bognar (1990) has also provided data on the vitamin loss during chilled storage of fruit and vegetables and the influence of chilling, storage and reheating conditions on prepared foods.

3.5.3 *Nutritional quality of specific* sous vide *foods*

(a) Meat-based products   Extensive work on meat has been done in France (Watier, 1988) and summarised by Watier and Belliot (1991). They studied the retention of the B vitamins in beef bourgignon, roast veal, roast lamb and roast fillet of pork (Watier, 1988). Their results are summarised in Table 3.4. Comparisons were made with values from the literature and it was concluded from this that the *sous vide* method preserved 'vitamins liable to oxidation better than traditional cooking, but this advantage was removed by storage and subsequent reheating'.

Metayer (1991) compared the retention of proteins, lipids and vitamins in *sous vide* products with traditionally prepared foods. It was concluded that the protein/lipid ratio was higher for traditionally cooked meats due to the difference in weight loss during cooking but still satisfactorily above the required ratio of 2. The accompanying sauce for *sous vide* dishes would bring this below the required ratio. Lipid retention was identical for roast and sautéed meats and vitamin levels (B_1, B_{12}, niacin) in meat and fish dishes cooked *sous vide*, stored at 3°C and reheated were equivalent to values in the literature for traditionally cooked dishes, but vitamin C was lost during storage and reheating.

The retention of certain fatty acids in *sous vide* processed seal meat

Table 3.4 Percentage retention of vitamins in *sous vide* processed meat and fish dishes after cooking, chilled storage for 21 days and reheating (adapted from Watier, 1988)

Vitamin	Percentage retention					
	Beef	Veal	Lamb	Pork	Salmon	Cod
Vitamin B_1	70	91	77	90	90	85
Vitamin B_2	100	52	100	100	100	63
Vitamin B_6	100	100	100	100	85	100
Vitamin PP*	93	93	91	88	95	100
Pantothenic acid	100	100	100	100	96	89
Vitamin B_{12}	87	100	100	100	92	72
Biotin	100	100	93	100	95	95
Vitamin A	–	–	–	–	78	67

*PP = Pellagra Preventative, vitamin B Complex, e.g. nicotinamide or nicotinic acid. Term not now in general use.

shepherd's pie was studied by Ghazala *et al.* (1996) for pasteurisation at 65, 70, 75, 80 or 85°C for 105, 60, 43, 35, or 30 minutes respectively. Processing at lower temperatures retained significantly more fatty acids than at the higher temperatures.

(b) Poultry-based products   During a study on the use of irradiation on *sous vide* products, the thiamin content of chicken breast was determined using HPLC (Shamsuzzaman *et al.*, 1992). It was found that *sous vide* cooking produced 98% retention based on the raw value (1.07 µg/g): increasing the electron beam treatment up to 2.9 kGy decreased this value to 86%: after 27 days storage at 2°C, these figures fell to 87 and 80% respectively.

Further work by Shamsuzzaman *et al.* (1995) on combined irradiation and *sous vide* processed chicken breast showed losses of between 23 and 46% in thiamin content during storage at 8°C, but any overall pattern was masked by the variability in the heat treatment received due to the variable product size.

Vitamins B_1, B_2 and C were compared in turkey breast for *sous vide*, modified atmosphere (MAP), cook–chill and cook–serve methods at all stages from raw to the point of consumption (Eriksen and Lassen, 1996). Both MAP and *sous vide* were found to offer better vitamin retention then cook–chill and cook–serve, even after 14 and 21 days storage.

(c) Fish-based products   The most extensive work has been done in France (Watier, 1988) and summarised by Watier and Belliot (1991). They studied the retention of the B vitamins in salmon and cod (Watier, 1988). Their results are summarised in Table 3.4. As in the work mentioned above on meat-based products, comparisons were made with values from the literature with similar conclusions.

(d) Vegetable-based products   The earliest work aimed specifically at *sous vide* food (Buckley, 1987) studied the stability of vitamin C in broccoli in a simulated *sous vide* system, finding that leaching and oxidation were avoided. The simulation consisted of cooking in a 'pressure steamer for 4 minutes'. This would provide a temperature above 100°C and so, strictly speaking, would not be considered as a *sous vide* process. However, the results showed 86% retention of vitamin C, compared with the raw product after 5 days chilled storage. Boiled or pressure-steamed broccoli showed retention figures of 28 and 62% respectively after 1 day of chilled storage, falling to 9.4 and 7.3% respectively after 5 days storage.

Again, extensive work has been done in France (Watier and Belliot, 1990) and summarised by Watier and Belliot (1991). They studied the retention of vitamin C, B_1 and folates in potatoes, carrots, green beans and cauliflower (Watier and Belliot, 1990). Their results are summarised in Table 3.5. As in the work mentioned above on meat-based products, comparisons were made with values from the literature with similar conclusions.

Smith and Fullum-Bouchard (1990) determined the vitamin C content using HPLC in the spinach component of chicken velouté prepared in cook–chill, *sous vide* (storage at 4°C) and cook–freeze systems (storage at –14°C) over 6 days, finding a decrease over the period for cook–chill samples and a decrease over the last half of storage for *sous vide* and cook–freeze samples.

Data on the retention of vitamins in *sous vide* foods has been quoted by Bognar *et al.* (1990). Although presented as figures relating to asparagus, broccoli in cream sauce, broccoli and chicken à la king produced by the *sous vide* method, the references cited (Kraxner, 1981; Erdmann and Klein, 1982; Ezell and Wilcox, 1959) provide no relevant data. For some data, it is likely that the authors intended to refer to Kossovitsas *et al.* (1973), who compared

Table 3.5 Percentage retention of vitamins in *sous vide* processed vegetables after cooking, chilled storage for 21 days and reheating (adapted from Watier and Belliot, 1990)

Vitamin	Percentage retention			
	Potatoes	Carrots	Green beans	Cauliflower
Carotene	–	89	100	–
Vitamin E	–	100	100	–
Vitamin C	63	86	40	69
Vitamin B_1	86	67	86	56
Vitamin B_2	100	100	71	86
Vitamin B_5	84	63	90	37
Vitamin B_6	80	100	79	63
Vitamin PP*	48	62	65	61
Vitamin B_9	100	100	85	100

*PP = Pellagra Preventative, vitamin B complex, e.g. nicotinamide or nicotinic acid. Term not now in general use.

the quality of prepared meals (broccoli in cream sauce, chicken à la king and cod in cream sauce) held in chilled and frozen storage. These data, however, were based on the Nacka method of preparation so cannot provide a reliable estimate for the *sous vide* method. Bognar *et al.* (1990) also quote thiamin losses of 20–30% in *sous vide* processed roast beef joints (Cooksey *et al.*, 1988) and 10% for vacuum-packed chicken and broccoli (Kraxner, 1981). The usefulness of the latter data is again doubtful.

Petersen (1993) studied the effect of *sous vide* processing, steaming and traditional boiling of broccoli florets on the retention of ascorbic acid, vitamin B_6 and folacin. Results showed that retention of ascorbic acid during processing at 100°C for 40 minutes was significantly reduced if the vacuum level used was 92% compared with 97% while vitamin B_6 was unaffected. Acknowledging that losses during storage and reheating also need to be taken into account, for the cooking stage, *sous vide* processing gave the best retention of the three vitamins studied, followed by steaming and boiling.

Goto *et al.* (1995) compared *sous vide* cooking with normal cooking for a range of dishes. For heating times of 30 seconds, 1, 2 and 3 minutes, the retention of total ascorbic acid content was higher for *sous vide* cooked vegetables (komatsuna leaf, snap bean pods, broccoli, pumpkin and potato) than those boiled in water.

The loss of active vitamin C in commercially frozen Brussels sprouts was compared for cook–serve and *sous vide* processes by Walker *et al.* (1996) using fluorimetric analysis. A range of cooking and warmholding times was studied showing a loss of 51% for the worst case; cooking for 30 minutes followed by 90 minutes of warmholding. For *sous vide*, the percentage loss was around 22% after cooking (9 minutes in steam to 75°C core temperature), chilling, 5 days chilled storage and reheating. However, the best cook–serve conditions of 15 minutes steaming followed by 30 minutes, giving a loss of 26.9%, were comparable with *sous vide* processed product reheated after 4 days storage, giving a 26.7% loss.

3.5.4 Nutritional quality of foods produced by methods similar to sous vide

(a) Poultry-based foods   Commercially produced turkey breast rolls prepared by the cook-in-bag process were vacuum packed and then cooked to an internal temperature of 71°C and stored for 87 days at 4°C. The only significant result was that amino nitrogen, non-protein nitrogen and pH did not change ($P > 0.05$) over the storage period (Smith and Alvarez, 1988).

(b) Vegetable-based foods   The method for processing ready-cooked vacuum-packed potatoes follows a similar procedure to *sous vide*, so that the nutritional results might be relevant. During the process, the vitamin C

content fell from 26.8 to 24.9 mg/100 g, a retention figure of 93% compared to a literature figure of 62% for boiled potatoes. Vitamin B_1 fell from 77.6 to 73 µg/100 g, a retention figure of 94% (Poulsen, 1978).

3.5.5 Summary of nutritional work

Like the sensory aspects of *sous vide* food, the range of techniques for determining nutrient levels, the inherent variation in raw materials and the types of studies make comparisons difficult. However, some points seem to emerge:

- The *sous vide* cooking process by itself can improve retention of vitamins and make these available for the consumer.
- This improvement may be lost by the time that chilling, chilled storage and reheating have taken place.

3.5.6 Relevance of comparisons made in nutritional studies of sous vide foods

The hidden benefits of the sometimes enhanced nutritional qualities of *sous vide* foods will have tangible benefits when used in particular foodservice environments such as hospitals and clinics, as discussed in another chapter. However, the present interest in 'healthy' eating and 'functional foods' by many consumers is certainly a marketing angle available for the exploitation and development of the market for *sous vide* products.

3.6 Overall conclusions

The choice facing any *sous vide* producer is deciding what chilled shelf life is actually needed. Using a heat treatment to guarantee a longer shelf life of, say, 21 days will unnecessarily degrade the sensory and nutritional qualities of the product if it would normally be used after 10 days of storage. Therefore selecting the shelf life could eliminate much experimentation.

Trying out all possible combinations of food products and heat treatments with replicates and sufficient samples while simultaneously controlling the inherent variation in the raw food materials in order to optimise the sensory, nutritional qualities of *sous vide* products and minimising microbiological risk, would be difficult to justify in terms of time and resources.

Experimental design can reduce the number of experiments necessary but methods based on computers may be more efficient. Computer programs have been devised which, using data on meal ingredients and the proposed processing conditions for the manufacture of chilled foods, will model the changes in temperature, microbial load and texture with time (Schellekens *et al.*, 1994). Other programs have used the neural network approach to process all available data to find underlying patterns and so infer likely product attributes for untried treatments. One example of this is for determining the effect of harvesting and processing methods on the quality of surimi. This produced a series of linear regression equations for predicting quality which could be approached through a number of decision points relating to time between harvesting of the fish and processing, temperature, moisture content, etc. (Peters *et al.*, 1996). Expert systems – using rules elicited and developed from interrogation of human experts and so answer queries through the application of logic – is another approach made possible by computers.

Other methods can be based on mathematical and statistical treatments: Xie *et al.* (1996) used fuzzy logic for unravelling the often contradictory information from assessors to study the organoleptic qualities of *sous vide* salmon, carrot and chicken products treated according to SVAC (1991) guidelines. This study attempted to clarify the variation between assessors, products and heat treatment to determine for each product the attribute which was most closely associated with the assessors' hedonic response. Other techniques such as response surface methodology, preference mapping and multidimensional scaling are reviewed extensively in Piggott (1986, 1988) but tend to be applied to single foods rather than trying to deal with the complexities and interactions of the food components found in *sous vide* dishes. Even then, the responses from trained assessors or untrained consumers are recorded in terms of rating attributes or acceptability, etc., on scales with appropriate anchor points. One scale which has been used to move away from this concept towards measuring the 'perfect' or 'ideal' point is the relative-to-ideal scale used for assessing the ideal salt concentration by Shepherd *et al.* (1984) in tomato soup and by Griffiths *et al.* (1984) in bread and tomato soup. Again single foods have been studied rather than meals.

Future developments aiming to optimise the quality of minimally processed foods such as *sous vide* meals will need to use methods which can accommodate more of what consumers experience. Food quality as perceived by the consumer is not something which can be broken down into neat components labelled as scores for 'taste', 'texture', 'odour' or measured quantities such as 'acidity' or 'fat content', etc. Claims for *sous vide* foods (Table 3.2) often involve emotional and psychological responses linked to many factors associated with the eating environment – especially in the restaurant (Pierson *et al.*, 1995; Reeve *et al.*, 1994), for example, decor (Bell *et al.*, 1994) and effort (Meiselman *et al.*, 1994). Lyman (1989) explored the association between preferences and the psychological stimuli of taste, odour, appearance, etc., linked to meanings and associations of food, noting that sensory properties can be relatively unimportant in determining food preferences. These preferences are built up through time and experience and form a perceptual judgement scale which is modified or extended as new foods or foods with exceptional qualities, such as *sous vide* products, are encountered.

The question then arises as to whether this type of factor can be accommodated in an objective methodology for optimisation and evaluation of the quality of foods such as *sous vide* products. The usual method of capturing information from consumers through questioning by interview or simple questionnaires is not difficult in a university training restaurant where customers are accustomed to participating in the training process (Reeve *et al.*, 1994). Trying to incorporate questions which can be used to assess emotional aspects towards the meal or the mood of the consumer may be possible but limited by the short time available. In a real eating environment, such as a restaurant, it would be difficult and unethical to wire up consumers to detect their physiological responses (brain activity, etc.) which indicate pleasure while eating, but with advances in electronics and sensing it may be possible, perhaps in the future, to do this remotely.

The full value of minimal processing methods such as *sous vide* for enhancing the acceptability of foods will only be confirmed when sensory analysts and product development technologists can use techniques which will take into account not only components of acceptability based on the sensory stimuli of food but also how close the food comes to the individual's perception of the ideal.

Acknowledgement

The author wishes to acknowledge, with thanks, the financial support of the Worshipful Company of Cooks, London.

References

ACMSF (Advisory Committee on the Microbiological Safety of Food) (1992) *Report on Vacuum Packaging and Associated Processes*, HMSO, London, 69 pp.

AOAC (Association of Analytical Chemists) (1984) *Official Methods of Analysis*. 14th edn, Washington, D.C., USA.

Amerine, M.A., Pangborn, R.M. and Roessler, E.B. (1965) *Principles of Sensory Evaluation*, Academic Press, London.

Ang, C.Y.W., Chang, C.M., Frey, A.E and Livingston, G.E. (1975) Effects of heating methods on vitamin retention in six fresh or frozen prepared food products. *Journal of Food Science*, **40**, 997–1003.

Ang, C.Y.W., Basillo, L.A., Cato, B.A. and Livingston, G.E. (1978) Riboflavin and thiamine retention in frozen beef-soy patties and frozen fried chicken heated by methods used in food service operations. *Journal of Food Science*, **43**, 1024–1025, 1027.

Anon. (1987a) Cooking in a vacuum. *Hospital Caterer*, January, 14.

Anon. (1987b) Sealed with a hiss. *Meat Processing*, July, 66.

Anon. (1993) Test van luchtledig bereide gerechten. *Test-Aankoop Magazine*, (352), 14–20.

Armstrong, G.A. (1996) Development and validation of a trained sensory panel for *sous vide* bolognaise testing, in *Proceedings of Second European Symposium on Sous Vide*, 10–12

April 1996, Alma University Restaurants/FAIR, University of Leuven, Belgium, pp. 271–280.

Bacon, F. (1990) Working to Rouxl. *Caterer & Hotelkeeper,* 22 February, 45–47.

Baird, B. (1990) *Sous vide*: What's all the excitement about? *Food Technology,* **44** (11), 92, 94, 96.

Bauler, M. (1990) *Sous vide*: The chef's viewpoint, in *Proceedings of a Conference – Sous vide in Australia* (ed. J. Sumner), Clunies Ross House, Melbourne, 14 November 1990, M&S Food Consultants: Albert Park, Victoria, pp. 8–16.

Bell, R., Meiselman, H.L., Pierson, B.J. and Reeve, W.G. (1994) The effects of adding an Italian theme to a restaurant on the perceived ethnicity, acceptability, and selection of foods. *Appetite,* **22**, 11–24.

Bender, A.E. (1987) Nutritional changes in food processing, in *Developments in Food Preservation – 4* (ed. S. Thorne), Elsevier Applied Science, London, pp. 1–34.

Bergslien, H. (1996) *Sous Vide* treatment of salmon (*Salmon salar*), in *Proceedings of Second European Symposium on Sous vide.* 10–12 April 1996, Alma University Restaurants/FAIR, University of Leuven, Belgium, pp. 281–291.

Bertagnoli, L. (1987) Pouches gain ground in foodservice. *Restaurants & Institutions,* **97** (10), 178–180, 184.

Bertelsen, G. and Juncher, D. (1996) Oxidative stability and sensory quality of *sous vide* cooked products, in *Proceedings of Second European Symposium on Sous Vide.* 10–12 April 1996, Alma University Restaurants/FAIR, University of Leuven, Belgium, pp. 133–145.

Bjorkman, A. and Delphin, K.A. (1966) Sweden's Nacka hospital food system centralizes food preparation and distribution. *Cornell Hotel & Restaurant Administration Quarterly,* **7** (3), 84–87.

Bognar, A. (1980) Nutritive value of chilled meals, in *Advances in Catering Technology* (ed. G. Glew), Applied Science, London, pp. 387–408.

Bognar, A. (1990) Vitamin status of chilled food, in *Processing and Quality of Foods: Vol. 3 – Chilled Foods – The Revolution in Freshness* (eds P. Zeuthen *et al.*), Elsevier Applied Science, London and New York, pp. 3.85–3.103.

Bognar, A., Bohling, H. and Fort, H. (1990) Nutrient retention in chilled foods, in *Chilled Foods – The State of the Art* (ed. T.R. Gormley), Elsevier Applied Science, London and New York, pp. 305–336.

British Nutrition Foundation. (1987) *Food Processing – A Nutritional Perspective.* Briefing Paper No. 11, 38 pp.

Buck, E.M., Hickey, A.M. and Rosenau, J. (1979) Low-temperature air oven vs. a water bath for the preparation of rare beef. *Journal of Food Science,* **44**, 1602–1605, 1611.

Buckley, C. (1987) Storage stability of vitamin C in a simulated *sous-vide* process. *Hotel and Catering Research Centre Laboratory Report* No. 238, Huddersfield Polytechnic, 2 pp.

Campbell, L. (1993) Swingin' with *Sous-Vide. Foodservice and Hospitality.* **26** (2), 88, 92, 96.

Cannon, J.E., Morgan, J.B., Schmidt, G.R., Delmore, R.J., Sofos, J.N., Smith, G.C. and Williams, S.N. (1995) Vacuum-Packaged Pre-cooked Pork from Hogs Fed Supplemental Vitamin E: Chemical, Shelf-Life and Sensory Properties. *Journal of Food Science,* **60** (6), 1179–1182.

Choain, F. and Noel, P. (1989) *Le Sous-Vide en Cuisine – Des fondements scientifiques aux applications pratiques,* Editions Jacques Lanore, Malalkoff, France, 207 pp.

Church, I.J. (1990) *An Introduction to 'Method Sous-Vide',* Department of Hospitality Management, Leeds Polytechnic, Leeds, UK, 46 pp.

Church, I.J. (1996) The sensory quality of chicken and potato products prepared using cook-chill and *sous vide* methods, in *Proceedings of Second European Symposium on Sous Vide,* 10–12 April 1996, Alma University Restaurants/FAIR, University of Leuven, Belgium, pp. 317–325.

Church, I.J. and Parsons, A.L. (1993) Review: *sous vide* cook-chill technology. *International Journal of Food Science and Technology,* **28** (6), 563–574.

Cooksey, D.K., Klein, B.P. and McKeith, F.K. (1988) Effect of packaging process on thiamine retention and other characteristics of pre-cooked vacuum-packaged beef roasts. *Proceedings of IFT Annual Meeting,* New Orleans, June 1988, Paper No. 653.

Cooksey, K., Klein, B.P. and McKeith, F.K. (1990) Thiamin retention and other characteristics of cooked beef loin roasts. *Journal of Food Science,* **55** (3), 863–864.

Coomes, S. (1994) *Sous vide* – closing the gap between truth and reality. *The Consultant*, Winter, 27–29, 58.

Creed, P.G., Reeve, W.G. and Pierson, B.J. (1993) The sensory quality of *sous vide* foods, in *Proceedings of the 1993 Food Preservation 2000 Conference: Integrating Processing, Packaging, and Consumer Research.* 19–21 October 1993, US Army Natick Research, Development and Engineering Center, Natick, Mass, USA, Vol. 1, pp. 459–472.

Creed, P.G. (1995) The sensory and nutritional quality of '*sous vide*' foods. *Food Control*, 6 (1), 45–52.

Creed, P.G. (1996) *The Sous Vide Method – An Annotated Bibliography*, Worshipful Company of Cooks Centre for Culinary Research, Bournemouth University, Poole, UK, 156 pp.

DoH (Department of Health) (1989) *Chilled and Frozen Foods*, Guidelines on Cook-Chill and Cook-Freeze Catering Systems, HMSO, London.

Dahl, C.A. and Matthews, M.E. (1980) Cook/Chill foodservice system with a microwave oven: thiamin content in portions of beef loaf after microwave heating. *Journal of Food Science*, 45, 608–612.

Daniels, D. (1988) *Sous vide* & CapKold... two approaches to cooking 'Under Vacuum'. *The Consultant*, 21 (2), Spring, 26–28.

De Liagre, C. (1985) Train fare. *Home and Garden*, 157, November, 202–205, 271–272.

Dinardo, M., Buck, E.M. and Clydesdale, F.M. (1984) Effect of extended cook times on certain physical and chemical characteristics of beef prepared in a waterbath. *Journal of Food Science*, 49, 844–848.

Dürr, P. (1979) Development of an odour profile to describe apple juice essences. *Lebensmittel-Wissenschaft und Technologie*, 12, 23–26.

Eitenmiller, R.R. (1990) Strengths and weaknesses of assessing vitamin content of food. *Journal of Food Quality*, 13, 7–20.

Endo, F. (1996) Personal communication. Wakayama University, Wakayama-shi, Japan.

Erdmann, J.W. and Klein, B.P (1982) Harvesting, processing and cooking influences on vitamin C in foods, in *Ascorbic Acid: Chemistry, Metabolism and Uses* (eds P.A. Seib and B.M. Tolber), Advances in Chemistry Series 200, American Chemical Society, pp. 499–532.

Eriksen, H. & Lassen, A. (1996) Vitamin retention in prepacked foods: comparison to traditional techniques, in *Proceedings of Second European Symposium on Sous Vide*, 10–12 April 1996, Alma University Restaurants/FAIR, University of Leuven, pp. 339–340.

Ezell, B.D. and Wilcox, M.S. (1959) Loss of vitamin C in fresh vegetables as related to wilting and temperature. *Journal of Agricultural Food Chemistry*, 7, 507–509.

Genigeorgis, C. (1993). Additional hurdles for '*sous vide*' products, in *Proceedings of First European Symposium – 'Sous Vide' Cooking*, 25–26 March 1993, Alma University Restaurants/FLAIR, University of Leuven, Belgium, 20 pp.

Ghazala, S. (1994) New packaging technology for seafood preservation: shelf-life extension and pathogen control, in *Fisheries Processing: Biotechnological Applications* (ed. A.M. Martin), Chapman & Hall, London, pp. 82–110.

Ghazala, S., Aucoin, J. and Alkanani, T. (1996) Pasteurization effect on fatty acid stability in a *sous vide* product containing seal meat (*Phoca groenlandica*). *Journal of Food Science*, 61 (3), 520–523.

Gittleson, B., Saltmarch, M., Cocotas, P. and McProud, L. (1992) Quantification of the physical, chemical and sensory modes of deterioration in *sous-vide* processed salmon. *Journal of Foodservice Systems*, 6, 209–232.

Glew, G. (1990) Pre-cooked chilled foods in catering, in *Processing and Quality of Foods: Vol. 3 – Chilled Foods – The Revolution in Freshness* (eds P. Zeuthen *et al.*), Elsevier Applied Science, London, pp. 3.31–3.41.

Goto, M., Hashimoto, K and Yamada, K. (1995) Difference of ascorbic acid content among some vegetables, texture in chicken and pork, and sensory evaluation score in some dishes between vacuum and ordinary cooking. *Nippon Shokuhin Kagaku Kogaku Kaishi*, 42 (1), 50–54.

Gould, G. (1996) Conclusions of ECFF Botulinum Working Party, in *Proceedings of Second European Symposium on Sous Vide*, 10–12 April 1996, Alma University Restaurants/FAIR, University of Leuven, Belgium, pp. 173–180.

Goussault, B. (1993) Produits cuits *sous vide* et plats cuisinés en atmosphère modifiée. *Industries Alimentaires et Agricoles (IAA)*, 110 (June), 443–445.

Gregory, J.F. (1983). Methods of vitamin assay for nutritional evaluation of food processing. *Food Technology*, **37** (1), 75–80.

Griffiths, R.P., Clifton, V.J. and Booth, D.A. (1984) Measurement of an individual's optimally preferred level of a food flavour, in *Progress In Flavour Research 1984* (ed. J. Adda), Proceedings of 4th Weurman Flavour Research Symposium, Dourdan, France, Elsevier Applied Science, London, pp. 81–90.

Hansen, T.B., Knøchel, S., Juncher, D. and Bertelsen, G. (1995) Storage characteristics of *sous vide* cooked roast beef. *International Journal of Food Science and Technology*, **30**, 365–378.

Hare, L. (1996) The development of rapid methods for the detection of folic acid and its derivates in food and their application to *sous vide* foods, in *Proceedings of Second European Symposium on Sous Vide*, 10–12 April 1996, Alma University Restaurants/FAIR, University of Leuven, Belgium, pp. 353–363.

Hattula, T. and Kiesvaara, M. (1996) Breakdown products of adenosine triphosphate in heated fishery products as an indicator of raw material freshness and of storage quality. *Lebensmittel Wissenschaft und Technologie*, **29** (1/2), 135–139.

Hunt, C. (1984) Nutrient losses in cook-freeze and cook-chill catering. *Human Nutrition: Applied Nutrition*, **38A**, 50–59.

Ivany, L. (1988) Perry promotes *sous vide* at SAFSR conference. *Restaurant Management*, **2** (6), 43.

Jones, S.L., Carr, T.L. and McKeith, F.K. (1987) Palatability and storage characteristics of pre-cooked pork roasts. *Journal of Food Science*, **52** (2), 279–281, 285.

Jonsson, L. and Danielsson, K. (1981) Vitamin retention in foods handled in foodservice systems. *Lebensmittel Wissenschaft und Technologie*, **14**, 94–96.

Kahn, L.N. and Livingston, G.E. (1970) Effect of heating methods on thiamin retention in fresh or frozen prepared foods. *Journal of Food Science*, **35**, 349–351.

Kalinowski, T. (1988) Ready, set, serve. *Canadian Hotel & Restaurant*, **66** (6), 24–26.

Khan, M.A., Klein, B.P and Lee, F.V. (1982) Thiamin content of freshly prepared and leftover Italian spaghetti served in a university cafeteria food service. *Journal of Food Science*, **47**, 2093–2094.

Kossovitsas, C., Navar, M., Chang, C.M. and Livingston, G.E. (1973) A comparison of chilled-holding versus frozen storage on quality and wholesomeness of some prepared foods. *Journal of Food Science*, **38**, 901–902.

Kraxner, U. (1981) *Qualität und Verhalten einiger Gemüsearten nach Lagerung in unterscheidlicher relativer Luftfeuchte*, Doctoral Thesis, Technical University, Munich, Germany, 141 pp.

Langley-Danysz, P. (1992) Est-il souhaitable d'allonger la DLC? Cuisson *sous vide*: le temps de la reflexion. *Revue Industries Agro-Alimentaires (RIA)*, (476), 38–41.

Leadbetter, S. (1989) *Sous-vide – A Technology Guide*, British Food Manufacturing Industries Research Association, Leatherhead, Surrey, UK, 22 pp.

Lefort, J., Sonnet, R. and Bulteau, M. (1993) Étude de la qualité bouchère des carcasses de porcs résistant différemment au stress. *Viandes et Produits Carnes*, **14** (5), 123–126.

Levine, J.B. and Rossant, J. (1987) 'Sous-vide' cooking: Haute cuisine in a pouch? *Business Week*, 13 July, 104–105.

Levy, P. (1986) A taste of technology. *Observer Magazine*, London, 2 March, 58–60.

Light, N. and Walker, A. (1990) *Cook-Chill Catering: Technology and Management*, Elsevier Applied Science, London, pp. 157–178.

Light, N., Hudson, P., Williams, R., Barrett, J. and Schafheitle, J.M. (1988) A pilot study on the use of *sous-vide* vacuum cooking as a production system for high quality foods in catering. *International Journal of Hospitality Management*, **7** (1), 21–27.

Lumley I.D. (1993) Vitamin analysis in foods, in *The Technology of Vitamins in Food* (ed. P. Berry Ottaway), Blackie Academic & Professional, Glasgow, UK, pp. 172–232.

Lyman B. (1989). *A Psychology of Food*, Van Nostrand Reinhold, New York, pp. 140–146.

Manser, S. (1988) Comparing notes. *Taste*, November, 16.

Martens, T. (1995) Current status of *Sous vide* in Europe, in *Principles of Modified-Atmosphere and Sous Vide Product Packaging* (eds J.M. Farber and K.L. Dodds), Technomic Publishing, Lancaster, USA, 37–68.

Mason, L.H., Church, I.J., Ledward, D.A. and Parsons, A.L. (1990) Review: The sensory quality of foods produced by conventional and enhanced cook-chill methods. *International Journal of Food Science and Technology*, **25**, 247–259.

Mayer, C.E. (1990) A computer can be a cook, but will it ever be a chef? Programs to establish the '*Sous-Vide*' process. *Washington Post*, 25 July, E1, E10.

McGuckian, A.T. (1969) The A.G.S. Food System – Chilled pasteurised food. *Cornell Hotel & Restaurant Administration Quarterly*, May, 87–92, 99.

Meiselman, H.L., Hedderley, D., Staddon, S.L., Pierson, B.J. and Symonds, C.R. (1994) The effect of effort on meal selection and meal acceptability in a student cafeteria. *Appetite*, **23**, 43–55.

Metayer, M. (1991) Qualité nutritionelle des produits *sous vide*. *Les Journées du Sous vide en Agro-Alimentaire*, ISVAC, 19–20 November 1991, Roanne, France, pp. 92–97.

Mielgaard, M.C., Dalgleish, C.E. and Clapperton, J.F. (1979). Beer flavour terminology. *Journal of the Institute of Brewing*, **85**, 38–52.

Ministère de l'Agriculture (1974) Réglementation des conditions d'hygiène relatives à la préparation, la conservation, la distribution et la vente des plats cuisinés à l'avance (Arrêté du 26 Juin 1974), *Journal Officiel de la Republique Française*, 16 Juillet 1974, 7397–7399.

Ministère de l'Agriculture (1988) *Prolongation de la dureé de vie des plats cuisinés à l'avance, modification du protocole permettant d'obtenir les autorisations.* (Note de Service DGAL/SVHA/N88/8106 du 31 Mai 1988), Service Vétérinaire d'Hygiéne Alimentaire, Paris, 8 pp.

Nandhasri, P. and Suksangpleng, S. (1986) Application of high performance liquid chromatography to determination of seven water-soluble vitamins in white sauce. *Journal of the Scientific Society of Thailand*, **12**, 111–118.

Nazaire, B. (1987) La cuisson *sous vide*: un mythe? *Revue Industries Agro-Alimentaires (RIA)*, (393), 28–29.

Nicholanco, S. and Matthews, M.E. (1978). Quality of beef stew in a hospital chill food-service system. *Journal of the American Dietetic Association*, **72**, 31–37.

Odake, S. (1996) Diffusion phenomena of substances in *sous vide* cooking, in *Proceedings of Second European Symposium on Sous Vide*. 10–12 April 1996, Alma University Restaurants/FAIR, University of Leuven, Belgium, pp. 367–368.

Peters, G., Morrissey, M.T., Sylvia, G. and Bolte, J. (1996) Linear regression, neural network and induction analysis to determine harvesting and processing effects on Surimi quality. *Journal of Food Science*, **61** (5), 876–880.

Petersen, M.A. (1993) Influence of *sous vide* processing, steaming and boiling on vitamin retention and sensory quality in broccoli florets. *Zeitschrift für Lebensmittel Untersuchung und Forschung*, **197** (4), 375–380.

Petit, R. (1990) Clean – *sous vide*. *Restaurant Hospitality*, **74** (11), 150.

Picoche, M. (1991) Incidence de la cuisson pasteurisation sur les qualités organoleptiques et microbiologiques des poissons et des legumes, in *Les Journées du Sous Vide en Agro-Alimentaire*, ISVAC, 19–20 November 1991, Roanne, France, pp. 133–142.

Pierson, B.J., Reeve, W.G. and Creed, P.G. (1995) 'The quality experience' in the food service industry. *Food Quality and Preference*, **6** (3), 209–212.

Piggott, J.R. (ed.) (1986). *Statistical Procedures in Food Research*, Elsevier Applied Science, London.

Piggott, J.R. (ed.). (1988) *Sensory Analysis of Foods*, 2nd edn, Elsevier Applied Science, London.

Polesello, A. and Rizzolo, A. (1986) Applications of HPLC to the determination of water-soluble vitamins in foods (a Review 1981–1985). *Journal of Micronutrient Analysis*, **2**, 153–187.

Poulsen, K.J. (1978) Vacuum-packaged cooked potatoes, in *How Ready are Ready-to-serve Foods?* (ed. K. Paulus), S. Karger, Basel, pp. 138–146.

Powers, J.J. (1988) Current practices and application of descriptive methods, in *Sensory Analysis of Foods*, (ed. J.R. Piggott), 2nd edn, Elsevier Applied Science, London, pp. 187–266.

Pralus, G. (1985) *La Cuisine Sous Vide – Une Histoire d'Amour – La Cuisine de l'An 2000*, published by author, Briennon, 42720 Pouilly-sous-Charlieu, France, 447 pp.

Pring, A. (1986) *Sous-vide*: Boon or threat? *Caterer & Hotelkeeper*, 24 July, 19.

Raffael, M. (1984) Revolution in the kitchen. *Caterer & Hotelkeeper*, 16 August, 34, 35.

Ready, C.A. (1971) Method of preparing and preserving ready to eat foods. US Patent 3,607,312, 4 pp.

Reeve, W.G., Creed, P.G. and Pierson, B.J. (1994) The restaurant – a laboratory for assessing meal acceptability. *International Journal of Contemporary Hospitality Management*, **6** (4), iv–vii.

Robson, C.P. and Collison, R. (1989) Pre-cooked chilled foods. *Catering and Health*, **1**, 151–163.

Rosinski, M.J., Barmore, C.R., Bridges, W.C. Jr, Dick, R.L. and Acton, J.C. (1989) Phosphate type and salt concentration effects on shear strength and packaging film adhesion to processed meat from a cook-in packaging system. *Journal of Food Science*, **54** (6), 1422–1425, 1430.

SVAC (*Sous vide* Advisory Committee) (1991) *Codes of Practice for Sous Vide Catering Systems*, SVAC, Tetbury, Gloucs., UK, 12 pp.

Sarli, T.A., Santoro, A., Anastasio, A. and Cortesi, M.L. (1993) Conservabilità di spigole ed orate sottoposte a cottura sottovuoto a basse temperature. *Industrie Alimentari*, **32** (316), 593–598.

Saxby, M.J. (1978) Applications of high-performance liquid chromatography in food analysis, in *Developments in Food Analysis Techniques – 1* (ed. R.D. King), Applied Science, London, pp. 125–153.

Scarpa, J. (1988) *Sous vide* – fresh technology. *Restaurant Business*, **87** (4), 136, 138, 140.

Schafheitle, J.M., Williams, R., Barrett, J., Hudson, P. and Light, N. (1986) Sous-vide: The way ahead?, in *The Future of Cook-chill Technology in Catering*. (ed. N. Light), Workshop at Dorset Institute of Higher Education, Poole, Dorset, UK, 17–18 April 1986.

Schafheitle, J.M. and Light, N.D. (1989a) Technical note: *Sous-vide* preparation and chilled storage of chicken ballotine. *International Journal of Food Science and Technology*, **24**, 199–205.

Schafheitle, J.M. and Light, N.D. (1989b) *Sous-vide* cooking and its application to cook-chill – what does the future hold? *Journal of Contemporary Hospitality Management*, **1**, 5–10.

Schafheitle, J.M. and Pierson, B.J. (1992) Unpublished data, Bournemouth University.

Schellekens, M. and Martens, T. (1992a) '*Sous Vide*' Cooking Part I: Scientific Literature Review, Publication No. EUR 15018 EN, Commission of the European Communities: Luxembourg, 185 pp.

Schellekens, M. and Martens, T. (1992b) '*Sous Vide*' Cooking Part II: Feedback from practice, *ibid.*, 54 pp.

Schellekens, M., Martens, T., Roberts, T.A., Mackey, B.M., Nicolaï, B.M., Van Impe, J.F. and De Baerdemaeker, J. (1994) Computer aided microbial safety design of food processes. *International Journal of Food Microbiology*, **24**, 1–9.

Schwarz, T. (1988) *Sous vide* guidelines. *Restaurant Business*, **87** (6), 10 April, 94, 96, 97.

Sellers, J. (1990) In the bag. *Foodservice Equipment & Suppliers Specialist*, **43** (3), 44, 46, 48.

Sessions, P. (1987) Sensible *sous vide*. *Hospital Caterer*, May/June, 14.

Shamaila, M., Powrie, W.D. and Skura, B.J. (1992) Sensory evaluation of strawberry fruit stored under modified atmosphere packaging (MAP) by quantitative descriptive analysis. *Journal of Food Science*, **57** (5), 1168–1172, 1184.

Shamsuzzaman, K., Chuaqui-Offermanns, N., Lucht, L., McDougall, T. and Borsa, J. (1992) Microbiological and other characteristics of chicken breast meat following electron-beam and *sous-vide* treatments. *Journal of Food Protection*, **55** (7), 528–533.

Shamsuzzaman, K., Lucht, L. and Chuaqui-Offermanns, N. (1995) Effects of combined electron-beam irradiation and *sous-vide* treatments on microbiological and other qualities of chicken breast meat. *Journal of Food Protection*, **58** (7), 497–501.

Sheard, M. and Church, I. (1992) *Sous Vide Cook–Chill*, Leisure and Consumer Studies, Leeds Polytechnic, Leeds, UK, 47 pp.

Shepherd, R., Farleigh, C.A., Land, D.G. and Franklin, J.G. (1984) Validity of a relative-to-ideal rating procedure compared to hedonic rating, in *Progress In Flavour Research 1984* (ed. J. Adda), Proceedings of 4th Weurman Flavour Research Symposium, Dourdan, France, Elsevier Applied Science, London, pp. 103–110.

Shortreed, G.W., Rickard, M.L., Swan, J.S. and Burtles, S.M. (1979) The flavour terminology of Scotch whisky. *Brewer's Guardian*, **11**, 55–62.

Smith, D.B. and Fullum-Bouchard, L. (1990) Comparative nutritional, sensory and microbiological quality of a cooked chicken menu item produced and stored by cook/chill, cook/freeze and *sous vide* cook/chill methods. Poster presented at *Canadian Dietetic Association Annual Conference*, 7 June 1990, 6 pp.

Smith, D.M. and Alvarez, V.B. (1988) Stability of vacuum cook-in-bag turkey breast rolls during refrigerated storage. *Journal of Food Science*, **53** (1), 46–48, 61.

Sornay, M. (1990) The *'sous vide'* system in catering, in *Chilled Foods – The Ongoing Debate* (eds T.R. Gormley and P. Zeuthen), Elsevier Applied Science, London, pp. 99–100.

Speirs, J.P. (1996) The development of microbiological assays for the detection of Vitamin B_1 (Thiamine), in *Proceedings of Second European Symposium on Sous Vide*, 10–12 April 1996, Alma University Restaurants/FAIR, University of Leuven, Belgium, pp. 369–377.

Stites, C.R., McKeith, F.K., Bechtel, P.J. and Carr, T.R. (1989) Palatability and storage characteristics of pre-cooked beef roasts. *Journal of Food Science*, **54** (1), 3–6.

Thomson, D.M.H. (ed.) (1988) *Food Acceptability*, Elsevier Applied Science, London.

Thorsell, U. (1996) Quality of meat prepared by the *sous vide* method, in *Proceedings of Second European Symposium on Sous Vide*, 10–12 April 1996, Alma University Restaurants/FAIR, University of Leuven, Belgium, pp. 387–388.

Tjomb, P. (1990) Cuisson *sous vide* soft ou hard. *Revue Industries Agro-Alimentaires (RIA)*, (441), 30–31.

Tjomb, P. (1992) Vers une optimisation des barèmes thermiques. Les petits plats mijotés à la portée de l'industrie. *Revue Industries Agro-Alimentaires (RIA)*, (480), 44, 46.

Tuan, S., Wyatt, J. and Anglemier, A.F. (1987) The effect of erythorbic acid on the determination of ascorbic acid levels in selected foods by HPLC and Spectrophotometry. *Journal of Micronutrient Analysis*, **3**, 211–228.

Turner, B.E. and Larick, D.K. (1996). Palatability of *sous vide* processed chicken breast. *Poultry Science*, **75**, 1056–1063.

Uribe-Saucedo, S.M. and Ryley, J. (1982) A comparative study of ascorbic acid and thiamine retention in pouch sterilisation and canning. *Proceedings of the Institute of Food Science and Technology*, **15** (3), 120–128.

Van Niekerk, P.J. (1988) Determination of Vitamins, in *HPLC in Food Analysis* (ed. E. Macrae), 2nd edn, Academic Press, London, pp. 133–184.

Varoquaux, P., Offant, P. and Varoquaux, F. (1995) Firmness, seed wholeness and water uptake during the cooking of lentils (*Lens culinaris* cv. *anicia*) for *'sous vide'* and catering preparations. *International Journal of Food Science & Technology*, **30** (1), 215–220.

Vedel, A., Charle, G., Charney, P. and Tourmeau, J. (1972) *B-1. Le Vin – définition, B-2. Caracteres organoleptiques*, Institut National des Appellations d'Origine des Vins et Eux-de-vie, Macon, France, cited by Powers, J.J. (1988), *op. cit.*

Walker, A., West, A. and Lawson, J. (1996) The nutritional quality of vegetables processed by cook-serve and cook-chill *sous vide* techniques, in *Proceedings of Second European Symposium on Sous Vide*, 10–12 April 1996, Alma University Restaurants/FAIR, University of Leuven, Belgium, pp. 391–397.

Watier, B. (1988) L'incidence des nouveaux procédés sur les teneurs en vitamines des aliments. *Information Diététique*, (3), 33–38.

Watier, B. and Belliot, J.P. (1990) Aspects vitaminiques des produits de la IVe gamme et de la Ve gamme. *Information Diététique*, (1), 42–48.

Watier, B. and Belliot, J.P. (1991) Vitamines et Technologie Industrielle Récente. *Cahiers de Nutrition et de Diététique*, **26** (1), 23–26.

Williams, A.A. (1975) The development of a vocabulary and profile assessment method for evaluating flavour contribution of cider and perry aroma constituents. *Journal of the Science of Food and Agriculture*, **26**, 567–582.

Xie, G., Keeling, H. and Sheard, M.A. (1996) Applying fuzzy mathematics to process and product development of *sous vide* foods, in *Culinary Arts and Sciences – Global and National Perspectives* (ed. J.S.A. Edwards), (Proceedings of ICCAS 96 – International Conference on Culinary Arts and Sciences, 25–28 June 1996, Bournemouth, UK), Computational Mechanics Publications, Southampton, UK, pp. 211–220.

Yoshimura, M., Shono, Y. and Yamauchi, N. (1995) Influence of vacuum packaging and processing temperature on the quality change of mitsuba. *Nippon Shokuhin Kogaku Kaishi*, **42** (8), 588–593.

4 The potential role of time–temperature integrators for process impact evaluation in the cook–chill chain

ANN VAN LOEY, TAMARA HAENTJENS and MARC HENDRICKX

4.1 Introduction

The impact of a treatment on a product safety or quality attribute depends on the rates of the reactions that affect this attribute, and on the time interval during which these reaction rates occur. The 'status' of a food, expressed in terms of its safety and/or quality, is determined by the effect of all reactions occurring in the product, integrated over the full history of the product until the moment of consumption. In Figure 4.1, the idea of the 'preservation reactor' is presented. This concept applies to the entire process of manipulating a food product – preparation and packaging, processing, distribution and storage – as well as to a single unit operation or processing step in this chain. The rates at which desired and undesired reactions (related to food safety and quality) take place are functions of both intrinsic (i.e. food-specific) properties and extrinsic (i.e. process-specific) factors (Maesmans *et al.*, 1990).

Sous vide cooking or vacuum cooking can be divided into two major processing steps, involving different temperature regimes. At first, raw materials are cooked under controlled conditions of temperature and time inside heat-stable vacuumed pouches. Afterwards, the product is quickly chilled and stored refrigerated until the moment of consumption after reheating. The severity of the heating process determines which vegetative microorganisms or spores can survive the cooking process. The subsequent chilling process and refrigerated storage has to prevent growth and germination of vegetative cells and spores still present in the food. As a consequence, the safety and quality status of a *sous vide* product as well as its shelf life is strongly dependent on its temperature exposure history, from production, through distribution and storage to consumption.

The *sous vide* Advisory Committee (SVAC), in their code of practice (1991), recognise three critical control points (CCPs) in the *sous vide* process: (i) cooking: time–temperature exposure, (ii) cooling rate, (iii) storage: time–temperature exposure. Temperature is one of the crucial extrinsic factors for guaranteeing safety and quality during the production and

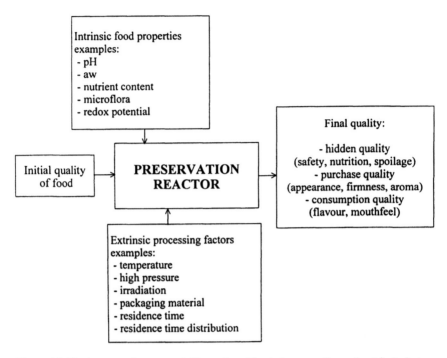

Figure 4.1 The 'preservation reactor'. Examples of intrinsic properties and extrinsic factors that can influence the quality of the product.

storage of *sous vide* products. As part of the approach to ensuring product safety and quality through temperature monitoring and control, attention has been focused on the potential use of time–temperature indicators (Selman, 1995).

The extent to which the information of a time–temperature indicator reflects a real time–temperature history depends on the type of indicator. Labuza and Taoukis (1990) proposed a three-category classification:

1. Critical temperature indicators (CTIs), which show exposure above (or below) a reference temperature. They merely indicate that the product was exposed to an undesirable temperature for a sufficient time to cause a change critical to its safety or quality.
2. Critical time–temperature integrators (CTTIs) which indicate a response that reflects the cumulative time–temperature exposure above a critical reference temperature. Their responses can be translated into an equivalent exposure time at the critical temperature.
3. Time–temperature integrators (TTIs) which give a continuous, temperature-dependent response. They integrate, in a single measurement, the full time–temperature history from the time of activation.

CTIs and CTTIs are partial history indicators (response over threshold) whereas TTIs are full history integrators (continuous response). The main focus of this chapter will be on full history TTIs for impact evaluation in the cook–chill chain.

In a first part, some general aspects of TTIs as modern process evaluation tools will be presented. Afterwards, a major focus will be on the potential use of integrators (i) to monitor the thermal efficacy of a cooking process and (ii) to monitor the chilling and refrigerated storage of food products. Similar approaches can, indeed, be taken for monitoring refrigerated distribution and storage of food products as for assessing thermal processes. TTIs to monitor lethal efficacy of a thermal treatment are based on reactions that occur at elevated temperatures whereas TTIs to monitor quality losses during low-temperature storage and distribution are based on reactions that occur at chilling temperatures.

4.2 General aspects on time–temperature integrators

4.2.1 Definition of a time–temperature integrator

A TTI can be defined as 'a small measuring device that shows a time–temperature dependent, easily, accurately and precisely measurable irreversible change that mimics the changes of a target attribute undergoing the same variable temperature exposure' (Taoukis and Labuza, 1989a, 1989b; Weng et al., 1991; De Cordt et al., 1992). The target attribute can be any safety or other quality attribute of interest such as microorganism (spore) inactivation, microbial growth, loss of a specific vitamin, texture, colour or residual shelf life. Of major interest are important hidden quality attributes (e.g. microbiological safety/risk analysis). A TTI mimics the evolution of a time–temperature related factor which is liable to affect the quality/safety of the foodstuff over the full range of temperatures likely to be experienced. The change occurring to the indicator during the exposure has to be irreversible in order to be able to quantify the impact of the exposure afterwards. However, after read-out of the monitoring system, it may in some cases be possible to reverse the change in order to reuse the indicator.

The major advantage of TTIs is the ability to quantify the integrated time–temperature impact on a target attribute without information on the actual temperature history of the product. By the above definition, a TTI is clearly defined without any reference to knowledge on the time–temperature profile.

It should be stressed that if temperature is not the only rate-determining factor, using only a TTI to monitor food safety or quality loss would result in error, as other factors that change with time can be critical. To evaluate such processes, a 'product history integrator' would be necessary, to mimic the behaviour of its target attribute exactly by responding in the very same way, to all of the changing intrinsic and extrinsic factors, as the quality attribute that it is designed for. However, the full product history integrator is currently only a concept and additional kinetic data and mathematical models are needed to describe the dependence of the reaction rates of desired and undesired reactions to all influencing factors, extrinsic as well as intrinsic.

4.2.2 Criteria of time–temperature integrators

From the definition of a TTI given above, several criteria that such a measuring device should meet can be formulated[1]:

- For convenience, the TTI has to be inexpensive, quickly and easily prepared, easy to recover, and give an accurate and user-friendly unambiguous read-out. The indicator may need to be stored and stabilised below a threshold temperature before use. It is essential that the start point of the life of the TTI, i.e. when it is activated, is known for certain, and that no reasonable possibility of pre-activation, partial activation or especially post-activation can occur.
- The TTI should experience the same time–temperature profile as the parameter under investigation. Ideally, the TTI should be incorporated into the food without disturbing heat transfer within the food. The presence of the TTI must not change the time–temperature profile of the food.
- The TTI should quantify the impact of the processing step on a target safety/quality attribute. This criterion results in a specific kinetic requirement: the basic kinetic requirement for a TTI to adequately mimic the target quality or safety attribute of interest and to give reliable information on the status of the product is that the temperature sensitivity of the rate constant, as expressed by the activation energy or z-value, of the TTI and the target quality (safety) attribute are equal.

There is no restriction on the absolute value of the reaction rate constant of a TTI system. However, depending on the analytical technique used to monitor the response of the TTI, the reaction rate of the chosen system will have to allow a detectable response to the temperature history. Because the aim of using a TTI is to calculate the process impact relying solely on the TTI response, it is necessary that the TTI response kinetics obey a rate

[1]A joint Ministry of Agriculture, Fisheries and Food (MAFF)/industry working party met during 1991 at the Campden and Chorleywood Food Research Association, and has completed a food industry specification to provide a basis for indicator manufacturers for chilled and frozen foods to design the performance of their indicators to meet the needs of the industry and at the same time to provide a basis for the users of such indicators to check the indicator performance against their requirements (George and Shaw, 1992).

equation that allows separation of the variables, so that a response function can be defined. Furthermore, the rate constant of TTI and target attribute have to obey the same law as to their temperature dependence. In general, this condition is fulfilled as mostly, one is dealing with the destruction of microorganisms, enzyme inactivation, chemical reactions, or physical phenomena whose rate constant or decimal reduction time obeys the Arrhenius law or the TDT model, respectively.

Although it was recognised early (Hayakawa, 1978) that any agent used for monitoring the lethal effect of heat processes should have the same activation energy as the target attribute, but could have a different rate constant, this basic requirement for proper functioning of a TTI has been frequently neglected or even ignored.

4.2.3 Classification of time–temperature integrators

TTIs can be classified in terms of working principle, type of response, origin, application in the food material and location in the food as shown in Figure 4.2 (Hendrickx et al., 1993).

Depending on the response property, TTIs can be subdivided into biological (microbiological and enzymatic), chemical and physical systems. For biological systems, the change in biological activity such as of microorganisms, their spores (viability), or enzymes (activity) due to the treatment is the basic working principle. Chemical systems are based on a purely chemical response towards temperature and time. The temperature sensing mechanism of existing physical TTIs is often based on diffusion.

Whenever the temperature sensitivity of the target quality or safety attribute can be described with a single activation energy (E_a), a single component TTI should be looked for or developed that has an identical E_a. However, when the activation energy of the TTI and target attribute differ, additional information on the temperature history is needed, in addition to the knowledge of the change in status of the TTI in order to predict the process impact on the target attribute. Mere correlation studies whereby the response of the TTI is statistically coupled to quality changes for a given set of time–temperature conditions without any matching of temperature dependency of rate constants, offer useful but limited information because the correlations found are valid only for the exact conditions tested. Extrapolation of the correlation equations to other temperatures or for fluctuating conditions could lead to errors. Alternatively, the use of multi-component TTIs has been suggested to quantify the change in status of a quality attribute from the reading of the impact of the process on a set of individual components each characterised by its own E_a but deviating from the E_a of the target quality attribute.

With respect to the origin of the TTI, extrinsic and intrinsic TTIs can be distinguished. An extrinsic TTI is added to the food, whereas intrinsic TTIs

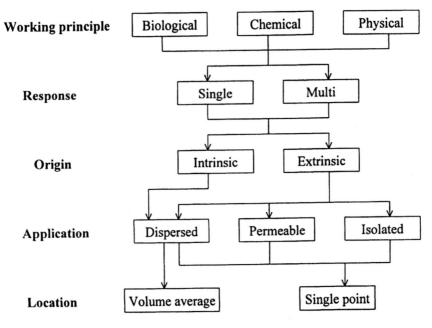

Figure 4.2 General classification of time–temperature integrators.

are intrinsically present in the food and represent the behaviour of a target index. With regard to the application of the TTI in the food product, three approaches can be distinguished: dispersed, permeable or isolated. In dispersed systems, the TTI (extrinsic or intrinsic) is homogeneously distributed throughout the food, allowing evaluation of the volume-average impact of a process. Those extrinsic TTIs that are not dispersed throughout the food may be permeable (permitting some diffusion of food components in the TTI) or isolated. All three approaches can be the basis for single-point evaluations of the process impact at specific locations within the food. Many indicators, in especially indicators for the chill chain, respond to temperatures on the outside of the pack, where there may be some thermal insulation between product and indicator. A TTI which reflects product temperature has far greater value and relevance than one which responds to the temperature on the outer surface of the pack.

4.2.4 Application scheme of a time–temperature integrator

The response of a TTI can be the measurement of the change in the status of a single component or of several individual components. If a single component TTI that has an activation energy equal to the one of the target attribute is at hand, a direct relation between the change in status of the TTI ($(X_t)_{TTI}$) and that of the target attribute it monitors ($(X_t)_{target}$) can be

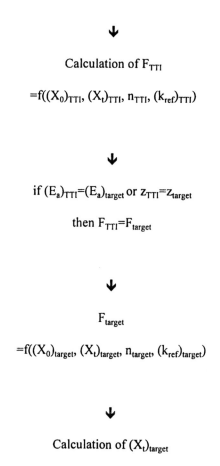

Measurement of $(X_0)_{TTI}$ and $(X_t)_{TTI}$

↓

Calculation of F_{TTI}

$$=f((X_0)_{TTI}, (X_t)_{TTI}, n_{TTI}, (k_{ref})_{TTI})$$

↓

if $(E_a)_{TTI}=(E_a)_{target}$ or $z_{TTI}=z_{target}$

then $F_{TTI}=F_{target}$

↓

F_{target}

$$=f((X_0)_{target}, (X_t)_{target}, n_{target}, (k_{ref})_{target})$$

↓

Calculation of $(X_t)_{target}$

Figure 4.3 Application scheme of a single component TTI to calculate the process impact on a target attribute (F_{target}) and/or the actual status of the target attribute after processing $(X_t)_{target}$.

proposed (Figure 4.3 – Van Loey *et al.*, 1996b). In this scheme, the equivalent time at reference temperature (F-value) is chosen as the concept to express the integrated time–temperature impact (see below).

A similar systematic approach using the concept of an equivalent temperature T_{eff} after a chosen process time, that allows the correlation of the response of a TTI $(X(t))$ to the remaining quality attribute $(A(t))$ of a food product, exposed to the same variable temperature conditions, was developed by Taoukis and Labuza (1989a) (Figure 4.4). It has been demonstrated

(Wells and Singh, 1988; Taoukis and Labuza, 1989a, 1989b; Hendrickx *et al.*, 1992) that exactly the same prerequisite for proper functioning of a TTI (i.e. equal activation energies of TTI and target attribute) is valid when the equivalent temperature at a chosen reference time concept is used to evaluate the quality retention in chilled foods and the impact of a heat treatment, provided of course that temperature is the only extrinsic factor responsible for changing the rate of safety or quality determining reactions. Again, if temperature is not the only rate determining factor, using a mere TTI to monitor food safety or quality loss would be in vain. An integrated quality monitor then should be used which mimics the behaviour of its target exactly by responding in the very same way to all changing intrinsic and extrinsic factors as the quality attribute it is designed for. In general for cold stored and chilled foods, and especially for the new generation of minimally processed and refrigerated foods (REPFEDs: Refrigerated Processed Foods with Extended Durability), information on the influence of other factors than temperature on the kinetics of quality attributes is scarce.

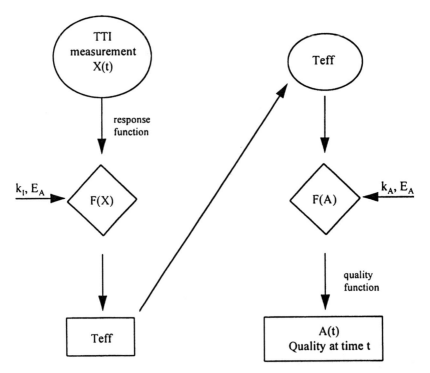

Figure 4.4 Schematic representation of the systematic approach for applying a TTI as a food quality monitor (redrawn from Taoukis and Labuza, 1989a).

4.3 Time–temperature integrators for the cook chain

4.3.1 Potential role of time–temperature integrators in the cook chain

To overcome the limitations of the most commonly used and well-elabo-rated techniques to evaluate thermal processes, namely the physical math-ematical method and the *in situ* method, TTIs have been and are being developed as alternative tools for thermal process evaluation. Hence, TTIs can be applied in the cook chain to monitor the severity of a heat treatment either in terms of food safety, or alternatively in terms of food quality. It should be stressed that, from the point of view of food safety, an overesti-mation of the actual processing impact by use of a TTI should be avoided, because it can lead to underprocessing of food products, which involves a risk for public health. Hence, process impact evaluation using a TTI should consistently and reliably provide a conservative estimate of the achieved microbiological safety level.

In the next paragraph, it is briefly described how the process impact can be determined based on the reading of a TTI. The concept can be applied for the evaluation of sterilisation as well as for pasteurisation processes.

The impact of a thermal treatment on a food attribute is usually quanti-fied using the concept of an 'equivalent time at reference temperature', referred to as the processing F-value or pasteurisation value (usually expressed in minutes). The processing value represents the equivalent time at a chosen constant reference temperature T_{ref} that would result in exactly the same impact on the specific quality attribute as the actual time–tem-perature variable profile to which the food (i.e. the attribute of interest) was subjected. The impact of a heat treatment on a product safety or quality attribute depends on the rates of the heat-induced reactions that affect this attribute, and on the time interval during which these reaction rates occur. Hence, mathematically the process value F is defined as the integral over time of the rate at each encountered temperature relative to the rate at the chosen reference temperature T_{ref} denoted as subscript (equation (4.1)).

$$^{E_a}F_{T_{ref}} = \int_0^t \frac{k}{k_{ref}} \, dt \qquad (4.1)$$

where F is the process-value, k the rate constant at T and k_{ref} the rate con-stant at reference temperature T_{ref}.

An analogue expression for the process value as in equation (4.1) can be obtained by use of the Thermal Death Time model (Bigelow, 1921):

$$^{z}F_{T_{ref}} = \int_0^t \frac{D_{ref}}{D} \, dt \qquad (4.2)$$

where F is the process value, D the decimal reduction time at T and D_{ref} the decimal reduction time at reference temperature T_{ref}.

The general expression of a first-order reaction ($n = 1$) can be written as equation (4.3) and of an nth-order reaction ($n \neq 1$) as equation (4.4):

$$\frac{dX}{dt} = -kX = -\frac{2.303}{D}X \qquad (4.3)$$

$$\frac{dX}{dt} = -kX^n \qquad (4.4)$$

Integration of these equations yields equations (4.5) and (4.6) respectively:

$$n = 1 \qquad \log\left(\frac{X}{X_0}\right) = -\int_0^t \frac{dt}{D} = -\int_0^t \frac{k}{2.303}\, dt \qquad (4.5)$$

$$n \neq 1 \qquad \frac{X^{1-n} - X_0^{1-n}}{1 - n} = -\int_0^t k\, dt \qquad (4.6)$$

so that the processing value F in case of a first-order heat inactivation can be written as equation (4.7) or, in the case of an nth-order ($n \neq 1$) inactivation, as equation (4.8):

$$n = 1 \qquad {}^{z(E_a)}F_{T_{ref}} = D_{ref}\log\left(\frac{X_0}{X}\right) = \frac{1}{k_{ref}}\ln\left(\frac{X_0}{X}\right) \qquad (4.7)$$

$$n \neq 1 \qquad {}^{E_a}F_{T_{ref}} = \frac{1}{k_{ref}}\left(\frac{X^{1-n} - X_0^{1-n}}{n - 1}\right) \qquad (4.8)$$

Hence, the use of a TTI to monitor the processing value is based on the response status of the indicator before (X_0) and after (X) thermal treatment, combined with its kinetics, using equation (4.7) or (4.8) depending on the order of the heat-induced reaction occurring to the indicator.

It can be seen that all TTIs are by definition *post factum* indicators of the impact of a thermal process, because the calculation of this process impact is based on the change in status of the TTI after thermal treatment, as compared with its initial status. Thus, TTIs are not suitable for on-line monitoring of the efficacy of thermal processes – which is feasible by direct time–temperature registration and consecutive integration of the product temperature history – but can be applied for at-line and/or off-line measurement of thermal efficacy. The time after thermal processing needed to calculate the process impact by read-out of a TTI depends upon the nature of the monitoring system: for example, microbiological assays are time-consuming (several days), whereas in general the evaluation of enzyme activity or quantification of a chemical compound is much faster (up to minutes).

4.3.2 State of the art of time–temperature integrators for the cook chain

Based on the classification described above (Figure 4.2), an overview of existing TTIs with indication of the z-value or activation energy, when available, is given in Table 4.1. As mentioned before, when no information is available on the temperature history of the product, a TTI can theoretically only be used to monitor those target aspects with equal z-values or activation energies. On practical grounds, however, the allowed difference in z-value between TTI and target attribute to ascertain a pre-set accuracy in process impact determination has been studied theoretically by Van Loey *et al.* (1995). A detailed description of the monitoring systems reported in Table 4.1 and their advantages and disadvantages has already been given by Maesmans (1993) and Hendrickx *et al.* (1995). To identify potential target attributes and their heat inactivation kinetics, we refer to databases and overview articles on the heat-inactivation kinetics of microorganisms or quality aspects (Norwig and Thompson, 1986; Villota and Hawkes, 1986; Pflug, 1987; Betts, 1992; Betts and Gaze, 1992; Villota and Hawkes, 1992). Globally spoken, based on heat-resistance data, potential growth temperature, oxygen tolerance or requirement of microorganisms, it can be stated that *Clostridia* and *Bacillus cereus* are the most important pathogens in *sous vide* products. Non-spore-forming pathogens of public health concern in *sous vide* processed products include *Salmonella*, *Staphylococcus*, *Aeromonas*, *Listeria* and *Yersinia*, while lactic acid bacteria are the main spoilage microorganisms (Schellekens and Martens, 1992).

With regard to the systems described in Table 4.1, several have been or are currently being applied for quantitative process evaluation in the research and development of (novel) thermal processing methods.

Commercially available indicators for thermal process validation have been summarised by Selman (1995) (Table 4.2).

Often, these commercially available process indicators consist of thermo-sensitive inks that change colour after exposure to a specific temperature profile and are limited to qualitative indications of the heating medium temperature: some only indicate whether a pre-set temperature has been reached and/or exceeded – e.g. 'Thermometer Strips' (3M Industrial Tapes and Adhesives, Manchester, UK), 'Colour-Therm' (Colour Therm, Surrey, UK), 'Celsistrip', 'Celsidot', 'Celsipoint' and 'Celsiclock' (Spirig Earnest, Germany). Others react after exposure for a fixed time period at a constant temperature – e.g. 'Autoclave Tape' (3M Industrial Tapes and Adhesives, Manchester, UK), 'TST' (Albert Browne Ltd, Leicester, UK), 'Cook-check' (PyMaH Corp., Flemington, NJ, USA). Still others integrate the full temperature history but can only be placed on the outside of a container – e.g. 'Sterigage' and 'Thermalog S' (PyMaH Corp., Flemington, NJ, USA). Therefore they respond only to heating medium temperature and provide no relevant information on the in-pack thermal efficacy of a process. These

Table 4.1 State of the art of TTIs for the cook chain

Origin	Application	Principle	E_a (kJ/mol) or z (°C)	Reference
Microbiological				
Extrinsic	Dispersed	*Bacillus stearothermophilus*	NR*	Burton *et al.* (1977)
		Bacillus coagulans	NR	Jones and Pflug (1981)
		Bacillus subtilis	NR	Pflug and Odlaug (1986)
		Clostridium sporogenes	NR	Pflug and Odlaug (1986)
		Inoculated Pack[a]		Yawger (1978)
	Permeable	*Bacillus stearothermophilus* in alginate	8.5°C	Bean *et al.* (1979)
		Bacillus stearothermophilus in alginate	11.4–11.8°C	Brown *et al.* (1984)
		Clostridium sporogenes in alginate	12.5–12.7°C	Brown *et al.* (1984)
		Bacillus stearothermophilus in alginate	9°C	Heppel (1985)
		Bacillus stearothermophilus in alginate	NR	Sastry *et al.* (1988)
		Clostridium sporogenes in a turkey cube	8.5°C	Segner *et al.* (1989)
		Bacillus stearothermophilus in polyacrylamide gel	11.7°C	Rönner (1990a, b)
		Bacillus stearothermophilus in alginate	8.8–9.1°C	Gaze *et al.* (1990)
	Isolated	*Bacillus anthracis* in perspex	60°C	Hunter (1972)
		Bacillus stearothermophilus in plastic	7.8–10°C	Pflug *et al.* (1980a, b)
		Bacillus stearothermophilus in glass bulb	10°C	Hersom and Shore (1981)
		Bacillus stearothermophilus in aluminium	NR	Rodriguez and Teixeira (1988)
Enzymic				
Intrinsic	Dispersed	Enzyme-linked immunosorbent assay for lactate dehydrogenase	NR	Smith (1995)
Extrinsic	Permeable	β-Galactosidase in alginate	NR	Matthiasson and Gudjonsson (1991)
		Lipase in alginate	NR	Matthiasson and Gudjonsson (1991)
		Nitrate reductase in alginate	NR	Matthiasson and Gudjonsson (1991)
		Immobilized amylase in alginate matrix	NR	Wunderlich (1995)

Isolated	Immobilized peroxidase in decanol	11.6°C	Weng et al. (1991)
	Immobilized peroxidase in dodecane	10.1°C	Weng et al. (1991)
	Imm. Bacillus licheniformis α-amylase	302 kJ/mol	De Cordt et al. (1992a)
	Bacillus amyloliquefaciens α-amylase + 83 wt% glycerol	701 kJ/mol	De Cordt (1994)
	Bacillys amyloliquefaciens α-amylase + 49 wt% glycerol + 31 wt% sucrose	560 kJ/mol	De Cordt (1994)
	Bacillus subtilis α-amylase (5–30 mg/ml)	8.4–12.8°C	Van Loey et al. (1996a)
	Bacillus subtilis α-amylase (200 mg/ml)	8.6°C	Van Loey et al. (1996a)
	Bacillus subtilis α-amylase (200 mg/ml) + trehalose (500 mg/ml)	6.2°C	Van Loey et al. (1996a)
	Bacillus amyloliq. α-amylase (200 mg/ml)	7.9°C	Van Loey et al. (1997a)
	Bacillus subtilis α-amylase (a_w = 0.76)	9.7°C	Van Loey et al. (1997b)
Chemical			
Intrinsic	Dispersed	Formation of 2,3-dihydro-3,5-dihydroxy-6-methyl-(4H)-pyran-4-one	96 kJ/mol	Kim and Taub (1993)
Extrinsic	Dispersed	Thiamine breakdown	26°C	Mulley et al. (1975a, b)
	Permeable	Maillard's reaction on paper disc	NR	Favetto et al. (1988, 1989)
	Isolated	Hydrolysis of dissacharides	18°C	Wen Chin (1977)
		Methylmethionine sulfonium breakdown	20–22.8°C	Berry et al. (1989)
		Acid hydrolysis of sucrose	94.6 kJ/mol	Sadeghi and Swartzel (1990)
		Destruction of Blue #2 at pH 11.3	58.2 kJ/mol	Sadeghi and Swartzel (1990)
		Destruction of Blue #2 at pH 9.5	74.5 kJ/mol	Sadeghi and Swartzel (1990)
Physical			
Extrinsic	Isolated	Thermal Memory Cell[b]	9.4–9.6°C	Swartzel et al. (1991)
		Thermalog S		Witonsky (1977)

[a]In inoculated pack studies, microorganisms relevant to the type of product and process are added; thus a single z-value cannot be given as this depends on the particular microorganisms used in a given study.

[b]The thermal memory cell is a multi-component TTI, based on the diffusion of at least two different ions, each with its own activation energy, in the insulator layer of a metal-insulator-semiconductor capacitor. Hence the activation energies that characterize the 'thermal memory cell' depend on the ions used.

*NR: not reported.

Table 4.2 Commercially available time–temperature process indicators/integrators for the cook chain (Selman, 1995)

Manufacturer	Trade name	Manufacturer	Trade name
• 3M Industrial Tapes and Adhesives	Autoclave Tape	• Redpoint	Spectratherm
	Thermometer Strips	• S.D. Special Coatings	Temperature Tabs
• Albert Browne Ltd	TST	• Spirg Earnest	Celsistrip
	Steriliser Control Tube		Celsidot
• Ashby Technical Products Ltd	ATP Irreversible T-Indicators		Celsipoint
• Cardinal Group	Easterday		Celsiclock
• Colour Therm	Colour-Therm		Integraph
• PyMaH Corporation	Cook-Chex	• SteriTec	Cross-checks
	SteriGage		Thermindex
	Thermalog S	• Thermindex Chemicals & Coatings Ltd	Pasteurisation Check
• Reatec AG	Reatec	• Thermographic Measurements Ltd	Thermax
• TLC Ltd	TLC 8		Autoclave Indicator

indicator systems are used extensively in the drug industry, for example, for the sterilisation of medical tools, but are less appropriate to monitor thermal processes in the food industry.

Because of the existing pressing need for new systems that allow quick, accurate and easy quantification of the thermal process impact, both in terms of food safety as in terms of food quality, the European Commission has, within the framework of AIR, supported the project 'The development of new Time–Temperature Integrators (product history indicators) for the quantification of thermal processes in terms of food safety and quality' (AIR1-CT92-0746).[2] The objective of the project was to solve these shortcomings through the development of new systems: chemical systems, protein-based systems and microbial systems.

4.4 Time–temperature integrators for the chill chain

4.4.1 Potential role of time–temperature integrators in the chill chain

Sous vide products are very dependent upon the maintenance of refrigeration temperature throughout the entire distribution chain as spoilage and pathogen growth is a function of both time and temperature. Storage, handling and distribution are being recognised as very important critical control points (CCP) in the chill chain. Ideally, chilled foods should be stored at the appropriate temperature during distribution and retail. During the storage and transportation phases of distribution, however, temperature conditions are often less than ideal, and temperature abuses occur. Temperature audits (Daniels, 1991) have shown significant refrigeration inadequacies at both retail and home level. Reaction rates that are related to quality-loss characteristics are generally strongly temperature dependent, even when all the other factors are controlled through effective packaging and maintained at the desirable levels. The accelerated quality losses due to the higher temperature exposure can result in unacceptable products before and at the retail level (Taoukis *et al.*, 1991). Hence, the safety and quality status and the shelf life of *sous vide* products depends on the temperature history of the product during storage, handling and transport. The commonly used open-date label to indicate end of shelf life does not tell anything about the actual temperature history (Sherlock *et al.*, 1991). If the food is temperature abused, an open date is meaningless and in fact a false sense of safety. The need for a tightly monitored distribution as well as for an alternative for open-date labelling has been repeatedly emphasised.

---

[2]Detailed information on the project can be obtained from the project co-ordinator: Prof. Marc Hendrickx, Laboratory of Food Technology, Kardinaal Mercierlaan 92, 3001 Heverlee, Belgium (e-mail: Marc.Hendrickx@agr.Kuleuven.ac.be).

Ideally, what is needed is a cost-effective way to individually monitor the temperature condition of food products through distribution to indicate their quality state and/or remaining shelf life or the potential that they are unsafe. This could lead to effective quality control during distribution, optimised stock rotation, and reduction of waste, as well as give some meaningful information on the remaining shelf life of the products. Effectively monitoring and controlling of CCPs during handling, distribution and storage is where TTIs can make significant contributions (Taoukis *et al.*, 1991).

TTIs can be used to monitor the temperature exposure of food products during distribution up to the time they are displayed at the supermarket. By being attached to individual cases or pallets, they can give a measure of the preceding temperature conditions at each receiving point. Additionally, it would allow targeting of responsibility and guarantee the producer and distributor that they deliver a properly handled product to the retailer, thus eliminating the possibility of unsubstantiated rejection claims by the latter. Because the temperature abuses in the distribution chain can vary from a few minutes to several hours, TTIs should be sensitive to short-time–temperature abuses. TTIs can also be used to indicate remaining shelf life or quality of food products, provided that the temperature sensitivity of the TTI and the targeted aspect is equal. TTIs have the potential for use both by the retailer to monitor handling practices throughout the distribution chain, and also by the consumer as a simple indicator of a product's freshness at the point of purchase. In addition, they could contribute towards increasing consumers' awareness of their own responsibility for keeping chilled foods at appropriate temperatures after purchase. TTIs can also be used for predicting food safety during refrigerated distribution and storage. It should be stressed that the use of TTIs for predicting potential pathogenic growth can be considered only after accurate and extensive modelling of the temperature dependence of the growth has been established. A statistically reliable and theoretically meaningful temperature/growth rate model for microorganisms causing spoilage or pathogenicity is needed (Fu *et al.*, 1991). Parameters that have to be taken into account include (i) an upper limit of the initial microbial population under a set quality control scheme based on HACCP and the food composition, (ii) the temperature behaviour of both lag phase and growth phase, including competition and any history effect, (iii) an upper limit for microbial load corresponding to the end of shelf life and (iv) the probability of pathogen growth and toxin production. Moreover, the design would have to be conservative, such that the TTI response might signal the food to be discarded when conditions for pathogenicity have not occurred (Taoukis *et al.*, 1991).

4.4.2 State of the art of time–temperature integrators for the chill chain

A variety of patents on indicators for the chill chain have been recorded and are summarised by Selman (1995). Among all of those patented, three types of full history TTIs are commercially available and are discussed below.

Monitor-Mark™ (3M Co., St Paul, Minn.): This is based on a time–temperature dependent diffusion of a blue dyed fatty acid ester through a porous wick. The measurable response, X, is the distance of the advancing easily visible diffusion front of the blue dye from the origin. Before use, the dye ester is separated from the wick by a barrier film so that no diffusion occurs. To activate the indicator, the barrier is pulled off and diffusion starts if the temperature is above the melting point of the ester. The melting point of the fatty acid ester used in a particular MonitorMark tag determines its response temperature. By varying the type and concentration of the ester, different melting temperatures and response life can be chosen. Thus, the indicator can be used either as a CTTI, with the critical temperature equal to the melting temperature of the ester, or as a TTI if the melting temperature is lower than the range of temperatures at which the food is stored. MonitorMark tags are available with response temperatures ranging from –17°C to 48°C and maximum running time of up to one year. Taoukis and Labuza (1989a, 1989b) showed that a plot of distance moved squared versus time gave a straight line which followed the Arrhenius relationship as a function of temperature. Because of the diffusion-based design, the activation energy of this tag ranges between 8 and 12 kcal/mol. Fu et al. (1991) have shown how the application of a simultaneous colour reaction along the track can be used to create new tags with activation energies up to 30 kcal/mol.

VITSAB®-TTI (Visual Indicator Tag Systems AB, Malmö, Sweden): This is based on a colour change caused by a pH decrease, due to a controlled enzymatic hydrolysis of a lipid substrate. Before activation, the lipase and the lipid substrate are in two separate compartments, one of which has a visible window. At activation, the barrier is broken, enzyme and substrate are mixed, the pH drops and the colour change starts as shown by the presence of an added pH indicator. The colour change from green to yellow can be visually recognised and compared to a colour band surrounding the window. Different combinations of enzyme-substrate types and concentrations can be used to give a variety of response life and temperature dependence. At the present time, four standard TTIs are being produced having activation energies between 13 and 47 kcal/mol.

Lifelines' Fresh-Scan (Lifelines Technology Inc., Morris Plains, NJ): This indicator is based on the solid-state polymerisation of thinly coated colourless acetylenic monomer that changes to a highly coloured polymer. The measurable change X is the reflectance which is measured with a laser optic wand and the data are stored in a hand-held computer. The indicator has two bar codes, one for the identification of the product and the other for identification of the indicator model. The indicators are active from the moment of production and have to be stored in the freezer before use where changes are minimal. It was found that the response follows typical first-order kinetics and that the tags have activation energies of about 20 to 24 kcal/mol (Taoukis and Labuza, 1989a, 1989b) but the catalyst can be varied to change this.

These three TTIs were the focus of both scientific and industrial trials with various perishable and semiperishable foods, including dairy, meat, fish, bakery, and fruit and vegetables products, summarised by Selman (1995). They may claim to satisfy most of the requirements for TTIs (Taoukis *et al.*, 1991). One important point must be highlighted and that is that all these indicators are placed on the outside of a package and therefore respond to the environmental temperature. The packaging itself may provide the food with some insulation to the environment and the food temperature will therefore lag behind any changes in outside temperature. Ideally, the indicator system should be inside the pack but with a response change visible externally.

A different category of TTIs is the consumer-readable TTI or consumer tag. These function on the same principles as the continuous-response ones but are designed to show a single end point, visually recognisable by consumers. These TTIs are used to assure the consumers that the products have been properly handled and indicate remaining shelf life. The use of consumer tags on perishable foods, not only gives the product a value-added profile, but also may quell the concerns of consumers who want high-quality convenience foods without worrying about the freshness of foods (Sherlock *et al.*, 1991). For example, the Lifelines' Fresh-Scan has been adapted to develop a consumer tag in a simple visual form, called Lifelines' Fresh-Check: a small circle of polymer is surrounded by a printed reference ring. The polymer, which starts out lightly coloured, gradually deepens in colour to reflect cumulative temperature exposure. Consumers may be advised on the pack not to consume the product if the polymer centre is darker than the reference ring, regardless of the use-by date. Also VITSAB developed a consumer readable indicator, configured in a 'bull's-eye' pattern with the outer ring being a reference colour and the centre circle showing a seemingly one-step change from green to yellow. These consumer TTIs have been successfully evaluated on their reliability under constant and variable temperature conditions and on the visual recognition of their end-point by panellists (Sherlock *et al.*, 1991). Sherlock and Labuza (1992) investigated the consumer perceptions of consumer tags on refrigerated dairy foods.

They concluded that the respondents of the consumer survey were particularly receptive to this concept, although consumer education about food spoilage will be necessary for a successful introduction of consumer tags.

Because the legal requirement for a 'best before' or 'use by' date on the pack will continue for the foreseeable future, consumer instructions on the pack need to clearly indicate action to be taken when there is conflict between end of product life indication as given by the 'best before' or 'use by' date and the TTI. Moreover, the TTI needs to be no less resistant to malpractice and tampering than is the printed date on the pack. The indicator or the package should self-indicate if removed from the product; at the same time, if removed it should damage the packaging in such a way that a fresh indicator cannot be applied without detection (Selman, 1995).

Despite the potentials of TTIs for the chill chain, their commercial application up to now has been limited. According to Taoukis et al. (1991), the main reasons are cost benefits, reliability and applicability. The indicator should be economically feasible. The cost is volume dependent, and since most indicators have not gone into mass production because of low demand, the cost remains high. The lack of reliability mainly dates back from the early studies on the use of TTIs. Early attempts to use TTIs as quality monitors were not well designed and thus unsuccessful as a result of poor understanding of the concepts involved. Most research publications have been ineffective in establishing how the TTI response can be used as a measure of food quality. As pointed out earlier, in most studies no meaningful mathematical modelling has been provided, so no extrapolations to other conditions could be made. Fortunately, recent publications have pointed out how mathematics can be used to model the tag response (Wells et al., 1987; Taoukis and Labuza, 1989a, 1989b; Wells and Singh, 1988). However, just like everything else, TTIs have some limitations. There are three sources of error in the estimated process impact: (i) the variation of the TTI response as the TTI tag ages, (ii) the statistical uncertainty in the kinetic inactivation parameters and (iii) in some cases, the difference in temperature sensitivity of the rate constants between the TTI and the target attribute.

4.5 Conclusion

The food industry as well as the consumer recognise a variety of benefits that can stem from the application of indicators both in the cook chain as well as in the chill chain. It is clear that there is a future for TTIs in thermal process evaluation and in monitoring and assurance of the distribution chain. At present, limited data for TTI responses are available in literature. The kinetic inactivation parameters (i.e. activation energy, reaction rate constant) of the indicators and of the target safety or quality aspects should be known to aid in the selection of relevant TTIs for different purposes.

Development and further improvement of different indicators are still in progress and will lead to new indicators that are more precisely designed to meet the needs of the food industry.

References

Bean, P., Dallyn, H. and Ranjith, H. (1979) The use of alginate spore beads in the investigation of ultra-high temperature processing. *J. Food Technol.*, 00, 281–294.

Berry, M.R., Singh, R.K. and Nelson, P.E. (1989) Kinetics of methylmethionine sulfonium in buffer solutions for estimating thermal treatment of liquids foods. *J. Food Proc. Preserv.*, 13, 475–488.

Betts, G.D. (1992) The microbiological safety of *sous vide* processing, in Technical Manual No. 39, Campden Chorleywood Food Research Association, UK, pp. 1–58.

Betts, G.D. and Gaze, J.E. (1992) Food pasteurization treatments. Technical Manual No. 27, Campden Chorleywood Food Research Association, UK

Bigelow, W.D. (1921) The logarithmic nature of thermal death time curves. *J. Infect. Dis.*, 29 (5), 528–536.

Brown, K.L, Ayres, C.A., Gaze, J.E. and Newman, M.E. (1984) Thermal destruction of bacterial spores immobilized in food/alginate particles. *Food Microbiol.*, 1, 187–198.

Burton, H., Perkin, A.G., Davies, H.L. and Underwood, H.M. (1977) Thermal death kinetics of *Bacillus stearothermophilus* spores at ultra high temperatures. III. Relationship between data from capillary tube experiments and from UHT sterilizers. *J. Food Technol.*, 12, 149–161.

Daniels, R.W. (1991) Applying HACCP to new-generation refrigerated foods at retail and beyond. *Food Technol.*, 45 (6), 122–124.

De Cordt, S., Hendrickx, M., Maesmans, G. and Tobback, P. (1992) Immobilized α-amylase from Bacillus licheniformis: a potential enzymic time-temperature integrator for thermal processing. *Int. J. Food Sci. Technol.*, 27, 661–673.

De Cordt, S. (1994) Feasibility of development of protein-based time-temperature-integrators for heat process evaluation. Ph.D. Thesis, Katholieke Universiteit te Leuven, Faculty of Agricultural and Applied Biological Sciences, Laboratory of Food Technology, Leuven, Belgium.

Favetto, G.J., Chirife, J., Scorza, O.C. and Hermida, C. (1988) Color-changing indicator to monitor the time-temperature history during cooking of meats. *J. Food Protect.*, 51 (7), 542–546.

Favetto, G.J., Chirife, J., Scorza, O.C. and Hermida, C. (1989) Time-temperature integrating indicator for monitoring the cooking process of packaged meats in the temperature range of 85–100°C. US Patent, 4,834,017.

Fu, B., Taoukis, P. and Labuza, T. (1991) Predictive microbiology for monitoring spoilage of dairy products with Time-Temperature Integrators. *J. Food Sci.*, 56 (5), 1209–1215.

Gaze, J.E., Spence, L.E., Brown, G.D. and Holdsworth, S.D. (1990) Microbiological assessment of process lethality using food/alginate particles. *Technical Memorandum No. 580*, Campden Chorleywood Food Research Association, pp. 1–47.

George, R.M. and Shaw, R. (1992) A food industry specification for defining the technical standards and procedures for the evaluation of temperature and time-temperature indicators. Technical Manual No. 35, Campden Chorleywood Food Research Association, UK.

Hayakawa, K.I. (1978) A critical review of mathematical procedures for determining proper heat sterilization processes. *Food Technol.*, 32 (3), 59–65.

Hendrickx, M., Saraiva, J., Lyssens, J., Oliveira, J. and Tobback, P. (1992) The influence of water activity on thermal stability of horseradish peroxidase. *Int. J. Food Sci. Technol.*, 27, 33–40.

Hendrickx, M., Maesmans, G., De Cordt, S., Noronha, J., Van Loey, A., Willockx, F. and Tobback, P. (1993) Advances in process modelling and assessment: the physical mathematical approach and product history integrators, in *Minimal Processing of Foods and Process Optimization* (eds R.P. Singh and F.A.R. Oliveira), CRC Press, pp. 315–335.

Hendrickx, M., Maesmans, G., De Cordt, S., Noronha, J., Van Loey, A. and Tobback, P. (1995) Evaluation of the integrated time-temperature effect in thermal processing of foods. *CRC Crit. Rev. Food Sci. Nutrit.*, **35** (3), 231–262.

Heppel, N.J. (1985) Measurement of the liquid-solid heat transfer coefficient during continuous sterilization of foodstuffs containing particles, in *Proceedings of IUFOST Symposium on Aseptic Processing and Packaging of Foods*, 9–12 September, Tylosand, Sweden, pp. 108–114.

Hersom, A.C. and Shore, D.T. (1981) Aseptic processing of foods comprising sauce and solids. *Food Technol.*, **35** (4), 53–62.

Hunter, G.M. (1972) Continuous sterilization of liquid media containing suspended particles. *Food Technol. Aust.*, April, 158–165.

Jones, A. and Pflug, I. (1981) *Bacillus coagulans*, FRR B666, as a potentional biological indicator organism. *J. Parenter. Sci. Technol.*, **53** (3), 82–87.

Kim, H.-J. and Taub, I.A. (1993) Intrinsic chemical markers for aseptic processing of particulate foods. *Food Technol.*, **47** (1), 91–99.

Labuza, T. and Taoukis, P. (1990) The relationship between processing and shelf life, in *Foods for the 90's* (ed. G. Birch), Elsevier Press, London, 73 pp.

Maesmans, G., Hendrickx, M., Weng, Z., Keteleer, A. and Tobback, P. (1990) Endpoint definition, determination and evaluation of thermal processes in food preservation. *Belgian J. Food Chem. Biotechnol.*, **45** (5), 179–192.

Maesmans, G. (1993) Possibilities and limitations of thermal process evaluation techniques based on Time-Temperature Integrators. Ph.D. Thesis, Katholieke Universiteit te Leuven, Faculty of Agricultural Sciences, Centre for Food Science and Technology, Leuven, Belgium.

Matthiasson, E. and Gudjonsson, T. (1991) *Enzymes as Thermal-Time Indicators in Food*, ITI, Reykjavik.

Mulley, A., Stumbo, C. and Hunting, W. (1975a) Kinetics of thiamine degradation by heat. A new method for studying reaction rates in model systems and food products at high temperatures. *J. Food Sci.*, **40** (5), 989–992.

Mulley, A., Stumbo, C. and Hunting, W. (1975b) Kinetics of thiamine degradation by heat. Effect of pH and form of the vitamin on its rate of destruction. *J. Food Sci.*, **40** (5), 993–996.

Norwig, J.F. and Thompson, D.R. (1986) Microbial population, enzyme and protein changes during processing, in *Physical and Chemical Properties of Food* (ed. M.R. Okos), ASAE, Michigan, USA, pp. 202–265.

Pflug, I.J. (1987) *Textbook for an Introduction Course in the Microbiology and Engineering of Sterilization Processes*, 6th edn, Environmental Sterilization Laboratory, Minneapolis, USA.

Pflug, I.J. and Odlaug, T.E. (1986) Biological indicators in the pharmaceutical and medical device industry. *J. Parenter. Sci. Technol.*, **40** (5), 242–248.

Pflug, I.J., Jones, A. and Blanchett, R. (1980a) Performance of bacterial spores in a carrier system in measuring the F_0-value delivered to cans of food heated in a Steritort. *J. Food Sci.*, **45** (4), 940–945.

Pflug, I.J., Smith, G., Holocomb, R. and Blanchett, R. (1980b) Measuring sterilizing values in containers of food using thermocouples and biological indicator units. *J. Food Protect.*, **43** (2), 119–123.

Rodriguez, A.C. and Teixeira, A.A. (1988) Heat transfer in hollow cylindrical rods used as bioindicator units for thermal process validation. *Trans. ASAE*, **31** (4), 1233–1236.

Rönner, U. (1990a) A new biological indicator for aseptic sterilization. *Food Technol. Int. Europe*, **90**, 43–46.

Rönner, U. (1990b) Bioindicator for control of sterility. *Food Lab. News*, **22** (6: 4), 51–54.

Sadeghi, F. and Swartzel, K.R. (1990) Time-temperature equivalence of discrete particles during thermal processing. *J. Food Sci.*, **55** (6), 1696–1698, 1739.

Sastry, S.K., Li, S.F., Patel, P., Konanayakam, M., Bafna, P., Doores, S. and Beelman, R.B. (1988) A bioindicator for verification of thermal processes for particulate foods. *J. Food Sci.*, **53** (3), 1528–1531.

Schellekens, M. and Martens, T. (1992) *Sous vide* Cooking – Part I: Scientific Literature Review. Publication No. EUR 15018 EN, by the Commission of the European Communities, Directorate General XII, Research and Development.

Segner, W.P., Ragusa, T.J., Marcus, C.L. and Soutter, E.A. (1989) Biological evaluation of a heat transfer simulation for sterilizing low-acid large particulate foods for aseptic packaging. *J. Food Proc. Preserv.*, **13**, 257–274.

Selman, J.D. (1995) Time-temperature indicators, in *Active Food Packaging* (ed. M.L. Rooney), Blackie, pp. 215–237.

Sherlock, M., Fu, B., Taoukis, P. and Labuza, T. (1991) A systematic evaluation of Time-Temperature Indicators for use as consumer tags. *J. Food Protect.*, **54** (11), 885–889.

Sherlock, M. and Labuza, T. (1992) Consumer perceptions of consumer Time-Temperature Indicators for use on refrigerated dairy foods. *J. Dairy Sci.*, **75**, 3167–3176.

Smith, D.M. (1995) Immunoassays in process control and speciation of meats. *Food Technol.*, **49** (2), 116–119.

SVAC (1991) *Code of Practice for* Sous vide *Catering Systems*, 12 pp.

Swartzel, K.R., Ganesan, S.G., Kuehn, R.T., Hamaker, R.W. and Sadhegi, F. (1991) Thermal memory cell and thermal system evaluation. US Patent 5,021,981.

Taoukis, P.S. and Labuza, T.P. (1989a) Applicability of time-temperature indicators as shelf life monitors of food products. *J. Food Sci.*, **54** (4), 783–788.

Taoukis, P.S. and Labuza, T.P. (1989b) Reliability of time-temperature indicators as food quality monitors under nonisothermal conditions. *J. Food Sci.*, **54** (4), 789–791.

Taoukis, P.S., Fu, B. and Labuza, T.P. (1991) Time-temperature indicators. *Food Technol.*, **45** (10), 70–82.

Van Loey, A., Ludikhuyze, L., Hendrickx, M., De Cordt, S. and Tobback, P. (1995) Theoretical consideration on the influence of the *z*-value of a single component time-temperature integrator on thermal process impact evaluation. *J. Food Protect.*, **58** (1), 39–48.

Van Loey, A., Ludikhuyze, L., Weemaes, C., Hendrickx, M., De Cordt, S. and Tobback, P. (1996a) Potential *Bacillus subtilis* α-amylase based time-temperature integrators to evaluate pasteurization processes. *J. Food Protect.*, **59** (3), 261–267.

Van Loey, A. Hendrickx, M., De Cordt, S. and Tobback, P. (1996b) Quantitative evaluation of thermal processes using time-temperature integrators. *Trends Food Sci. Technol.*, **7** (1), 16–26.

Van Loey, A., Arthawan, A., Hendrickx, M., Haentjens, T. and Tobback, P. (1997a) The development and use of an α-amylase based Time-temperature Integrator to evaluate in pack pasteurisation processes. *Lebens. Wissen. Technol.*, **30**, 94–100.

Van Loey, A., Haentjens, T., Hendrickx, M., and Tobback, P. (1997b) The development of an enzymic time-temperature integrator to assess thermal efficacy of sterilization of low-acid canned foods. Accepted for publication in *Food Biotechnology*.

Villota, R. and Hawkes, J.G. (1986) Kinetics of nutrients and organoleptic changes in foods during processing, in *Physical and Chemical Properties of Food* (ed. M.R. Okos), ASAE, Michigan, USA, pp. 266–366.

Villota, R. and Hawkes, J.G. (1992) Reaction kinetics in food systems, in *Handbook of Food Engineering* (eds D.R. Heldman and D.B. Lund), Marcel Dekker, NY, USA, pp. 39–144.

Wells, J.H., Singh, R.P. and Nobel, A.C. (1987) A graphical interpretation of time-temperature related quality changes in frozen foods. *J. Food Sci.*, **52**, 436–444.

Wells, J.H. and Singh, R.P. (1988) Application of time-temperature-indicators in monitoring changes in quality attributes of perishable and semiperishable foods. *J. Food Sci.*, **53**, 148–156.

Wen Chin, L. (1977) Disaccharide hydrolysis as a predictive measurement for the efficacy of heat sterilization in canned foods. Ph.D. thesis, Department of Food Science and Nutrition, University of Massachusetts, Amhers, Massachusetts, USA.

Weng, Z., Hendrickx, M., Maesmans, G. and Tobback, P. (1991) Immobilized peroxidase: a potential bioindicator for evaluation of thermal processes. *J. Food Sci.*, **56** (2), 567–570.

Witonsky, R.J. (1977) A new tool for the validation of the sterilization of parenterals. *Bull. Parenter. Drug Assoc.*, **11** (6), 274–281.

Wunderlich, J. (1995) Temperature bio-indicators based on immobilized enzyme for the controlling of continuous HTST heating of particulate foods and other heating processing, in *Proceedings of New Shelf Life Technologies and Safety Assessments*, VTT Symposium, Helsinki, Finland (eds R. Ahvenainen, T. Mattila-Sandholm and T. Ohlsson), 197 pp.

Yawger, E.S. (1978) Bacteriological evaluation for thermal process design. *Food Technol.*, **32** (6), 59–62.

5 Computer-integrated manufacture of *sous vide* products: the ALMA case study

TOON MARTENS and BART NICOLAÏ

5.1 Introduction

The superior quality of *sous vide* products is based on the better retention of aromas and nutrients during cooking in the bag and the product specific minimal heat treatment. The shelf life of the cooked products is extended by the heat treatment and this offers the possibility to optimise the production and distribution costs of pre-cooked products. Modern information technology combined with recent developments in food technology plays an important role in the development of the *sous vide* business to guarantee the safety and the quality of these products with a limited shelf life, optimise production scheduling, minimise production and distribution cost, have a fast feedback from retailers and even customers, etc.

5.2 Computer-integrated manufacturing

5.2.1 The CIM model

The computer-integrated manufacturing (CIM) model is a model that integrates the data flows between the different activities of a company: research and development, production, logistics, marketing, maintenance, etc. (Figure 5.1). In many companies, especially in the small- and medium-sized companies, the different pieces of automation are not integrated and many parts of the CIM model do not exist. The CIM model is an essential part of the factory of the future (Scheer, 1988).

5.3 Production planning and control

Material requirements planning (MRP) is developed in the mechanical jobshop environment (Figure 5.2). A formal plan is calculated to know when raw materials, subassemblies, etc., have to be available to deliver the product Just-in-Time (JIT). MRP takes account of stocks, capacities, delivery and production time periods.

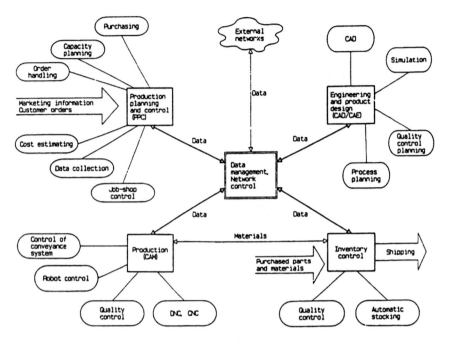

Figure 5.1 CIM integration model.

In the late 1980s ALMA tried to implement different MRP softwares. In that time, very few packages were available on the PC platform and more than 10 hours were needed to recalculate the plan. In foodservice a very high flexibility is needed and, for this reason, MRP could not be fully implemented. Many specific requirements typically for the food industry: biological variability of raw materials, limited shelf life of products, conversion problems between metrics (litres, kg, packs), etc., were not available in the standard MRP software.

To have a reproducible quality of *sous vide* products it is very important to use the right quantities of ingredients, especially the spices. Because the products are cooked in the bag and stored during several days or even weeks, *sous vide* recipes are quite different from traditional recipes and during *sous vide* cooking no adjustment can be made. For this reason a separation was made between the weighing of ingredients and the *sous vide* cooking area. The computer calculates the right amount of ingredients from the bill of material. Since all *sous vide* products are produced for JIT delivery, no products are produced as stock. Many of the concepts of JIT are very useful in the design of a *sous vide* production system. The JIT strategy is often formulated with zeros and one: 'zero defects, zero set-up time, zero inventory, zero lead time, lot size of 1'.

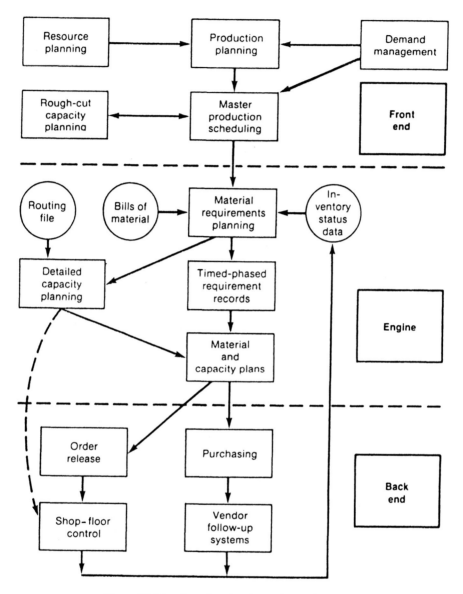

Figure 5.2 Manufacturing planning and control system.

The critical success factors to implement JIT are:

1. Order and cleanliness: In a traditional kitchen a lot of equipment is available and there is no clear routing of products and people. A *sous vide* system requires a high degree of hygiene. In France this is called *'l'ultra propre'*.

2. Zero defects: In a *sous vide* system, defects of raw materials and operators cannot be corrected. A JIT quality policy does not concentrate on the inspection of the output, but on product and process design. The Hazard Analysis Critical Control Points (HACCP) approach is based on the same principles. To guarantee zero defects, a continuous control and monitoring of the process is needed. In the ALMA *sous vide* system the heat treatment of the *sous vide* cooking is downloaded from the computer with the recipe database and automatically executed by the process computer that controls the cooking process.
3. Uniform production load: JIT tries to synchronise the production system with the needs of the customer. In a JIT environment the production output is not determined by the available capacity but by the needs of the customer. Since production volumes are fluctuating, people and machines have to be multifunctional and the machinery simple in operation. In ALMA, all the necessary information to execute a *sous vide* preparation is on the computer screen and a lot of tasks are executed automatically by the computer.
4. Redefinition of layout: In a JIT environment there are no buffers between different JIT cells and the products are moving fast through the production cycle. Very few traditional kitchen layouts allow this fast flow of products and for this reason in many cases it is much more efficient to build a new facility with limited floor space. The cost of a new production facility for 8000 meals/day was about 25% cheaper than rebuilding the existing 30-year-old kitchen. The capacity of the *sous vide* kitchen could be 20 000 meals per day if two shifts were organised.
5. Networking with suppliers: A JIT production system is only possible with the help of the suppliers. This means that instead of multiple sourcing a co-makership relation has to be developed. We experienced this in the start-up of the *sous vide* plant. In the period from 1990–1991, it was hard to find suppliers that could deliver meat-chicken-turkey products with the right microbiological quality. Most of the data are now exchanged by fax because very few suppliers are connected to the Internet. For the near future we hope to exchange data from production planning with our suppliers, Electronic Data Interchange (EDI) are now available.

5.4 Computer-aided design (CAD)

5.4.1 Challenge testing versus modelling

The design of safe *sous vide* products is traditionally based on 'challenge tests'. The food is inoculated with various pathogens and their potential to grow under conditions that resemble processing, storage distribution and retailing conditions is then evaluated experimentally (Gorris, 1994). The challenge tests themselves are simple but tedious, particularly if different alternative processing (different pack sizes) and distribution (retail, catering) have to be evaluated or if the effects of process deviations have to be established (Nicolaï, 1996).

To have better control of the temperature distribution during *sous vide* processing, equipment design has to be improved. Computational Fluid Dynamics (CFD) software is a powerful tool to study temperature distribution in ovens, cooling equipment, retail display cabinets, etc. Most of the equipment used in the catering and retail sector is very badly designed but also in legislation one can find safety guidelines that are impossible to achieve or have a high cost with a minimal effect on safety. ChefCad (Schellekens *et al.*, 1994) is the first expert system that tried to integrate knowledge on heat transfer, growth and inactivation kinetics, texture kinetics, etc., to design safe *sous vide* products on a computer screen.

5.4.2 Predictive microbiology

It is now apparent that in many instances microbial growth is largely determined by a relatively small number of factors, e.g. temperature, pH, salt, organic acid and gas composition. Microbiological evaluation of the ever-changing range of products, processes, and storage conditions currently in the marketplace by traditional methods is impossible. Most of the *sous vide* producers have a wide variety of products. In ALMA more than 1000 different recipes are made on a yearly basis. The concept 'predictive microbiology' was proposed to evaluate the safety of products based on the main controlling factors. Most of the work in predictive microbiology was done on the growth of pathogenic microorganisms. For inactivation the traditional models and data from pasteurisation and sterilisation literature are used.

Joint efforts in the UK resulted in a PC Software Food Micromodel. A more intelligent program was developed by Wijtzes (1996). The decision support system allows determination of which microorganisms could grow in a certain food product.

All the above software does not yet include models which can be applied under time varying conditions of temperature and pH. But also, the inactivation models have to be reviewed. The safety concepts developed for sterilised products do not hold for *sous vide* products. For some *sous vide* products very low heating temperatures and very long processing times are applied to yield a very soft texture. Some microorganisms can repair the effects of the heat treatment, but they can also adapt to conditions of sublethal heating by synthesis of heat shock proteins.

The data in Table 5.1 were obtained within the EU project AIR2-CT93-1519: the microbial safety and quality of foods processed by the *sous vide* method as a method of commercial catering.

Table 5.1 Effect of heat shock on the thermal inactivation of *E. coli* 0157 E30228

Experiment	D^{55} value (mins)	
	Control	Heat shocked
E30228 NS +	4.03	5.87
E30228 NS +	3.20	5.62
E30480 NS +	4.49	5.42

Shoulders and tailing can be very important, and thermal death calculations based on the traditional Bigelow model will give a wrong idea of the safety of these products cooked at low temperature for a long time. Van Impe *et al.* (1997) developed a model of growth and inactivation, based on the Gompertz model, that can take account of shoulder and tailing during heating. So far, almost exclusively, deterministic models have been developed. Recently, models appeared which aimed to predict not only the mean, but also the variability of the microbial response (Nicolaï, 1995b; Whiting and Buchanan, 1996).

The modelling approach has been extended to risk-assessment models, which take account of the initial distribution of microorganisms on foods and the probability of an infectious dose (Whiting, 1997).

These new modelling approaches and data on inactivation during low temperature–long time heat treatment put the traditional safety guidelines into question. Food technologists have successfully applied the mathematical method based on the 12D concept in sterilisation. But for minimally heated products, the adapted traditional approach – 6 decimal reductions of non-proteolytic *Clostridium botulinum* which has been proposed (Peck and Stringer, 1996) or the pasteurisation based on *Enterococcus faecalis* – cannot be used as the general safety guidelines for *sous vide* products. A new concerted action, 'Harmonisation of safety criteria for minimally processed foods', has been started to discuss relevant safety criteria (http://www.harmony.alma.be).

5.4.3 Software for analysis of heat transfer in foods

Heat transfer in foods can occur through different mechanisms: heat conduction, convective heat transfer, combined heat and moisture transfer and thermal radiation.

To calculate conduction heat transfer in different shapes of foods several finite element packages are now commercially available, e.g. Patran p3/Thermal (PDA Engineering, Costa Mesa, USA), MSC/Nastran (Los Angeles, USA), Algor (Pittsburgh, USA), Elfen (Rockfield Software Ltd, Swansea, UK), Ansys (Houston, USA), Samcef (Samcef, Liège, Belgium), SAP90 (Computers and Structures, Ltd, Berkeley, USA).

The finite element method has been used for heat conduction analysis of foods with complicated geometrical shapes such as chicken legs (De Baerdemaeker *et al.*, 1977), a baby food jar (Naveh *et al.*, 1983), broccoli stalks (Jiang *et al.*, 1987) and tomatoes (Pan and Bhowmik, 1991). A typical application to ready-to-eat foods was presented by Nicolaï *et al.* (1995) who considered the reheating process of ready-made lasagne in a combination oven. The authors found a good agreement between predicted and calculated temperatures under steam and wet air conditions. However, when dry air at elevated temperatures was applied as heating agent, the agreement was poor. This was attributed to moisture losses which decreased the thickness of the lasagne to more than 15%. Note that, because of the presence of the packaging material, in *sous vide* type foods no moisture losses should be expected.

Xie and Sheard (1996) used the Cranck Nicholson finite difference method to locate the least-lethality point in a *sous vide* product pouch pasteurised in a combination oven. For pouches with a depth less than 30 mm, heated at temperatures, the error in lethality estimation associated with the variation in the location of the least-lethality point is negligible.

Computer simulation was used to design the cooling system for the central production of ALMA. A generic high moisture product was considered ($k = 0.6$ W/m °C, $\rho = 1000$ kg/m^3, $c = 4184$ J/kg °C). The product is packaged in brick-like plastic bags with dimensions 5.6 cm $\times$ 32.8 cm $\times$ 18.2 cm and is initially at 2°C. At $t = 0$ the bag is put in a steamer and heated with saturated steam at 99°C until a core temperature of 65°C is reached. Subsequently, the product is cooled using three alternative methods: (i) by spraying cold water (2°C, $h = 1000$ W/m^2 °C) on the top and vertical sides of the product and by ventilating the bottom side with cold air ($h = 10$ W/m^2 °C, $T_\infty = 2$°C); (ii) by ventilating all sides with cold air; and (iii) by spraying cold water at 10°C ($h = 1000$ W/m^2 °C) on the top and vertical sides for 20 minutes prior to ventilation of all sides with cold air. The finite element package ANSYS vs 5.2 (Houston, USA, tel. +1-412-746-3304, http://www.ansys.com/) was applied. The finite element grid for the problem is shown in Figure 5.3. Observe that a two-dimensional finite element model is sufficient, as one dimension of the product is large with respect to the other two.

The temperature in the centre of the product corresponding to the different cooling methods is shown in Figure 5.4. The cooling time necessary to achieve a centre temperature of 10°C is equal to 170, 460 and 412 min, respectively.

Considering the guidelines for temperature requirements in the catering sector as summarised in the EU FERCO-FORCE program (Anon., 1994) (Table 5.2), it is clear that none of the norms is achieved for this process. The process engineers at ALMA therefore decided to reduce the thickness of the package to 3 cm and to apply cooling procedure (iii). The resulting

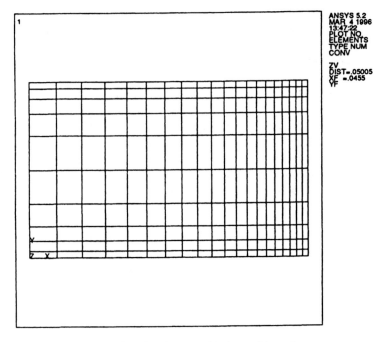

Figure 5.3 Finite element grid of *sous vide* product.

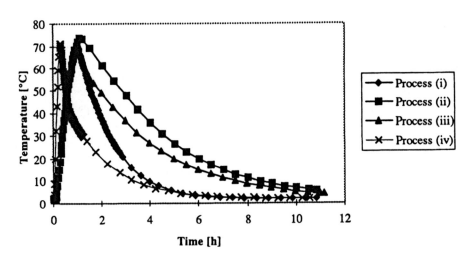

Figure 5.4 Temperature in the centre of the product for the different processes.

Table 5.2 Guidelines for catering companies in different EU countries (FERCO-FORCE, 1994)

Country	Core temperature	Cooling requirements	Storage temperature
France	65°C	10°C < 2h	–
Belgium	65°C	10°C < 2h	–
Spain	–	3°C as fast as possible	–
Portugal	–	10°C < 2h	< 3°C
Germany	65°C	7°C < 4h	–
Ireland	65°C	–	–
Italy	65°C	8°C < 3h	–
Holland	65°C	15°C < 2h	–
		7°C < 5h	

temperature course (process (iv)) is shown in Figure 5.4. The target core temperature of 65°C is achieved much more quickly (15 min), and during the cooling phase core temperatures of 15 and 10°C are achieved after approximately 2 and 3 h, respectively, which is at least compliant to the Dutch guidelines. It is clear that the simulations indicate that the target cooling velocities as established in the guidelines are very stringent, and the majority of them cannot be satisfied in practice. A careful reconsideration of the norms is therefore mandatory, taking into account the possibility and extent of pathogen growth during the cooling phase.

A second problem studied with computer simulation was the temperature distribution in a convection oven. Sheard and Rodger (1993) showed

Table 5.3 Range of heating times (minutes) from 20°C to 75°C core temperature in standardised *sous vide* packs (Sheard and Rodger, 1993)

Oven	Size	Fastest	Slowest	Difference
A Lower oven	3-grid			
Shelves 2, 4, 6		17.0	40.5	23.5
Shelves 3, 5, 7		16.0	38.0	22.0
A Upper oven	3-grid			
Shelves 2, 4, 6		21.0	51.5	30.5
Shelves 3, 5, 7		22.0	51.0	29.0
B	3-grid			
Shelves 2, 4, 6		22.5	40.5	18.0
Shelves 3, 5, 7		31.0	53.0	22.0
C	6-grid	9.0	26.5	17.5
D	6-grid	21.5	45.5	24.0
E	6-grid	15.0	35.0	20.5
F	10-grid	21.5	63.5	42.0
G	10-grid	20.0	42.0	22.0
H	10-grid	15.5	47.0	31.5
I	10-grid	31.5	85.5	54.0
J	10-grid	20.0	52.0	32.0

that there was a large difference (more than 3 times as long) to obtain a core temperature of 75°C in standardised *sous vide* packs at different locations of a convection oven (Table 5.3). In the framework of the EU AIR2 project 92-1519, Verboven *et al.* (1995, 1996) applied CFD to study the temperature distribution in a convection oven (http://www.agr.kuleu-ven.ac. be/aee/amc/research/ process/process.htm).

A schematic of the horizontal cross-section of the appliance is shown in Figure 5.5. The air is heated by electrical resistors in the back of the oven, and is circulated as shown by means of a fan. The food products are placed upon a wheel rack consisting of 10 gastronorm layers. The air flow in the oven was modelled using the CFD package FLOW3D (CFDS, AEA, Oxfordshire, UK). In Figure 5.6, the velocity vectors in three different horizontal cross-sections of the oven are shown. Observe that symmetry with respect to the vertical plane through the middle of the oven was assumed. A reasonable agreement has been obtained between calculated and measured results.

A complete 3D model was developed, which included the shape of the rack. The dimensions of the side walls have a great influence on the flow patterns inside the oven. Foster and James (1996) used CFD to study the temperature distribution in a refrigerated display cabinet that is currently used in supermarkets for *sous vide* and other chilled food products. Small changes in the design of the retail display cabinet had a major effect on the chilling efficiency.

Simulation routines for radiation heat transfer (microwaves) are currently being developed in the EU-project FAIR-1192: optimal control of microwave combination ovens for the food heating.

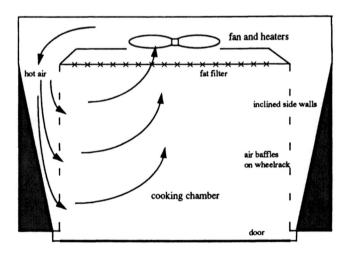

Figure 5.5 Cross-section of the *sous vide* appliance.

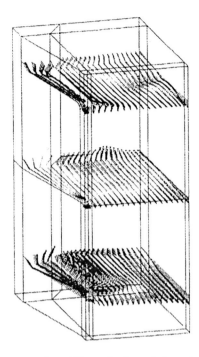

Figure 5.6 Velocity vectors at three different horizontal planes in the *sous vide* appliance.

5.4.4 Integrated packages for product design

The preparation of *sous vide* meals involves a complicated chain of heating and cooling processes. While the above software packages can be used to investigate several important issues such as heat transfer and microbial kinetics, there is need for general computer-aided safety design packages which integrate all important subprocesses that affect the quality and safety of the food. An early example of an integrated approach to process the optimisation of sterilisation processes was presented by Teixeira *et al.* (1969) who suggested that the heat conduction in cans could be numerically calculated by means of the finite difference method, and used the computed centre temperature as an input for the calculation of the process lethality by numerical integration. As a further improvement, the use of time-varying retort temperature profiles was considered by Teixeira *et al.* (1975) in order to maximise the retention of thiamine while safeguarding the required process value. This eventually led to the STERILMATE software package for the computer-aided design of sterilisation processes (Kim *et al.*, 1993). More elaborate computer-aided optimisation procedures have been described by Banga *et al.* (1991) and Silva *et al.* (1992). Commercially available packages for sterilisation are now available, e.g. TPRO for

Windows (Norback, Ley and Associates, Middleton, USA, http://www. norbackley.com/).

A major development in computer-aided process design was the CookSim package (Race and Povey, 1990). This package is essentially a knowledge-based system which guides the user towards a safe thermal process design by automatically solving the mathematical models underlying the heat transfer process and the associated microbial kinetics. The CookSim package consists of four components. The *knowledge base decision support component* forms the central part of the system. It provides facilities to browse and edit data and tools to display simulation results, compare processes and bacterial destruction rates. The knowledge base is built on top of the rule-based expert system shell ART. Rules for simulation model selection and thermal process optimisation are included. The object-oriented *database component* contains the recipe, consisting of the food, the package and the process in the CookSim terminology. The food is a particularly complicated object, which consists of ingredients in different proportions. Each ingredient has various parameters such as its physical state, microbial and nutritional target parameters. The inheritance mechanism is effectively used to avoid unnecessary duplication of data. The *simulation component* includes a finite difference procedure for conduction heat transfer analysis, and algorithms to calculate the thermal inactivation. The *user interface component* is graphically oriented and provides an easy way for the user to interact with the software. The authors also envisaged several techniques for process optimisation, including neural nets, genetic algorithms, heuristics and simulated annealing.

A related package for the computer-aided design of complicated recipes consisting of consecutive heating/cooling steps is called ChefCad (Schellekens *et al.*, 1994; Nicolaï *et al.*, 1994a, b) and was developed in the course of EU FLAIR project AGRF-CT91-0047. The system consists of several modules. The *data and knowledge base* contains the declarative and procedural knowledge of the system. The declarative knowledge encompasses all the data in the system, including the current recipe, a list of food ingredients (the complete food database is in the system), species of microorganisms and the parameters of their growth/inactivation models, equipment types such as ovens and refrigerators, etc. The procedural knowledge base contains finite element routines for the numerical solution of 2D heat conduction problems, an automatic finite element grid generator, routines to calculate the thermophysical properties from the chemical composition of the food, routines to calculate the surface heat transfer coefficient of the heating/cooling fluid, differential equation solvers for the microbial growth/inactivation and texture changes. The *inference engine* is the core of the system. It is a part of the programming environment and contains procedural knowledge for making logical inferences but is not immediately accessible to the programmer. The inference engine processes the user requests which arrive through the user interface. The necessary declarative data are fetched from the data and knowledge base, and passed to the calculation routines which are then fired. The calculation results are then transferred back to the user interface for visualisation. Also, a microbial safety diagnosis of the recipe is made by inferencing appropriate rules. The *user interface* is obviously the most visible part of the system. It is graphically oriented (X-Windows/Motif) and the user can interact with the system by means of graphical widgets such as push-buttons and pull-down menus. The time course of important process variables such as the food centre temperature, the microbial load and the texture can be visualised easily. The main window of the package is shown in Figure 5.7.

In an ongoing EU project, PECO CT93-0240, software is being developed for the prediction of the thermal behaviour of foods in different processing conditions. To this end the existing program COSTHERM for the prediction of thermophysical properties is being extended. It is also intended to select useful correlation formulas for surface heat transfer coefficients and to develop and validate user-friendly 1D and 2D solvers.

In chemical engineering, software packages are developed that describe a whole plant (FLOWTRAN, PROCESS, ASPEN, GEMS, etc.) and some specific applications in the food industry have been described (Petersen and

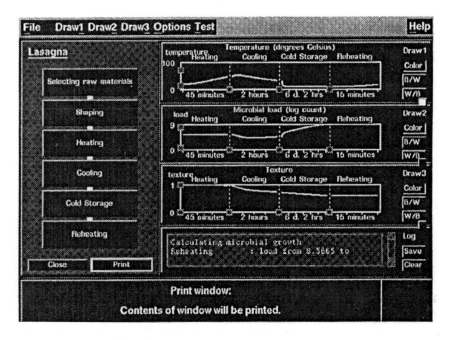

Figure 5.7 Main window of the ChefCad package.

Drown, 1995). There is still a long way to go before software packages are developed which describe a full model of a *sous vide* plant.

5.4.5 Computer-aided layout

Layout is a critical step in the planning of a *sous vide* plant. Layout is critical to provide a high level of hygiene and have an efficient flow of materials. The CRAFT computer-aided layout routine (Tompkins and White, 1984) has been used to generate layout alternatives. Other software packages for computer-aided layout are COFAD, CORELAP, ALDEP and PLANET (Tompkins and White, 1984). CRAFT and COFAD require quantitative flow inputs, PLANET accepts either quantitative or qualitative flow inputs and CORELAP and ALDEP require qualitative flow inputs.

Layouts may be generated by improving existing layouts or by constructing the layout from scratch. Because of the many possible combinations and hygiene rules, one normally starts with a basic layout that is improved. Basic layout schemes can be found in 'L'usine agro-alimentaire' (CRITT IAA, 1992). Four basic layout schemes are described. We used the CRAFT software to generate layout alternatives.

5.5 Computer-aided production management

5.5.1 Scheduling

For the production of the menu components of 8000–12 000 meals per day, 15 cooking kettles were available with a volume of 250 up to 600 litres. In the new *sous vide* plant only 5 kettles were available: 3 of 150 litres and 2 of 350 litres; one kettle of 350 litres is a pressure cooking kettle that was intended to be used for ragouts. Most of the kettles are used to prepare the sauce that has to be added to a meat product or vegetable that is cooked in a steaming cabinet afterwards in a bay. A scheduling program was developed. The program analyses the orders of a time horizon within the shelf life limits. The total amounts have to be split into batch sizes that fit in a kettle of 150 litres or 350 litres. The minimum production volume is 50 litres. To split the batches, two rules are applied: first, for a certain batch only the minimum number of kettles may be used and, second, the allocation of the kettles has to be done in such a way that the unused capacity of kettles is minimised. To reduce the complexity of the scheduling problem, a heuristic was introduced which stated that when the packaging machine (=bottleneck) became available a new batch had to be ready. Initially, the kettles were heavily used, since a lot of preparations were traditionally cooked and not filled. Recipe development of *sous vide* products is very time consuming. Gradually more and more products were actually *sous vide* cooked and

Table 5.4 Architecture of automation of ALMA central *sous vide* production

Level	Function	Hardware	Software
1	Administration Long-term planning	} PCs	OS VINES-networking DOS + UNIX Application: MRP, scheduling
2	Production control		
3	Monitoring Operating	FERRANTI Micro PMS	OS: real time OSC Application: batch control software
4	Measuring control	} PLC SIEMENS S5	Step 5 Program
5	Sensors Activators		

now scheduling is no longer a problem. The problem is now the availability of the steaming cabinets.

5.5.2 Computer-aided recipe handling

Kettles and steaming cabinets are controlled by a Siemens S5-135U PLC (Table 5.4).

The parameters for the machine control are downloaded from a real time process computer FERRANTI micro PMS. This system will now be replaced by a Windows NT machine with INTOUCH monitoring software. The operator interface for recipe handling is an ORACLE program developed by ALMA. The operations that have to be performed by the operator and process computer (indicated by an asterisk (*) on Table 5.5) are part of the recipe. The heat treatment, necessary to guarantee the safety of the *sous vide* product, can only be changed in exceptional cases and any change to the standard recipe is logged.

The operators have a complete overview of the production schedule of one day and of the status of every batch. Different operators can work different batches simultaneously, which increases productivity enormously. All batches of cooked products for 8000–12 000 meals per day are handled by three operators.

5.5.3 Computer-aided quality management

For *sous vide* products microbial safety is the first quality parameter that has to be guaranteed.

The application of the HACCP methodology is obliged by the new EU directive on food hygiene (COM (91) 525, 14/06/93). HACCP encompasses the identification of hazards associated with the production, distribution

Table 5.5 Example of the definition of a recipe on PMS: stew from Bengalen

Step number	Action	Number of ingredient	Parameters		Wait for step	
1	Add	1	100%	–	0	0
2	Heat*	2	150°C	10 min	1	0
3	Stir*	4	20 RPM	0 min	1	0
4	Stir*	4	0 RPM	0 min	2	0
5	Add	1	100%	–	4	0
6	Heat*	2	150°C	15 min	5	0
7	Stir*	4	20 RPM	15 min	5	0
8	Add	1	100%	–	6	7
9	Heat*	2	150°C	15 min	8	0
10	Stir*	4	20 RPM	15 min	8	0
11	Add	1	100%	–	9	10
12	Add	1	100%	–	11	0
13	Add	1	100%	–	12	0
14	Add	1	100%	–	13	0
15	Add	1	100%	–	14	0
16	Add	1	100%	–	15	0
17	Add	1	100%	–	16	0
18	Cold water*	7	100%	–	17	0
19	Heat*	2	100°C	100 min	18	0
20	Stir*	4	20 RPM	0 min	18	0
21	Heat*	2	100°C	60 min	19	0
22	Heat*	2	100°C	0 min	21	0
23	Stir*	4	0 RPM	0 min	22	0
24	Add	1	100%	–	23	0
25	Heat*	2	100°C	5 min	24	0
26	Stir*	4	20 RPM	5 min	24	0
27	Add	1	100%	–	25	26
28	Add	1	100%	–	27	0
29	Add	1	100%	–	28	0
30	Heat*	2	100°C	100 min	29	0
31	Heat*	2	100°C	2 min	30	0
32	Empty	9	30 RPM	–	31	0

*Operations which have to be performed by the operator and process computer.

and particular use of each food product, and the assessment of their severity and risk by well-documented and verifiable means. It also prescribes the determination of the actions which must be performed to control identified hazards at critical control points (CCPs), the monitoring of the criteria that indicate whether the CCPs are under control, the preparation of corrective actions if control is lost, and final verification to ensure that the HACCP system is efficient.

The application of the HACCP methodology requires a lot of specific knowledge, and involves a considerable administrative effort. Several attempts have been made to automate the HACCP procedure.

A hypertext-based system, Mirfak, was described by Laporte *et al.* (1993). Information concerning microbiological analyses, foodstuff processing and explanation of the HACCP system is incorporated in the system. The system also includes rules and standards for a good hygienic design,

miscellaneous information such as the characteristics of growth and destruction of the main microorganisms, and a list of generic components (e.g. production and storage apparatus, and premises) of a sandwich production
factory. Several browsing tools (scanning, exploration, search, navigation
memory and a directory) are available. The provision of inferencing procedures is envisaged for future releases.

TNO Voeding (Zeist, the Netherlands, http://www.tno.nl/) now markets
a MS-Windows-based software package for computerised HACCP analysis, called *FIST* HACCP (formerly ProQ). It offers support in recording,
processing and reporting the complete HACCP protocol in an easy step-by-
step approach. First, *company data* have to be defined, including the personnel, the process steps and the production documents which are common
for all subsequent HACCP analyses. In a next step, the actual HACCP
analysis can be accomplished by entering the HACCP data, including the
people who are involved in a particular process, the documents which
describe the terms of reference, the product description and the intended
use of the product; and the production process. The latter can be represented visually under the form of flow charts. Subsequently, the user can
enter for each process step the potential hazards and the preventive actions
to control these hazards. A decision tree then assists the user to determine
whether a given process step is a CCP. If so, the corresponding CCP data
can be entered and will be displayed in the flow chart. The continuous monitoring of the HACCP implementation is guaranteed by defining appropriate verification procedures. A wide range of reporting facilities is provided
as well. *FIST* HACCP 2.4 can also be extended with optional modules. The
FIST Risk module allows the user to specify probabilities associated with
production process hazards and assess the overall risk in terms of a risk
index. With *FIST* Connections, documents of various formats including
work instructions, drawings and photographs of machines, final product
descriptions, etc., can be integrated in *FIST* HACCP. Product formulations
can be specified by means of the module *FIST* Specifications. Derived properties can be defined and calculated using a formula editor. Also available
are a quality control (*FIST* Quality) and a verification (*FIST* Verification)
module.

Another comparable package is doHACCP (Norback, Ley and Associates, Middleton, USA, http://www.norbackley.com/).

HACCP will be implemented in ALMA according to the ISO 9001 standard. The HACCP plan of ALMA focuses on temperature control and
training. The *sous vide* cooking process is controlled by a process computer
and the refrigerated storage is now equipped with a permanent alarming
system that is directly connected to the maintenance company.

The quality handbook that will be developed will be available on the
Internet to guarantee that every member of the staff can consult the latest
version. Also the work instructions will be available on the internal web.

Any change to a document will be automatically mailed to the distribution list of that document.

Quality problems will be entered into a database with an indication of the responsible person so that follow-up is guaranteed. More general questions and problems can be posted to an internal ALMA discussion list.

5.6 Computer-aided customer information and feedback

In 1995, ALMA started a website (http://www.alma.kuleuven.ac.be/) to inform the students and the staff of the university on the menus for the next two weeks with information on the ingredients and nutritional value. Also, general information on ALMA and nutrition are available.

Customers can post their remarks and questions to a newsgroup (kuleuven.alma on newsserver news.kulnet.kuleuven.ac.be). Information on the past research work at ALMA is available on the website http://www.harmony.alma.be.

5.7 Conclusions

Information technology plays an important role in modern *sous vide* production and distribution. In the production of the *sous vide* menu components there are more computers working than people. During the night, when nobody is in the factory, the computer cooks long time–low temperature batches that are automatically cooled. The *sous vide* factory of the future is already partly existing, but to implement the full possibilities of *sous vide* technology and information technology, we have to go a long way because of our limited resources.

References

Anon. (1994) FERCO-FORCE Code Européen de bonnes pratiques en matières d'hygiene pour la restauration collective.
Banga, J.R., Perez-Martin, J.M., Gallardo, J.M. and Casares, J.J. (1991) Optimization of the thermal processing of conduction-heated canned foods: study of several objective functions. *Journal of Food Engineering*, **14**, 25.
De Baerdemaeker, J., Singh, R.P. and Segerlind, L.J. (1977) Modelling heat transfer in foods using the finite element method. *Journal of Food Process Engineering*, **1**, 37–50.
CRITT IAA. (1992) *L'usine agro-alimentaire*, Editions RIA, France Agricole, Paris, France.
Drown, D.C. & Peterson, J.N. (1995) Computer aided design in the food processing industry, Part 2. Applications of process flowsheeting. *Journal of Food Technology*, **20**, 407–417.
Foster, A.M. and James, S.J. (1996) Using CFD in the design of food cooking, cooling and display plant equipment, in *Second European Symposium on Sous Vide Proceedings*, ALMA, Leuven, Belgium, pp. 43–57.

Gorris, L.G.M. (1994) Improvement of the safety and quality of refrigerated ready-to-eat foods using novel mild preservation techniques, in *Minimal Processing of Foods and Process Optimization* (eds R.P. Singh and F.A. Oliveira), pp. 57–73.

Jiang, H., Thompson, D.R. and Morey, R.V. (1987) Finite element model of temperature distribution in broccoli stalks during forced-air precooling. *Transactions of the ASAE*, **30**(5), 1473–1477.

Kim, Teixeira, A.A., Bichier, J. and Tavares, M. (1993) *STERILMATE: software for designing and evaluating thermal sterilization processes*, ASAE Paper No. 93–4051, American Society of Agricultural Engineers, St Joseph, Michigan, USA.

Laporte, E., Muratet, G., Cerf, O. and Bourseau P. (1993) Development of a knowledge base for the impovement of hygiene in the food industry, in *Proceedings of the AIFA Conference on Artificial Intelligence for Agriculture and Food*, 27–29 October, Nîmes, France, 79 pp.

Naveh, D., Kopelman, I.J. and Pflug, I.J. (1983) The finite element method in thermal processing of foods. *Journal of Food Science*, **48**, 1086–1093.

Nicolaï, B.M., Van Impe, J.F. and Schellekens, M. (1994a) Application of expert systems technology to the preparation of minimally processed foods: a case study. *A Benelux Quarterly Journal of Automatic Control*, **35**, 50.

Nicolaï B.M., Obbels, W., Schellekens, M., Verlinden, B., Martens, T. and De Baerdemaeker, J. (1994b) Computational aspects of a computer aided design package for the preparation of cook–chill foods, in *Proceedings of the Food Processing and Automation Conference III*, ASAE, St Joseph, Orlando, USA, 190 pp.

Nicolaï, B.M., Van Impe, J.F., Martens, T. and De Baerdemaeker, J. (1995a) A probabilistic model for microbial growth during cold storage of foods, in *Proceedings of the 19th International Congress IIR/IIF August 1995, volume 1*, IIR/IIF Institut International du Froid, Paris, France, pp. 232–239.

Nicolaï, B.M., Van Den Broek, P., Schellekens, M., De Roeck, G., Martens, T. and De Baerdemaeker J. (1995b) Finite element analysis of heat conduction in lasagna during thermal processing. *International Journal of Food Science and Technology*, **30**, 347–363.

Nicolaï, B.M. and Van Impe, J.F. (1996) Predictive food microbiology: a probabilistic approach. *Mathematics and Computers in Simulation*, **42**, 287–292.

Pan, J.C. and Bhowmik, S.R. (1991) The finite element analysis of transient heat transfer in fresh tomatoes during cooling. *Transactions of the ASAE*, **34**(3), 972–976.

Peck, M.W. and Stringer, S.C. (1996) *Clostridium botulinum*: mild preservation techniques, in *Proceedings of Second European Symposium on sous vide*, ALMA, Leuven, Belgium, pp. 181–197.

Race, P. and Povey, M.J. (1990) CookSim: a knowledge based system for the thermal processing of food, in *Expert Systems and Their Applications*, Avignon, France, pp. 115.

Scheer, A.W. (1988) *CIM: Computer Integrated Manufacturing*, volume XI, Springer, Berlin.

Schellekens, M., Martens, T., Roberts, T.A., Mackey, B.M., Nicolaï, B.M., Van Impe, J.F. and De Baerdemaeker, J. (1994) Computer aided microbial safety design of food processes. *International Journal of Food Microbiology*, **24**, 1.

Sheard, M.A. and Rodger, C. (1993) Optimum heat treatments for 'sous vide' cook–chill products, in *Proceedings of First European 'Sous Vide' Cooking Symposium*, Leuven, Belgium, pp. 117–126.

Silva, C., Hendrickx, M., Oliveira, M. and Tobback, P. (1992) Critical evaluation of commonly used objective functions to optimize overall quality and nutrient retention of heat-preserved foods. *Journal of Food Engineering*, **17**, 241.

Teixeira, G.E., Dixon, J.R., Zahradnik, J.W. and Zinsmeister G.E. (1969) Computer optimization of nutrient retention in thermal processing of conduction-heated foods. *Food Technology*, **23**(6), 137.

Teixeira, G.E., Zinsmeister G.E. and Zahradnik, J.W. (1975) Computer simulation of variable retort control and container geometry as a possible means of improving thimine retention in thermally processed foods. *Journal of Food Science*, **40**, 656.

Tompkins, J.A. and White, J.A. (1984) *Facilities Planning*, John Wiley, New York, USA.

Van Impe, J.F. *et al.* (1997) *Dynamic modelling of the microbial evolution*. Report of Final Project Meeting EU Project FAIR CT93–1519.

Verboven, P., Nicolaï, B.M. and De Baerdemaeker, J. (1995) Design of a two-dimensional CFD model of the dry air flow in a forced convection oven, in *Proceedings of the CFDS User Group Meeting*, Windsor, UK, 27–28 June 1995.

Verboven, P., Nicolaï, B.M. and De Baerdemaeker, J. (1996) Modelling of forced convection heating of foods in a commercial oven using CFD, presented at the 1996 IChemE Research Event, Leeds, UK, 2–3 April 1996.

Vollmann, T.E., Berry, W.L. and Whybark, D.C. (1988) *Manufacturing Planning and Control Systems*, Irwin Inc., Homewood, Illinois, USA

Whiting, R.C. and Buchanan, B.L. (1996) Integrating predictive microbiology and microbial risk assessment to enhance the development of HACCP programs, in *Proceedings of the 2nd International Conference on Predictive Microbiology*, Tasmania, Australia, 18–22 February 1996.

Whiting, R.C. (1997) Microbial database building: what have we learned? *Food Technology*, **51**, 82–87.

Wijtzes, T. (1996) *Modelling the microbial quality and safety of foods*, Proefschrift Landbouwuniversiteit Wageningen, Wageningen, The Netherlands.

Xie, G. and Sheard, M.A. (1996) Location of the least-lethality point in conduction-heated *sous vide* pouches pasteurised in a combination oven, in *Proceedings of Second European Symposium on Sous Vide*, ALMA, Leuven, Belgium, pp. 399–411.

6 Critical factors affecting the safety of minimally processed chilled foods

GAIL D. BETTS

6.1 Minimal-processing techniques

Over the past two decades there has been an increase in the demand for minimally processed chilled foods, e.g. *sous vide* products, which contain little or no chemical preservatives and are given a mild heat treatment in order to achieve a fresh-cooked taste.

Such products rely on the combination of minimal-processing and storage under controlled chill conditions to prevent growth of pathogenic organisms and achieve microbiological safety, unlike conventional thermally processed products which rely on thermal destruction of any pathogens present.

There is therefore an inherent potential for survival and growth of many pathogenic organisms in these products if any part of the combined preservation system breaks down. The critical stages of minimally processed chilled food production are discussed along with the organisms of particular concern and guidelines are given for the process parameters required to achieve product safety.

Minimally processed foods are produced using a range of modern technologies which aim to change the food as little as possible and at the same time endow it with a shelf life sufficient for its transport from the producer to the consumer (Ohlsson, 1994).

There are many different minimal-processing techniques such as *sous vide* cooking, modified atmosphere packaging and non-thermal processing systems including ultrasound or high-pressure treatments which can be applied at various stages throughout the manufacturing process of a food. All these techniques have a common goal of achieving food products which contain little or no added preservatives, and are exposed to mild heat treatments in order to achieve a fresh-cooked taste.

As a consequence of this, minimal-processing techniques do not always inactivate microorganisms but rely on a combination of preservation techniques and storage under controlled chill temperatures to prevent growth of pathogenic and spoilage organisms to achieve microbiological safety and stability (Leistner and Gorris, 1995).

There is, therefore, an inherent potential for survival and growth of pathogenic microorganisms in these products if any part of the combined

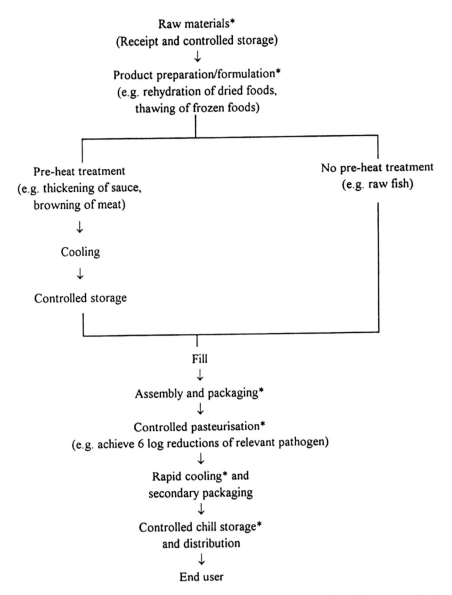

Raw materials*
(Receipt and controlled storage)
↓
Product preparation/formulation*
(e.g. rehydration of dried foods,
thawing of frozen foods)

Pre-heat treatment
(e.g. thickening of sauce,
browning of meat)
↓
Cooling
↓
Controlled storage

No pre-heat treatment
(e.g. raw fish)

Fill
↓
Assembly and packaging*
↓
Controlled pasteurisation*
(e.g. achieve 6 log reductions of relevant pathogen)
↓
Rapid cooling* and
secondary packaging
↓
Controlled chill storage*
and distribution
↓
End user

Figure 6.1 Flow diagram for production, storage and distribution of minimally processed food containing a heat treatment, e.g. *sous vide*. The asterisk (*) indicates critical microbiological safety factors.

preservation system fails. The critical factors affecting the safety of minimally processed chilled foods will depend to some extent on the specific techniques used, although it is possible to divide minimally processed foods into two generic groups: (i) those containing a pasteurisation step, e.g. *sous vide* products; and (ii) those which do not contain a kill step, e.g. minimally processed vegetables.

The main stages affecting the safety of these two generic groups are shown in Figures 6.1 and 6.2. The critical factors are discussed in further detail later in the chapter.

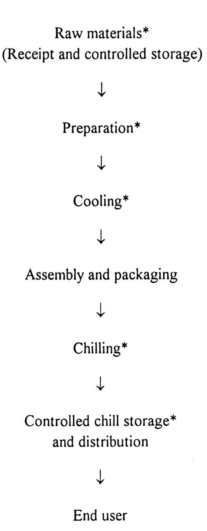

Figure 6.2 Flow diagram for production, storage and distribution of minimally processed food not receiving a heat treatment, e.g. fresh vegetables. The asterisk (*) indicates critical microbiological safety factors.

6.2 Microorganisms of concern

Minimally processed chilled foods without a heat step rely almost exclusively on chill temperature and storage time as the critical factors controlling safety, while those including a heat treatment rely on a combination of mild pasteurisation to destroy some of the pathogens present as well as chill temperature and storage time to prevent the growth of any surviving pathogens.

The organisms of most concern to minimally processed chilled foods are those which are able to survive a pasteurisation process, if applied, and grow at chill temperature conditions. In order to assess the safety concerns with these organisms, it is essential to determine their growth and survival limits and thus ensure that their growth is prevented.

There are two main groups of organisms which should be considered in relation to minimally processed foods: psychrotrophs and mesophiles.

Psychrotrophic organisms have a minimum growth temperature of <0 to 5°C, an optimum growth temperature of 25 to 30°C and a maximum growth temperature of >30°C. The group of psychrotrophic organisms includes some spoilage organisms, e.g. *Pseudomonas* spp., but also contains some human pathogens such as *Listeria monocytogenes, Yersinia enterocolitica, Aeromonas hydrophila* and non-proteolytic *Clostridium botulinum* types B and E.

While psychrotrophic bacteria grow optimally at 25 to 30°C they are able to grow, albeit slowly, at good refrigeration temperatures, i.e. <5°C. In extended shelf life foods which are vacuum packaged, there may be sufficient time for these organisms to adjust to the chill environment and grow. As chilled storage may be insufficient to inhibit growth of psychrotrophic pathogens, it is important to ensure that any pathogenic or spoilage organisms of particular concern are destroyed during processing or that the shelf life is limited, e.g. to less than 10 days (see Section 6.3.1).

Mesophilic pathogens generally have a minimum growth temperature of approximately 10°C, an optimum growth temperature of 30 to 37°C, and a maximum growth temperature of 35 to 45°C. The group of mesophilic organisms contains many human pathogens, e.g. proteolytic *C. botulinum, C. perfringens, Bacillus cereus, Salmonella* spp. and *Staphylococcus aureus*, and should be of little concern to foods stored at good refrigeration temperatures, i.e. < 5°C. However, mesophilic pathogens may be of concern to foods which are temperature abused during chill storage/distribution to temperatures above 10°C. Such temperatures are commonly seen throughout the retail distribution chain; for example, Rose (1986) reported air temperatures of –5.0 to +13.8°C within retail chill cabinets and counters with product temperatures within the range –8.0 to +18.4°C. Similar data were reported by Conner *et al.* (1989) for domestic refrigerators, where 20% of those examined were found to have maximum operating temperatures above 10°C. More recently, Willcocx *et al.* (1994a) showed that the temperature of some Belgian retail display cabinets had a weekly mean temperature as high as 12.3°C with the highest daily temperature measured at 16.0°C. In these temperature conditions any mesophilic organisms present in minimally processed chilled foods could grow and cause poisoning or spoilage.

Table 6.1 lists some of the important organisms of concern to minimally processed chilled foods. Of these organisms, the psychrotrophic pathogens *L. monocytogenes* and psychrotrophic *C. botulinum* are considered to be of most concern and will be discussed in further detail.

6.2.1 *Psychrotrophic* Clostridium botulinum

Psychrotrophic *C. botulinum* is the organism of most concern to minimally processed chilled foods with an extended shelf life as it may be able to grow during storage and produce a powerful neurotoxin which, if ingested, would cause botulism. In addition, this organism may be able to survive the mild heat treatments given to these types of foods.

(a) Low-temperature growth   Several studies have been done on the growth of psychrotrophic strains of *C. botulinum* from the 1960s to the present day, which illustrate the time taken for toxin production in a variety of laboratory media and food products (Table 6.2). The majority of these studies have been concerned with growth of *C. botulinum* type E, particularly in association with fish products, for example crab, cod, shrimp, salmon and trout. The time taken for toxin to be produced is dependent on product type and packaging format and can occur very rapidly.

Garcia *et al.* (1987) illustrated that toxin was produced in modified atmosphere packaged salmon fillets in 6 to 12 days at 8°C depending on the spore inoculum and packaging system used. For example, in vacuum-packaged salmon fillets inoculated with 10^4 *C. botulinum* per gram, toxin was produced within 6 days at 8°C and within 3 days at 12°C.

Further workers (Garcia and Genigeorgis, 1987) showed that, at 4°C, toxin was produced in vacuum-packaged salmon fillets within 15 days.

Some studies have reported on the growth of *C. botulinum* in meat and poultry products. For example, Schmidt *et al.* (1961) showed that type E produced toxin within 19 days at 6°C and 31 days at 3.3°C in beef stew.

Similar data were reported by Crandall *et al.* (1994) who found that non-proteolytic type B spores were able to produce toxin by 31 days at 4°C and by 6 days at 10°C for *sous vide* cooked beef and gravy.

In poultry products, toxin has been produced within 28 days at 8°C in a chicken product (Brown and Gaze, 1990) and within 8 days at 8°C for cooked turkey (Genigeorgis *et al.*, 1991).

Table 6.1 Extremes of temperature, a_w, pH and salt concentration permitting growth of food poisoning bacteria of potential concern to minimally processed chilled foods

Organism	Minimum temperature (°C)	Minimum a_w	Minimum pH	Maximum NaCl (% w/v)	Aerobic/Anaerobic respiration
Salmonella spp.	4.0	0.94	4.5	4.0	Facultative
L. monocytogenes	−0.4	0.92	4.3	12.0	Facultative
A. hydrophila	−0.1	–[b]	4.0	4.0	Facultative
Y. enterocolitica	−1.0	0.96	4.2	7.0	Facultative
S. aureus	6.7[a]	0.86	4.0	7.5	Facultative
V. parahaemolyticus	5.0	0.94	4.8	8.0	Facultative
B. cereus	4.0	0.91	4.3	–[b]	Facultative
Psychrotrophic C. botulinum (B, E and F)	3.3	0.97	5.0	5.0	Anaerobic
E. coli O157:H7 and other VTEC	7.0	0.95	4.0+	–[b]	Facultative

[a]No evidence of toxin production at this temperature.
[b]Data not available.

The above data represent approximate values for these growth limits under otherwise optimal conditions. Exact values will vary depending on particular strain of microorganisms and food composition.
(Adapted from Betts, 1996)

Limited studies have been done on the growth of *C. botulinum* in vegetable products. Brown and Gaze (1990) found that *C. botulinum* types E and B did not grow in cooked carrot homogenate stored at 3, 5 or 8°C for 12 weeks. Similarly, Petran *et al.* (1995) found that *C. botulinum* non-proteolytic type B was unable to grow in lettuce and shredded cabbage stored at 4.4 or 12.7°C for up to 28 days. Carlin and Peck (1995) evaluated the growth of non-proteolytic types E, B and F in cooked vegetables and found that growth occurred in 13 of 28 puréed vegetables, although the incubation temperature used in these studies was the optimum for growth (30°C).

Until recently it was generally accepted that 3.3°C was the lowest temperature permitting growth of non-proteolytic strains of *C. botulinum* (Schmidt *et al.*, 1961; Anon., 1992a). However, recent work (Graham *et al.*, 1997) has provided evidence to suggest that the minimum growth temperature for some strains of psychrotrophic *C. botulinum* may be lower than 3.3°C. In their studies, the growth of non-proteolytic types B, E and F at 2, 3, 4, 5, 8 and 10°C was evaluated in laboratory media. The mean temperatures from three experiments at 3°C were 3.15, 3.07 and 2.95°C and growth of non-proteolytic *C. botulinum* occurred in each experiment by 5 to 7 weeks.

It is apparent that non-proteolytic *C. botulinum* has the ability to grow under a range of chill temperatures and in a variety of minimally processed chilled foods. Growth and toxin production at refrigeration temperatures may occur in foods which appear to be organoleptically acceptable (Conner *et al.*, 1989) and is of particular concern for refrigerated foods of extended durability, as the extended storage period may allow growth of this organism even at good refrigeration temperatures (Notermans *et al.*, 1990).

It can be concluded that control of the growth of *C. botulinum* is one, if not the major, critical factor affecting the safety of minimally processed foods. This can be done by destroying the organism during the heat treatment or inhibiting its growth by product formulation or restricted shelf life. These points are discussed in further detail in Section 6.3.2.

(b) Heat resistance   The heat resistance of psychrotrophic *C. botulinum* in various laboratory buffers and foods is shown in Table 6.3. The majority of reported literature is focused on fish products and *C. botulinum* type E, although some data are available for non-proteolytic type B which appears to have a greater resistance to heat. Scott and Bernard (1982) studied the heat resistance of *C. botulinum* types B and E in phosphate buffer and found that while the type E studied had a D-value of 0.33 min at 82.2°C, the $D_{82.2}$ of the B strain test ranged from 1.49 to a maximum of 32.3 min which is unusually high.

The heat resistance of non-proteolytic *C. botulinum* in food products is similar to that in phosphate buffer. In fish products, the values reported ranged from a D_{80} of 0.78 min in oyster homogenate (Bucknavage *et al.*,

Table 6.2 Time to toxin production of psychrotrophic *Clostridium botulinum* at chill temperatures

Temperature of storage (°C)	Time to produce toxin (days)	*C. botulinum* type	Growth medium	Reference
3.3	31	E	Beef stew	Schmidt et al. (1961)
3.3	129	B	Broth	Eklund (1993)
4	30	E	Broth	Solomon et al. (1977)
4	31	B	*Sous vide* beef	Crandall et al. (1994)
4	18–21	E	Cod	Post et al. (1985)
4	20	E	Broth	Solomon et al. (1982)
4	33	B	Broth	Eklund (1993)
4.4	55	E	Crab meat	Cockey and Tatro (1974)
4.4	70	B	Cod	Brown and Gaze (1990)
5	42	E	Cod	" " "
5	84	B	Chicken	" " "
5	15	E	Herring	Cann et al. (1965)
5	27	B	Broth	Eklund (1993)
5.6	19	E	Beef stew	Schmidt et al. (1961)
6	120	E	Chopped meat (3.5% NaCl)	Abrahamson et al. (1986)
8	28	B & E	Tagliatelle with meat	Notermans et al. (1990)
8	12	B & E	Salmon fillets under MAP	Garcia et al. (1987)
	9			
	6			
8	15	B	Broth	Solomon et al. (1982)
8	28–56	E	Chicken	Brown and Gaze (1990)
8	42–56	B	Chicken	" " "
8	28–56	B	Cod	" " "
8	21	E	Cod	" " "

8	8–14	E & B	Cooked turkey	Genigeorgis et al. (1991)
8	8–20	E & B	Cooked turkey (1.47% brine)	"
8	10–24	E & B	Cooked turkey (2.2% brine)	"
8	9–12	E & B	Rockfish under MAP	Ikawa and Genigeorgis (1987)
8	8–20	E	Cod	Post et al. (1985)
10	8	E	Fish fillets	Huss (1981)
10	9	E	Vacuum-packed potatoes	Notermans et al. (1990)
10	8	E	Crab meat	Cockey and Tatro (1974)
10	7	E	Herring	Taylor et al. (1990)
	7	E	Smoked mackerel	"
10	6	E	Cod	"
10	6	B	Sous vide beef	Crandall et al. (1994)
	6–8	E	Shrimp	Lerke and Farber (1971)
10	6–8	E	Cod	Cann and Taylor (1979)
10	30	E	Trout 2–5% NaCl	Lerke (1973)
	7	E	Crab pH 7.2	"
	9	E	Crab pH 5.68	"
	12	E	Crab pH 5.35	
	14	E	Crab pH 5.23	
11	6	E	Beef stew	Schmidt et al. (1961)
12	14	E	Crab meat	Solomon et al. (1977)
12	12	E	VP fish	Lilly and Kautter (1990)
12	6–14	E	Cod	Post et al. (1995)
12	6	E	Broth	Solomon et al. (1982)
12	4–13	E & B	Cooked turkey	Genigeorgis et al. (1991)
	7–11	E & B	Cooked turkey (1.47% brine)	
	9–13	E & B	Cooked turkey (2.2% brine)	
12	3–6	E & B	Rockfish under MAP	Ikawa and Genigeorgis (1987)

Table 6.3 Heat resistance of psychrotrophic *Clostridium botulinum*

C. botulinum type	Heating medium	Heating temperature (°C)	D value (min)	Reference
Non-proteolytic B (4 strains)	Phosphate buffer (pH 7.0)	82.2	1.49–32.3	Scott and Bernard (1982)
E	Phosphate buffer (pH 7.0)	82.2	0.33	Scott and Bernard (1982)
E (2 strains)	Phosphate buffer (pH 7.0)	80.0	3.3	Ohye and Scott (1957)
F	Phosphate buffer (pH 7.0)	82.2	0.25–0.84	Lynt *et al.* (1979)
E	Aqueous suspension	80	0.33–1.25	Roberts and Ingram (1965)
B	Phosphate buffer (pH 7.0)	85	100*	Peck *et al.* (1993)
E	Phosphate buffer (pH 7.0)	85	46*	Peck *et al.* (1993)
E (5 strains)	Blue crab meat	82.2	0.49–0.74	Lynt *et al.* (1977)
F	Crab meat	85	0.53	Lynt *et al.* (1977)
E	Crab meat	82.2	1.9	Bohrer *et al.* (1973)
E	Oyster homogenate	80	0.78	Bucknavage *et al.* (1990)
E	Oyster homogenate	82.2	0.07–0.43	Tui-Jyi and Kuang (1992)
E	Clam liquor	82.2	0.20	Licciardello (1983)

	Food	Temp	D-value	Reference
E	Tuna in oil	76.7	40.90	Bohrer et al. (1973)
		80	10.50	
		82.2	6.60	
E	White fish	82.2	2.21	Crisley et al. (1968)
E	Sardines in tomato sauce	82.2	6.60	Bohrer et al. (1973)
E	Corn in brine	76.7	11.2	Bohrer et al. (1973)
		80	3.2	
		82.2	1.3	
E	Shrimp	80	1.9	Bohrer et al. (1973)
		82.2	1.3	
E	Cod	80	15.1	Gaze and Brown (1990)
		90	0.79	
B	Cod	80	18.30	Gaze and Brown (1990)
		90	1.10	
E	Carrot	80	4.33	Gaze and Brown (1990)
		90	0.48	
B	Carrot	86	4.24	Gaze and Brown (1990)
		90	0.43	
B	Turkey (1% NaCl)	90	1.1	Juneja and Eblen (1995)
B	Buffer	90	0.46	Juneja et al. (1995)
	Turkey	90	0.8	

*Recovery medium contained added lysozyme.

1990) to 18.3 min for type B in cod (Gaze and Brown, 1990). The presence of oil or fat in the heating menstruum appears to affect the thermal resistance characteristics of *C. botulinum*; for example, at 82.2°C the D-values for various fish products ranged from 0.07 min in oyster homogenate (Tui-Jyi and Kuang, 1992) to 2.21 min in white fish chubs (Crisley *et al.*, 1968) whereas for a high fat product such as tuna in oil, the D-value was 6.6 min (Bohrer *et al.*, 1973).

Of the literature reported to date (Table 6.3), the highest heat resistance values appear to be for type B in fish or poultry. Gaze and Brown (1990) reported the heat resistance of type B in cod at 90°C to be 1.1 min. Based on these data, they recommended that *sous vide* products be heated for a minimum of 10 min at 90°C or equivalent in order to achieve a 6 log reduction in numbers of psychrotrophic *C. botulinum*. This has since been recommended by the ACMSF (Anon., 1992a) for all vacuum-packaged (VP) and modified atmosphere-packaged (MAP) foods.

Recent work (Stringer *et al.*, 1997) has suggested that this process time and temperature may not always be sufficient to achieve the 6D reduction in numbers of psychrotrophic *C. botulinum*. In work by these authors, spores of non-proteolytic B, E and F strains were heated in laboratory media for between 0 and 60 min at 90°C and incubated at 5, 10 or 30°C in the absence or presence of lysozyme. When the samples were incubated at 5°C for 23 weeks, a process of 1 min at 90°C ensured that a 6D reduction was achieved. However, at 10°C a process of 60 min at 90°C was required to ensure a 6D reduction in the presence of lysozyme.

These data emphasise the importance of good chill temperatures during storage of minimally processed foods and illustrate that for certain products which may contain high levels of lysozyme, other safety factors in addition to heating temperature should be considered.

(c) Effect of MA packaging As *C. botulinum* is anaerobic and can thrive in modified atmosphere (MA) conditions, there is little to be gained by using modified atmospheres to control its growth. However, there is a commonly held belief that this organism does not grow in the presence of atmospheric oxygen. Various studies have shown that oxygen levels up to 4.4% are needed to inhibit growth of this organism and in fact in many air-packed products the redox potential in parts of the product is sufficiently low to allow growth and toxin production by *C. botulinum*.

For example, Post *et al.* (1985) found that inoculated cod fillets stored at 12°C in various atmospheres became toxic most rapidly in 100% N_2, then air, then 100% CO_2, then vacuum. Eklund (1993) reported that the use of 60% $CO_2/25\%$ O_2 /15% N_2 and 90% $CO_2/10\%$ N_2 delayed the production of *C. botulinum* toxin at 10°C compared to air-packed samples, although by 10 days the inhibitory effect had been overcome and all samples were toxic. These authors considered that CO_2 does not increase the rate of toxin production by *C. botulinum*, but does inhibit the growth of spoilage organisms at 10 and 25°C and can therefore allow development of toxin before spoilage occurs.

There is little data to show any inhibitory effects of MA gases alone on growth of *C. botulinum*, in fact there are data to show that low levels of CO_2 stimulate the germination of *C. botulinum* spores.

Foegeding and Busta (1983) studied the germination rate and percentage of germinated *C. botulinum* spores under CO_2 alone, N_2 alone, CO_2 plus hydrogen and N_2 plus hydrogen. For all three strains of *C. botulinum* tested, CO_2, either alone or in combination with hydrogen, enhanced the germination rate while germination in the absence of CO_2 was low.

It can be seen that inclusion of O_2 in MA packs does not necessarily inhibit the growth of *C. botulinum* and should be avoided as a specific antimicrobial factor to inhibit growth of this organism. In order to ensure safety of MAP/VP minimally processed chilled foods, the critical safety factors discussed in this chapter should be used.

6.2.2 Listeria monocytogenes

L. monocytogenes is of particular importance to short shelf life minimally processed chilled foods because it is more heat resistant than other vegetative pathogens and is able to grow rapidly at good refrigeration temperatures.

(a) Low-temperature growth   The ability of *L. monocytogenes* to grow at low temperatures is well documented, and the lowest recorded temperature at which growth has been reported in laboratory media is –0.1 to –0.4°C (Walker *et al.*, 1990).

Similar growth kinetics have been reported in a variety of foods. For example, in smoked cod, it was reported that *L. monocytogenes* grew at 3°C in vacuum packages or 100% CO_2, but was unable to grow at –1.5°C in either packaging type within 42 days (Bell *et al.*, 1995). Further studies in cold-smoked salmon (Hudson and Mott, 1993) showed that *L. monocytogenes* grew in both aerobic and vacuum packages at 5 and 10°C. At the higher temperature, the lag phase and growth rate were comparable for both packaging types, while at 5°C the lag phase was extended in vacuum-packaged salmon; however, once growth was initiated, the growth rate was similar.

In some products, the growth of *L. monocytogenes* does not occur as readily at low temperatures. For example, in cooked, uncured turkey loaf stored under vacuum at 3°C (Ingham and Tautorus, 1991), there was no increase in numbers over a 15 day storage period, while Degnan *et al.* (1992) found there was a decrease in levels of *L. monocytogenes* on vacuum-packaged wieners stored at 4°C for 72 days.

L. monocytogenes has also been shown to grow at low temperatures on minimally processed vegetable products. Carlin *et al.* (1996) evaluated the growth of this organism on minimally processed fresh endive at 3 and 10°C in various modified atmospheres containing between 10 and 50% CO_2. Growth occurred at 3°C for all MA conditions tested although the increase was low during 10 days storage at 3°C (between 0.3 and 1.5 logs). At 10°C rapid growth occurred by 7 days storage in all atmospheres tested, with the highest level of CO_2 (5%) resulting in the greatest increase in numbers.

(b) Heat resistance The heat resistance of *L. monocytogenes* is generally higher than that of other vegetative pathogens and it is therefore important that the process for minimally processed chilled foods with a short shelf life is sufficient to inactivate this organism.

Gaze *et al.* (1989) studied the heat resistance of *L. monocytogenes* Scott A and NCTC 11994 in chicken, carrot and steak over the temperature range 60 to 70°C. The highest heat resistance value found was 0.27 min for NCTC 11994 in carrot at 70°C. It was therefore concluded that a heat treatment of 2 min at 70°C was required to achieve a 6 log reduction in numbers of this organism. This process recommendation has since been given in UK Government Guidelines for Cook–Chill and Cook–Freeze Systems (Anon., 1989).

Many authors have reported on the heat resistance of *L. monocytogenes* in a variety of food products including pork, ham, beef, salmon and cod (Table 6.4).

It can be seen (Table 6.4) that the D-values reported for a range of products fall within or below the range reported by Gaze *et al.* (1989) and therefore it is concluded that a minimum process of 70°C for 2 min (or equivalent) should achieve a minimum 6D kill of *L. monocytogenes* in short shelf life minimally processed foods.

(c) Effect of MA packaging There has been considerable research over the past 5 years on the growth of *L. monocytogenes* in a range of MAP food products. Although some of the data are contradictory, there appears to be a general consensus that high levels of CO_2, i.e. >70%, are required in the absence of O_2, to inhibit this organism. For example, in turkey sandwiches (Farber and Daley, 1994) growth of *L. monocytogenes* was inhibited by 70% CO_2 at both 4 and 10°C.

Hart *et al.* (1991) studied the growth of *L. monocytogenes* on chicken meat packaged under 30 and 100% CO_2 and stored at 1, 6 and 15°C. At 1°C, no growth was observed in any atmosphere while at 6°C, growth was observed in air-packed controls but was inhibited by MA, particularly 100% CO_2. At 15°C the growth rate was similar in all gaseous environments.

Further work on raw minced chicken found that growth of *L. monocytogenes* was inhibited at 4, 10 and 27°C in a MA containing 75% CO_2/25% N_2

Table 6.4 Heat resistance of *Listeria monocytogenes*

Heating medium	Heating temperature (°C)	D value (min)	Reference
Ground pork	60	1.14–1.7	Ollinger-Snyder *et al.*, 1995
Ham	60	0.97–3.48	Carlier *et al.*, 1996
Cook-chill roast beef and gravy	65	0.56–0.88	Grant and Patterson, 1995
	70	0.37	
Sous vide beef	64	1.40–1.7	Hansen and Knøchel, 1996
Sous vide cod	65	0.27	Embarek and Huss, 1993
Sous vide salmon	65	1.18	
Green shell mussels	62	1.85	Bremer and Osborne, 1995
Physiological saline	60	0.72–3.1	Sörquist, 1994

while the addition of 5% O_2 (72.5% CO_2/22.5% N_2/5% O_2) removed the inhibitory effect and growth of *L. monocytogenes* was similar to that in air-packed controls (Wimpfheimer *et al.*, 1990).

Avery *et al.* (1994) found that growth of *L. monocytogenes* occurred in VP beef at 5 and 10°C whereas no growth was observed under 100% CO_2 at either temperature. Similar data were reported for lamb products where growth was inhibited by 100% CO_2 at 5°C but occurred in 80% CO_2 (Sheridan *et al.*, 1995).

In contrast, Fang and Lin (1994) studied cooked pork inoculated with *L. monocytogenes* and stored at 4 and 20°C in 100% CO_2, 80% CO_2/20% air or 100% air and showed that growth occurred under all conditions at both temperatures.

In conclusion, *L. monocytogenes* is able to grow in chill products under a range of MAP conditions, and other factors, e.g. thermal processing, should be used where the level of CO_2 used is not sufficiently inhibitory.

6.2.3 Other pathogenic organisms

For minimally processed chilled foods which receive a pasteurisation step, there should be minimal risk from the growth of other pathogens as the heat resistance of these pathogens is lower than that for *L. monocytogenes* and a process of 70°C for 2 min should achieve a 6 log reduction in numbers. However, for products which do not receive any heat treatment, or for those which are assembled after processing, there is the potential for the presence and growth of other pathogenic organisms. Inhibitory factors such as MAP and chilled temperatures will therefore be critical safety factors in controlling the growth of these pathogens.

Yersinia enterocolitica is a psychrotrophic pathogen which can grow in a range of MAP conditions, although the inhibitory effects of CO_2 are highly dependent on other factors such as product, storage temperature and competitive organisms. In some cases growth of this organism was inhibited by as little as 20% CO_2, whereas in other cases growth occurred, albeit at a reduced rate, in 100% CO_2. For example, Gill and Reichel (1989) found that *Y. enterocolitica* was able to grow in 100% CO_2 on high pH ground beef at 5 and 10°C, whereas Hudson *et al.* (1994) reported that *Y. enterocolitica* grew on sliced roast beef at –1.5°C in VP samples but not in 100% CO_2 packaged meat.

Other workers (Kleinlein and Untermann, 1990) found that a relatively low level of CO_2 was able to inhibit the growth of *Y. enterocolitica* at low temperatures. At temperatures of 1 and 4°C the growth of *Y. enterocolitica* in minced beef was reduced in a mixture of 20% CO_2/80% O_2 compared to air-packed controls. At 10°C there was only slight inhibition of growth, whereas at 15°C the growth was similar for MAP and air-packed samples.

Aeromonas hydrophila is a psychrotrophic organism able to grow both aerobically and in the absence of oxygen. Thus there is the potential for this organism to grow in MAP or VP products stored at chill temperatures.

The levels of CO_2 required to inhibit this organism are very dependent on product type and temperature. A level of 10% CO_2 was not sufficient to have an anti-microbial effect on *A. hydrophila* inoculated onto broccoli and stored at 4 and 15°C (Berrang *et al.*, 1989) while a level of 30% was able to reduce the growth of *A. hydrophila* in cooked mince and surimi products at 5 and 13°C (Ingham and Potter, 1988).

Hudson *et al.* (1994) found that growth of *A. hydrophila* was reduced on cooked roast beef packaged in 100% CO_2 compared to VP. At –1.5°C there was no growth of this organism in CO_2, while there was an increase of over 10^4 per gram in VP product after 1000 hours. At 3°C there was an increase in numbers of *A. hydrophila* under MAP conditions of approximately 10^3 per gram although the growth rate and final numbers achieved were lower than those observed under VP.

Salmonella species, in common with other members of the Enterobacteriaceae family, are facultative anaerobes and their growth is not necessarily inhibited by the removal of oxygen. There are, however, data to illustrate that inclusion of CO_2 in the gaseous atmosphere can inhibit the growth of *Salmonella* species, particularly when used in combination with chilled storage.

Siliker and Wolfe (1980) found that growth of *Salmonella* on ground beef was substantially reduced at 10°C in a MA containing 60% CO_2 compared to the air-packed control. Similar results were found by Luitens *et al.* (1982) who found little growth of *Salmonella* inoculated onto beef in VP or MA conditions of 60% CO_2/40% O_2 when stored at 20°C, while an increase of 10^3 was observed in air-packed controls.

Temperature combined with storage atmosphere is important in the inhibition of *Salmonella*, particularly close to the minimum growth temperature. Ingham *et al.* (1990) found that *Salmonella* inoculated onto crab meat did not grow at 7°C in air or in a MAP containing 50% CO_2/10% O_2/balance air. Growth occurred at 11°C and was approximately 50% lower for MA samples compared to the air controls.

Hintlian and Hotchkiss (1987) found that *S. typhimurium* was able to grow at 12.8°C on cooked beef packaged in 75% CO_2 containing 0, 2, 5, 10 or 25% O_2 although the growth rate and final numbers reached were lower than in air-packed controls.

It has been shown that many pathogenic microorganisms are able to survive pasteurisation treatments and/or grow in VP or MAP foods stored at chill temperature. It is essential that the risks from these organisms are controlled during product development and manufacture.

6.3 Critical factors

The following stages in the production of minimally processed foods are critical to the safety of these products.

• Shelf life determination
• Product formulation
• Raw materials
• Preparation
• Assembly and packaging
• Pasteurisation
• Cooling
• Chilled storage and distribution
• Temperature control.

6.3.1 Shelf life determination

While many of the pathogens of concern to minimally processed foods can grow at chill temperatures, this growth is relatively slow. It is therefore possible to minimise the risk with respect to these organisms by restricting the shelf life and thus the likelihood of growth. It has been stated that *C. botulinum* is the organism of most concern to minimally processed VP or MAP foods (Anon., 1992a) and that the shelf life of these foods should be based on the potential for growth of this organism.

Storage of minimally processed chilled foods at 3°C or less will prevent the growth of *C. botulinum. Sous vide* products are stored at <3°C throughout storage and distribution and maintenance of this temperature is critical to the safety of these foods. If the chill temperature is greater than 3°C then a number of other factors are critical to the safety of these products.

For products with a short shelf life, i.e. 10 days or less at chill temperatures of <8°C, there is unlikely to be any growth of *C. botulinum*; therefore, there are no specific recommended controlling factors for foods based on inhibition of this organism (Anon., 1992a). Figure 6.3 shows the time to toxin production for *C. botulinum* in a range of foods. It can be seen that, below 8°C, there is no toxin production within 10 days. In a few cases toxin was produced at 8°C between 6 and 10 days. In such cases, the products were fish or poultry products or the inoculum level used was high. For fish products, the risks of growth and toxin production may be slightly higher and where risk assessment indicates that a product may be a significant source of *C. botulinum*, extra control measures in addition to temperature control should be given to short shelf life products, e.g. limiting shelf life or reducing the storage temperature still further (Anon., 1992a).

For long shelf life, minimally processed food, there is the possibility that psychrotrophic *C. botulinum* could grow and produce toxin if the organism was present. This possibility is increased as the storage temperature is increased or the storage time lengthened. Therefore, for long shelf life products, >10 days at 8°C, a series of controlling factors are recommended in Section 6.3.2.

6.3.2 Production formulation

The microbiological safety of minimally processed chilled foods with a shelf life of greater than 10 days at chill temperatures, i.e. ≤8°C, should be

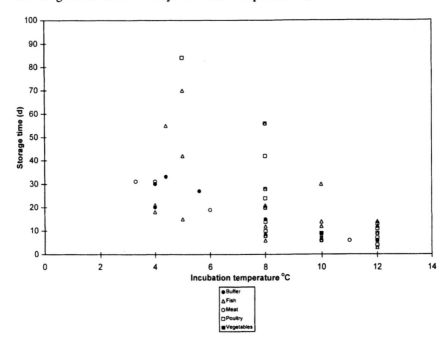

Figure 6.3 Time to toxin production for psychotrophic *Clostridium botulinum* stored in various menstrua at refrigerated temperatures.

controlled by one or more of the following factors (Anon., 1992a; Betts, 1996):

(i) minimum heat treatment of 90°C for 10 min or equivalent (see Section 6.3.6)
(ii) pH of 5 or less throughout the food
(iii) salt level (aqueous phase) of 3.5% throughout the food
(iv) a_w (water activity) of 0.97 or less throughout the food
(v) a combination of heat and preservative factors which can be shown to consistently prevent growth and toxin production by *C. botulinum*.

One or more of these factors should be achieved in the food every time it is produced and must be achieved consistently and uniformly throughout the product.

It is also possible to use a lower level of factors (i) to (iv) in a food to achieve a combined preservation effect or use additional preservatives such as nitrite. Where a combination of factors is used, it is necessary to illustrate that the preservation system chosen can consistently prevent growth and toxin production by psychrotrophic *C. botulinum*.

pH. With respect to the pH, the level of acid in a minimally processed food, e.g. *sous vide* meal, can be used as a controlling factor to prevent the growth of microorganisms and a pH of 5.0 or below throughout all parts of a food is sufficient to inhibit the growth of psychrotrophic *C. botulinum* in combination with chill temperatures (Anon., 1992a).

In some multi-component foods there may be a variation in pH within the product due to diffusion and mixing limitations and if pH is the critical safety factor in use then it is important that the pH throughout all parts of the final product meets the target value.

Salt content. A level of 3.5% (w/w) salt throughout the aqueous phase of a food stored at below 10°C has been reported to be sufficient to inhibit the growth of *C. botulinum* (Anon., 1992a).

The percentage of salt in the aqueous phase of a product can be calculated from the salt content (grams of NaCl present in 100 g of product) and the moisture content (grams of water per 100 g of product) using the following calculation.

$$\frac{NaCl}{NaCl \text{ content} + \text{moisutre content}} \times 100$$

The examples shown in Table 6.5 illustrate the salt content needed to give a minimum salt concentration of 3.5% in the aqueous phase for a range of water contents (Betts, 1996).

If the salt level is used as the critical safety factor, it is important that the

Table 6.5 Amount of salt required to achieve 3.5% aqueous phase in the final product

Water (% w/w)	Required salt (% w/w)
50	1.82
55	2.00
60	2.18
65	2.36
70	2.54
75	2.73
80	2.91

concentration throughout all parts of the final product meets the target value.

Water activity. With regard to water activity (a_w), the ACMSF (Anon., 1992a) stated that for foods with salt or other solutes (e.g. sugars, etc.) as the main a_w depressant, a level of 0.97 should be achieved throughout all parts of that food to inhibit growth of psychrotrophic *C. botulinum*. This level is not sufficiently low to prevent growth of other pathogens of potential concern (Table 6.1) although provided the heat treatment is a minimum of 70°C/2 min, all vegetative pathogens should be inactivated.

In some multi-component foods there may be a variation in a_w within the product. Where a_w is used either alone or in combination with other factors to control the safety of a food it is important that it is controlled to ensure that it is at the target level.

6.3.3 Raw materials

The particular microbiological hazards associated with raw materials are the presence and growth of foodborne pathogens and the formation of heat stable toxins by toxigenic bacteria.

In order to control the presence of undesirable microorganisms, it is recommended that microbiological specifications are set for all raw materials and packaging materials to ensure that the microbiological quality of the raw materials is acceptable.

There should be good separation of different types of raw materials, e.g. raw and cooked components during transport and storage, to minimise the risk of contamination of the more sensitive cooked ingredients from raw ingredients with a high microbial loading. There should also be good hygienic design of storage areas to enable adequate cleaning to reduce the risk of environmental contamination.

The storage and handling of the raw materials is the first stage in the manufacturing process which is critical to the safety of minimally processed chilled foods. Correct procedures at this stage will restrict microbial contamination and growth. Ineffective control may allow the numbers of microorganisms to reach excessively high levels which will affect the adequacy of safety factors further down the processing line. In order to reduce the growth of microorganisms present on raw materials there should be good stock rotation and temperature control during storage is essential. The storage time and temperature must be appropriate to minimise bacterial growth and toxin production.

The control of storage temperature is particularly important with respect to production of toxin by vegetative organisms. For example *Staphylococcus aureus* produces enterotoxins which are fairly resistant to heat; a D-value of 73.7 min at 90°C has been reported for *S. aureus* enterotoxin A (Modi *et al.*, 1990). While the recommended pasteurisation process for minimally processed chilled foods (10 min at 90°C) will eliminate *S. aureus* cells, it is unlikely to eliminate the preformed toxin. If the conditions during storage were such that *S. aureus* could grow and produce toxin it is possible that the toxin would remain active and be present in the final product.

Psychrotrophic and proteolytic *C. botulinum* toxins are also fairly resistant to heat. Sakaguchi (1979) recommended heating for 30 min at 80°C or boiling for a few minutes in order to inactivate preformed toxins. Similar data are reported by Woodburn *et al.* (1979) who recommended treatments of 20 min at 79°C or 5 min at 85°C. Conditions for raw material storage should be controlled so that the production of all bacterial toxins is prevented.

6.3.4 Preparation

The preparation of raw materials is also critical to the safety of minimally processed chilled foods, as failure to control preparation procedures may lead to significant microbial growth. There should be good temperature control in preparation areas to comply with legal temperature specifications and the temperature of the raw materials should be kept within the target temperature range, e.g. < 5°C.

Control of times and temperatures used for the pre-cooking and subsequent cooling of intermediate ingredients is important to minimise growth of bacteria. All intermediate product not for immediate use should be chilled rapidly to below 5°C and should be kept at this temperature until required.

There should be careful and appropriate handling of raw ingredients to minimise the potential for cross contamination, e.g. utensils should not be used on both raw and cooked materials.

There should be controlled thawing of frozen ingredients and controlled rehydration of dried ingredients: the time and temperature of these procedures should minimise the risks of microbial growth.

It is critical that the product formulation in terms of a_w, salt or pH levels is achieved during preparation, particularly if the formulation is being used as the critical safety factor to inhibit growth of *C. botulinum* (see Section 6.3.3). If *C. botulinum* is not the target microorganism, for example in short shelf life products, then the product formulation can still be important to inhibit other pathogens. Table 6.1 gives the minimum or maximum levels of various factors permitting the growth of some pathogenic bacteria and these data can help in product design.

Another microbiological hazard associated with prepared raw materials is survival of microorganisms during subsequent cooking due to excessively large particulates. Specifications should be set for portion sizes of solid ingredients, e.g. chicken breasts or steak pieces. Failure to control portion sizes could lead to inadequate pasteurisation due to poor heat penetration and thus survival of microorganisms.

6.3.5 Assembly and packaging

The microbiological hazards associated with the assembly and packaging of minimally processed chilled foods are microbial survival during subsequent pasteurisation, microbial contamination and microbial growth.

With respect to foods which are pasteurised under vacuum, i.e. *sous vide* products, the product fill temperature is important as any variations may affect the vacuum intensity and this, in turn, may lead to variations in pasteurisation value achieved. The temperature of product during filling should be monitored to ensure compliance with specifications.

The seal integrity of all VP and MAP products should be monitored. Any failure of the seals may lead to post-process contamination and loss of atmosphere.

In addition, for minimally processed foods which are packed in a modified atmosphere after preparation or cooking, it is crucial to ensure that the correct gaseous mixture is achieved.

Many short shelf life minimally processed chilled foods rely on MAP and, in particular, CO_2 to inhibit the growth of pathogens which may be present. This is a critical safety factor for this category of product and the levels of CO_2 required to inhibit pathogenic growth must be established (see Section 6.2.3) and adhered to.

6.3.6 Pasteurisation

It has already been stated that due to the potential for growth of psychrotrophic *C. botulinum* at chill temperatures, the processes for minimally processed chilled foods should be aimed at achieving destruction of this organism.

From the data of Gaze and Brown (1990) it was shown that a treatment of 7 min at 90°C would achieve a 10^6-fold reduction in numbers of *C. botulinum*. Similar processes can be found in the literature for appropriate time/temperature treatments for *sous vide* foods; for example Notermans *et al.* (1990) state that a process of 4 min at 90°C should reduce non-proteolytic *C. botulinum* by a factor of 10^5.

French legislation for *sous vide* foods (Anon., 1988), based on the thermal characteristics of *Enterococcus faecalis*, recommends a process of 100 min at 70°C for products with a 21 day shelf life or 1000 min at 70°C for products with a 42 day shelf life. These processes are equivalent in lethality to 1 minute and 10 min at 90°C respectively.

In conclusion it can be recommended that due to the potential for rapid toxin production by psychrotrophic *C. botulinum* at abuse chill temperatures, all minimally processed chilled foods with an extended shelf life of greater than 10 days should be pasteurised to adequate lethality to achieve a 10^6-fold reduction in numbers of this organism; current knowledge suggests that this can be achieved by heating to a minimum of 90°C for 7 min. It is therefore recommended that a process of 10 min at 90°C or equivalent (Table 6.6) is used to allow an added safety margin and that these products are stored at good chill temperature throughout the shelf life. For *sous vide* products in particular, a storage temperature of 0 to 3°C should be used.

It has been suggested that *sous vide* products with a short shelf life of 1 to 10 days can be processed to a minimum of 100 min at 70°C and should be stored at 0 to 3°C throughout the shelf life (Betts, 1992). However, since then it has been recommended (Betts, 1996) that as there is minimal potential for growth of *C. botulinum* in short shelf life products there is no necessity to give a process recommendation based on this organism.

Table 6.6 Alternative time/temperature combinations to achieve the equivalent of 90°C for 10 minutes[a]

Process temperature (°C)	Time (min)
75	464
76	359
77	278
78	215
79	167
80	129
81	100
82	77
83	60
84	46
85	36
86	28
87	22
88	17
89	13
90	10

[a]These data have been calculated using a z-value of 9°C and a reference temperature of 80°C in accordance with the ACMSF recommendations (Anon., 1992a).

Therefore the minimum treatment that should be given to short shelf life minimally processed chilled foods is 70°C for 2 min or equivalent (Table 6.7). This is sufficient to achieve a 10^6-fold reduction of *L. monocytogenes* (Gaze *et al.*, 1989) and should ensure safety from this organism and other vegetative pathogens. This process was originally designed for products with a maximum shelf life of 5 days (Anon., 1989) and it is therefore likely that a higher process than this may be needed to inactivate other spoilage organisms that may be able to grow at chill temperatures and spoil the product if stored for up to 10 days.

The pasteurisation process is one of the most important critical control points of the *sous vide* process as the failure to achieve the correct lethality may lead to survival of infectious and toxigenic microorganisms capable of growth at chill temperatures. There are several control steps which should be taken to ensure that the correct lethality is achieved: the design of the cooking vessel used, the choice of cooking times and temperatures, and the orientation and spacing of product within the cooker (Betts, 1992).

Of particular importance to the safety of the product is the determination of the correct cooking times and temperatures required to achieve the processes given above in the product core for each product type, as the process requirements will vary depending on product density, heat penetration rates, particulate size, etc. This was demonstrated by Ghazala *et al.* (1995) who used a computer model system to simulate the thermal process requirements for *sous vide* spaghetti/meat sauce and rice/salmon products to achieve a 10^5-fold reduction in *Enterococcus faecium*.

Table 6.7 Alternative time/temperature combinations to achieve the equivalent of 70°C for 2 minutes[a]

Process temperature (°C)	Time (min)
60	43.5
61	31.8
62	23.3
63	17.1
64	12.7
65	9.3
66	6.8
67	5.0
68	3.7
69	2.7
70	2.0
71	1.5
72	1.1
73	0.8
74	0.6
75	0.5

[a]Calculated using a z-value of 7.5°C as recommended in Gaze *et al.* (1989).

They found that the spaghetti product needed to be heated for 127 min at 60°C or 36 min at 75°C whereas the rice/salmon product needed 163 min at 60°C and 42 min at 75°C to achieve equivalent 5D processes.

6.3.7 Cooling

The hazard associated with the cooling of minimally processed chilled foods is the growth of surviving spore-formers: the rate of cooling should be sufficiently fast to prevent the growth of these microorganisms. It has been recommended that for cook–chill products the cooling process should begin within 30 min of the end of cooking and should reach a core temperature of 0 to +3°C within a further 90 min (Anon., 1989).

Rapid cooling is essential to control the growth of any spore-forming bacteria which survive the heat process and, in particular, psychrotrophic C. botulinum which can grow at temperatures down to 3.3°C.

Other recommendations have been given for VP and MAP chill foods. For example, Betts (1996) states that there should be an appropriate cooling procedure designed to reduce the product to the target chill temperature as soon as possible and, where practicable, this should be to 5°C or below. For individual portion sized packs, i.e. approximately 350 g, it should be possible to achieve the target temperature within 90 min.

If this cannot be achieved in, for example, large joints of meat, the food should pass through the temperature range 50 to 10°C in the shortest possible time. If this occurs within 4 hours then it is unlikely that the risks from C. botulinum will be increased.

Correct cooling following pasteurisation is a critical safety factor to control microbial growth in minimally processed foods. The product core temperature should be recorded on removal from the cooling process to ensure that correct temperature reduction has been achieved.

Hot product should not be put in the same cooler or storage chiller as 'cooled' product, as it will raise the temperature of the chiller and lengthen the cooling process.

It is important that the maximum capacity of the cooling equipment is not exceeded and that chillers are appropriately loaded, e.g. in accordance with manufacturers' recommendations.

Cooling should be carried out in a hygienic area such that the risk of recontamination with pathogenic organisms is minimised. For example, accumulation of condensate on equipment and other surfaces should be avoided as this may be a significant source of cross-contamination.

6.3.8 Chilled storage and distribution

The microbiological hazards associated with the storage and distribution of minimally processed chilled foods are microbial contamination and growth. The products must be stored at a temperature of 3 to 8°C (or 0 to 3°C for *sous vide* foods); the temperature should be monitored and recorded where possible. There should be good stock rotation to ensure that packs are not kept for longer than the specified amount of time and the effectiveness of the procedure should be regularly monitored. The storage area should be hygienically designed to allow efficient cleaning and the chiller must be able to maintain the target temperature when fully stocked.

The design of loading bays, distribution vehicles and storage areas in distribution centres should be such that the specific storage temperature is not exceeded. Where possible a temperature of +3°C or less should be maintained to minimise microbial growth. The temperature of the product on despatch, in vehicles, in loading bays and at point of delivery to customers should be monitored and recorded.

6.3.9 Temperature control

The control of product temperature is one of the most critical factors in the safety of minimally processed VP and MAP foods, from receipt and storage of perishable food ingredients to distribution and end use of products, particularly for those products which do not include a heating step. For this reason it is necessary to re-emphasise where temperature control is important.

The particular target temperatures recommended for each manufacturing stage are based on scientific information on growth and death characteristics of relevant bacteria, particularly *C. botulinum* and *L. monocytogenes*, and should be adhered to.

The stages where temperature control is important in manufacture of minimally processed chilled foods are listed below:

- Transport of perishable or frozen raw materials.
- Unloading of perishable or frozen raw materials.
- Storage of perishable or frozen raw materials.
- Food preparation and assembly.
- Pre-treatment cooking stages of raw materials, e.g. browning of meats and subsequent cooling rate.
- Storage of prepared raw materials.
- Temperature and time achieved during pasteurisation or cooking stage.
- Temperature profile during cooling.
- Temperature of product at point of packaging, since if the temperature is too high, condensation may form on the packaging film and cause an increase in water activity.
- Storage temperature during distribution.

Due to the importance of temperature control, it is recommended that temperatures are monitored throughout production. A wide variety of devices are available for use in monitoring temperature. These should be capable of measurement to ±0.5°C and should be calibrated against a reference thermometer traceable to National Standards, at regular intervals, e.g. every 6 to 12 months.

Where the temperature of any manufacturing stage deviates from the target and tolerances, then immediate corrective action should be taken to restore the target temperature and review the safety of the food which has been produced, handled or stored during the period when the temperature was outside the tolerance.

6.4 Application of HACCP

The areas highlighted in this chapter are the most likely critical safety factors in the manufacture of minimally processed chilled foods. However, it is important that the specific critical control points are identified for each product formulation and this can best be achieved by the application of Hazard Analysis Critical Control Points (HACCP). Schellekens (1996) covers the use of HACCP in her review of new research issues in *sous vide* cooking and reports that the European Chilled Food Federation specifies that safety of chilled prepared foods should be ensured by applying HACCP principles.

In the USA there is a similar approach with the FDA recommendations for *sous vide* processing, which require that *sous vide* products are produced and distributed with a HACCP approach. Currently in the European Union it is a requirement under the directive on the hygiene of foodstuffs (Anon., 1993) for all food companies to use a system of hazard analysis which should be based on the principles of HACCP, although the use of HACCP *per se* is not mandatory.

There are several articles which describe the use of HACCP for minimally processed foods.

Smith (1990) gives a useful description of the application of HACCP to ensure the microbiological safety of *sous vide* meat and pasta products and identifies raw material quality, time/temperature relationships, packaging control and the use of additional barriers such as pH and reduction of a_w in formulated products as some of the critical control points.

Betts (1992) evaluated the microbiology safety of *sous vide* processing and describes in detail the particular microbiological hazards and controls associated with all stages of this process; Snyder (1995) describes the application of HACCP for MAP and *sous vide* products and gives examples of chemical and physical hazards as well as the microbiological hazards described in this chapter.

If the safety of minimally processed food is to be ensured in the future, then it is critical that all potential hazards for each product are identified and controlled using a HACCP approach.

6.5 Future trends

The developments of future safety factors for minimally processed foods are two-fold. Currently, there are no definitive recommendations for the processing of minimally processed foods, although there is a general consensus throughout the EU that the process of 10 min at 90°C should be achieved for long shelf life products (Table 6.8). There is currently a European Union-funded project considering the harmonisation of safety criteria for minimally processed foods (EU concerted action FAIR CT 96-1-2-1996–1999) which is aiming to address the issue of standardising safety criteria such as heat treatment, product formulation and storage conditions.

In the USA the process requirements are not so restrictive; for example, a 10^4-fold reduction of *L. monocytogenes* must be achieved; however, there is the requirement to show that the products are inhibitory to the growth of *C. botulinum* under good storage and abuse temperatures (Farber, 1995).

The second area likely to see developments is the use of novel processing techniques. The introduction to this chapter mentioned some of the potential novel processing techniques such as high pressure, ultrasound and electric fields. These technologies are well suited to minimally processed foods as they all allow the thermal input to be reduced while still maintaining the level of microbial inactivation required. Earnshaw *et al.* (1995) described the potential for using these techniques and stressed that the success of these approaches will rely on an in-depth knowledge of the physiological basis of their modes of action. This, in turn, would allow design of optimised preservation systems and provide evidence to regulatory bodies to enable authorisation of novel technology in the food-manufacturing industry.

As the inactivation of microbial pathogens will remain one of the critical safety factors in the processing of minimally processed chilled foods for many years to come, there is a requirement for further research into a variety of food products, using several strains of the important pathogens of concern, e.g. *L. monocytogenes* and *C. botulinum*, in order that confidence in the effects of non-thermal techniques can be conclusively established.

Table 6.8 Recommended process parameters for *sous vide* and minimally processed chilled foods

Reference	Pasteurisation	Cooling	Storage temp (°C)	Distribution temp (°C)	Shelf life
DOH Guidelines (Anon., 1989)	Core temperature of 70°/2 min should be achieved.	The temperature of the food should be reduced to 0 to +3°C within 90 min. Chilling should begin within 30 min of end of cooking.	0 to +3°C	0 to +3°C	5 days including day of cooking and consumption.
French Regulations (Anon., 1988)	Dependent on proposed shelf life. The following temperature and timescale recommended. *Temp.* (°C) *Time* (min) *Shelf life* 70 40 6 days 70 100 21 days 70 1000 42 days	The core temperature should be reduced to < +10°C within 120 min.	0 to +3°C	0 to +3°C	6, up to 21 or up to 42 days depending on pasteurisation process achieved.
Gagnon *et al.* (1990)	Not stated. Pasteurisation process should destroy defined level of identified target microorganism. Reference is made to French regulations.	Core temperature of ≤ +4°C to be achieved within 120 min.	−1°C to +4°C	−1°C to +4°C	Not stated. Reference is made to French legislation as an example of shelf life.
Dutch Belgian Guidelines (Anon., 1992b)	90°C for 10 min (*sous vide* products); 90°C for 10 min plus 70°C for 2 min – for products which are cooked, assembled, VP and pasteurised.	Not stated.	0 to 5°C	0 to 5°C	10 to 42 days.
ACMSF (Anon., 1992a)	90°C for 10 min.	Not stated.	< 8°C	< 8°C	>10 days (maximum not stated).
Betts (1996)	90°C for 10 min (long shelf life) 70°C for 2 min (short shelf life)	<5°C within 120 min (provisions given for large joints).	< 8°C (0 to 3°C for *sous vide* products).	< 8°C (0 to 3°C for *sous vide* products).	>10 days (long shelf life). 10 days or less (short shelf life).

References

Abrahamson, K., Gullmer, B. and Molin, N. (1986) The effect of temperature on toxin formation and toxin stability of *Clostridium botulinum* type E in different environments. *Canadian Journal of Microbiology*, **12**, 385–393.

Anon. (1988) Prolongation de la durée de vie des plâts cuisineé à l'avance, modification du protocole permettant d'obtenir les autorisations. Note de Service DGAL/SVHAIN 881 No. 8106 du 31 Mai 1988. République Française Ministère de l'Agriculture.

Anon. (1989) *Chilled and Frozen, Guidelines on Cook–Chill and Cook–Freeze Systems*. ISBN 0 1132 1161 9. Department of Health, HMSO, London.

Anon. (1992a) *Report on Vacuum Packaging and Associated Processes*. Advisory Committee for the Microbiological Safety of Foods. HMSO, London.

Anon. (1992b) *Draft Code for the Production, Distribution and Sale of Chilled Long Life, Pasteurised Meals*. Belgian Dutch Chilled Meals Working Group, The Netherlands.

Anon. (1993) Council Directive 93/43/EEC. On the hygiene of foodstuffs. Official Journal of the European Communities L175/1. 19/7/93.

Avery, S.M., Hudson, J.A. and Penney, N. (1994) Inhibition of *Listeria monocytogenes* on normal ultimate pH beef (pH 5.3–5.5) at abusive storage temperatures by saturated carbon dioxide controlled packaging. *Journal of Food Protection*, **57**(4), 331.

Bell, G., Penney, N. and Moorhead, S.M. (1995) Growth of the psychrotrophic pathogens *Aeromonas hydrophila*, *Listeria monocytogenes* and *Yersinia enterocolitica* on smoked blue cod packed under vacuum or carbon dioxide. *International Journal of Food Science and Technology*, **30**, 515–521.

Berrang, M.E., Bracket, R.E. and Beuchat, L.R. (1989) *Aeromonas hydrophila* on fresh vegetables stored under a controlled atmosphere. *Applied and Environmental Microbiology*, **55**, 2167–2171.

Betts, G.D. (1992) The microbiological safety of *sous vide* processing. *Technical Manual No. 39*. Campden & Chorleywood Food Research Association, Chipping Campden, Glos., UK.

Betts, G.D. (1996) Code of practice for the manufacture of vacuum and modified atmosphere packaged chilled foods with particular regard to the risks of botulism. *CCFRA Guideline No. 11*. Campden & Chorleywood Food Research Association, Chipping Campden, Glos., UK.

Bohrer, C.W., Denny, C.B. and Yao, M.G. (1973) Thermal destruction of type E *Clostridium botulinum*. *Final Report on RF 4603*. National Canners Association Research Foundation, Washington, DC.

Bremer, P.J. and Osborne, C.M. (1995) Thermal death times of *Listeria monocytogenes* in green shell mussels prepared for not smoking. *Journal of Food Protection*, **58** (6), 604–608.

Brown, G.D. and Gaze, J.E. (1990) Determination of the growth potential of *Clostridium botulinum* Types E and non-proteolytic B in *sous vide* products at low temperatures. *Technical Memorandum No. 593*. Campden & Chorleywood Food Research Association, Chipping Campden, Glos., UK.

Bucknavage, M.W., Pierson, M.D., Hackney, C.R. and Bishop, J.R. (1990) Thermal inactivation of *Clostridium botulinum* type E spores in oyster homogenate at minimal processing temperatures. *Journal of Food Science*, **55** (2), 372–373.

Cann, D.C. and Taylor, L.Y. (1979) The control of the botulism hazard in hot-smoked trout and mackerel. *Journal of Food Technology*, **14**, 123–129.

Cann, D.C., Wilson, B.B., Hobbs, G. and Shewan, J.M. (1965) The growth and toxin production of *Clostridium botulinum* type E in certain vacuum packed fish. *Journal of Applied Bacteriology*, **28** (3), 431–436.

Carlier, V., Augustin, J.C. and Rozier, J. (1996) Heat resistance of *Listeria monocytogenes* Phagovar (2389/2425,3274/2671/47/108/340): D- and z-values in ham. *Journal of Food Protection*, **59** (6), 588–591.

Carlin, F., Nguyen-the, C., Da Silva, A.A. and Cochet, C. (1996) Effects of carbon dioxide on the fate of *Listeria monocytogenes*, of aerobic bacteria and on the development of spoilage in minimally processed fresh endive. *International Journal of Food Microbiology*, **32**, 159–172.

Carlin, F. and Peck, M.W. (1995) Growth and toxin production by non-proteolytic and proteolytic *Clostridium botulinum* in cooked vegetables. *Letters in Applied Microbiology*, **20**, 152–156.

Cockey, R.R. and Tatro, M.C. (1974) Survival studies with spores of *Clostridium botulinum* type E in pasteurised meat of the blue crab, Calinectes sapidus. *Applied Microbiology*, **27**, 629–622.

Conner, D.E., Scott, C.M. and Bernard, D.T. (1989) Potential *Clostridium botulinum* hazards associated with extended shelf-life refrigerated foods : A review. *Journal of Food Safety*, **10**, 151–153.

Crandall, A.D., Winkowski, K. and Montville, T.J. (1994) Inability of *Pediococcus pentosaceus* to inhibit *Clostridium botulinum* in *sous vide* beef with gravy at 4 and 10°C. *Journal of Food Protection*, **57** (2), 104–107.

Crisley, F.D., Peeler, J.T., Angellotti, R. and Hall, H.E. (1968) Thermal resistance of spores of 5 strains of *Clostridium botulinum* type E in ground whitefish chubs. *Journal of Food Science*, **33** (4), 411–416.

Degnan, A.J., Yousef, A.E. and Luchansky, J.B. (1992) Use of *Pediococcus acidilactici* to control *Listeria monocytogenes* in temperature-abused vacuum-packaged wieners. *Journal of Food Protection*, **55** (2), 98–103.

Earnshaw, R.G., Appleyard, J. and Hurst, R.M. (1995) Understanding physical inactivation processes: combined preservation opportunities using heat, ultrasound and pressure. *International Journal of Food Microbiology*, **28**, 197–219.

Eklund, M.W. (1993) Control in fishery products, in Clostridium botulinum. *Ecology and Control in Foods* (eds A.H.W. Hauschild and K.L. Dodds), Marcel Dekker, pp. 209–232.

Eklund, M.W., Wieler, D.I. and Poysky, F.T. (1967) Outgrowth and toxin production of non-proteolytic type B *Clostridium botulinum* at 3.3 to 5.6°C. *Journal of Applied Bacteriology*, **83** (4), 1461–1462.

Embarek, P.K.B. and Huss, H.H. (1993) Heat resistance of *Listeria monocytogenes* in vacuum-packaged pasteurised fish fillets. *International Journal of Food Microbiology*, **20**, 85–95.

Fang, T.J. and Lin, L. (1994) Growth of *Listeria monocytogenes* and *Pseudomonas fragi* on cooked pork in a modified atmosphere packaging nisin combination system. *Journal of Food Protection*, **57** (6), 479–485.

Farber, J.M. (1995) Regulations and guidelines regarding the manufacture and sale of MAP and *sous vide* products, in *Principles of Modified Atmosphere and* sous vide *Product Packaging* (eds J.M. Farber and K.L. Dodds), Technomic Publishing Company, Pennsylvania.

Farber, J.M. and Daley, E. (1994) Fate of *Listeria monocytogenes* on modified atmosphere packaged turkey roll slices. *Journal of Food Protection*, **57** (12), 1098–1100.

Foegeding, P.M. and Busta, F.F. (1983) Effect of carbon dioxide, nitrogen and hydrogen gases on germination of *Clostridium botulinum* spores. *Journal of Food Protection*, **46** (11), 987–989.

Gagnon, B. (1990) *Canadian Code of Recommended Manufacturing Practices for Pasteurised/Modified Atmosphere Packed/Refrigerated Food*, Agri-Food Safety Division, Agriculture Canada, Ottawa, Canada.

Garcia, G. and Genigeorgis, C. (1987) Quantitative evaluation of *Clostridium botulinum* non-proteolytic types B, E and F growth risk in fresh salmon tissue homogenates stored under modified atmospheres. *Journal of Food Protection*, **50** (5), 390–397.

Garcia, G.W., Genigeorgis, C. and Lindroth, S. (1987) Risk of growth and toxin production by *Clostridium botulinum* non-proteolytic types B, E and F in salmon fillets stored under modified atmospheres at low and abused temperatures. *Journal of Food Protection*, **50** (4), 330–336.

Gaze, J.E. and Brown, G.D. (1990) Determination of the heat resistance of a strain of non-proteolytic *Clostridium botulinum* type B and a strain of type E, heated in cod and carrot over the temperature range 70 to 92°C. *Technical Memorandum No. 592*, Campden & Chorleywood Food Research Association, Chipping Campden, Glos., UK.

Gaze, J.E., Brown, G.D., Gaskell, D.E. and Banks, J.G. (1989) Heat resistance of *L. monocytogenes* in homogenates of chicken, beef, steak and carrot. *Food Microbiology*, **6**, 251–259.

Genigeorgis, C.A., Meng, J. and Baker, D. (1991) Behaviour of non-proteolytic *Clostridium botulinum* type B and E spores in cooked turkey and modelling lag phase and probability of toxigenesis. *Journal of Food Science*, **56** (2), 373–379.

Gill, C.O. and Reichel, M.P. (1989) Growth of the cold-tolerant pathogens *Yersinia enterocolitica, Aeromonas hydrophila* and *Listeria monocytogenes* on high-pH beef packaged under vacuum or carbon dioxide. *Food Microbiology*, **6**, 223–230.

Ghazala, S., Ramaswamy, H.S., Smith, J.P. and Simpson, M.V. (1995) Thermal process simulations for *sous vide* processing of fish and meat foods. *Food Research International*, **28** (2), 117–122.

Graham, A.F., Mason, D.R., Maxwell, F.J. and Peck, M.W. (1997) Effect of pH and NaCl on growth from spores of non-proteolytic *Clostridium botulinum* at chill temperatures. *Letters in Applied Microbiology*, **24**, 95–100.

Grant, I.R. and Patterson, M.F. (1995) Effect of gamma radiation and heating on the destruction of *Listeria monocytogenes* and *Salmonella typhimurium* in cook-chill roast beef and gravy. International *Journal of Food Microbiology*, **27**, 117–128.

Hansen, T.B. and Knøchel, S. (1996) Thermal inactivation of *Listeria monocytogenes* during rapid and slow heating in *sous vide* cooked beef. *Letters in Applied Microbiology*, **22**, 425–428.

Hart, C.D., Mead, G.C. and Norris, A.D. (1991) Effects of gaseous environments and temperature on the storage behaviour of *Listeria monocytogenes* on chicken breast meat. *Journal of Applied Bacteriology*, **70**, 40–46.

Hintlian, C.B. and Hotchkiss, J.H. (1987) Comparative growth of spoilage and pathogenic organisms on modified atmosphere-packaged cooked beef. *Journal of Food Protection*, **50** (3), 218–223.

Hudson, J.A. and Mott, S.J. (1993) Growth of *Listeria monocytogenes*, *Aeromonas hydrophila* and *Yersinia enterocolitica* on cold-smoked salmon under refrigeration and mild temperature abuse. *Food Microbiology*, **10**, 61–68.

Hudson, J.A., Mott, S.J. and Penney, N. (1994) Growth of *Listeria monocytogenes*, *Aeromonas hydrophila* and *Yersinia enterocolitica* on vacuum and saturated carbon dioxide controlled atmosphere packaged sliced roast beef. *Journal of Food Protection*, **57** (3), 204–208.

Huss, H.M. (1981) Clostridium botulinum *type E and Botulism*. Technological Laboratory, Ministry of Fisheries Technical University, Lyngby, Denmark.

Ikawa, J.Y. and Genigeorgis, C. (1987) Probability of growth and toxin production by non-proteolytic *Clostridium botulinum* in rockfish fillets stored under modified atmospheres. *International Journal of Food Microbiology*, **4**, 167–181.

Ingham, S.C., Alford, R.A. and McCown, A.P. (1990) Comparative growth rates of *Salmonella typhimurium* and *Pseudomonas fragi* cooked crab meat stored under air and modified atmosphere. *Journal of Food Protection*, **53** (7), 566–567.

Ingham, S.C. and Potter, N.N. (1988) Growth of *Aeromonas hydrophila* and *Pseudomonas fragi* on mince and surimis made from Atlantic pollock and stored under air or modified atmosphere. *Journal of Food Protection*, **51**, 966.

Ingham, S.C. and Tautorus, C.L. (1991) Survival of *Salmonella typhimurium*, *Listeria monocytogenes* and indicator bacteria on cooked uncured turkey loaf stored under vacuum at 3°C. *Journal of Food Safety*, **11**, 285–292.

Juneja, V.K. and Eblen, B.S. (1995) Influence of sodium chloride on thermal inactivation and recovery of non-proteolytic *Clostridium botulinum* type B strain KAP spores. *Journal of Food Protection*, **58** (7), 813–816.

Juneja, V.K., Eblen, B.S., Marmer, B.S., Williams, A.C., Palumbo, S.A. and Miller, A.J. (1995) Thermal resistance of non-proteolytic type B and type E *Clostridium botulinum* spores in phosphate buffer and turkey slurry. *Journal of Food Protection*, **58** (7), 758–763.

Kleinlein, N. and Untermann, F. (1990) Growth of pathogenic *Yersinia enterocolitica* strains in minced meat with and without protective gas with consideration of the competitive background flora. International *Journal of Food Microbiology*, **10**, 65–72.

Lambert, A.D., Smith, J.P., and Dodds, J.L. (1991) Effect of initial O_2 and CO_2 and low-dose irradiation on toxin production by *Clostridium botulinum* in MAP fresh pork. *Journal of Food Protection*, **54** (12), 939–944.

Leistner, L. and Gorris, L.G.M. (1995) Food preservation by hurdle technology. *Trends in Food Science and Technology*, **6**, 41–45.

Lerke, P. (1973) Evaluation of potential risk of botulism from sea food cocktails. *Applied Microbiology*, **25** (5), 807–810.

Lerke, P. and Farber, L. (1971) Heat pasteurisation of crab and shrimp from the pacific coast of the United States: public health aspects. *Journal of Food Science*, **36**, 277–279.

Licciardello, J.J. (1983) Botulism and heat processed seafoods. *Marine Fisheries Review*, **45** (2), 1–7.

Lilly, J.T. and Kautter, D.A. (1990) Decomposition. Outgrowth of naturally occurring *Clostridium botulinum* in vacuum-packaged fresh fish. *Journal of the Association of Official Analytical Chemists*, **73** (2), 211–212.

Luitens, L.S., Marchello, J.A. and Dryden, F.D. (1982) Growth of *Salmonella typhimurium* and mesophilic organisms on beef steak as influenced by type of packaging. *Journal of Food Protection*, **45**, 263.

Lynt, R.K., Kautter, D.A. and Soloman, H.M. (1979) Heat resistance of non-proteolytic *Clostridium botulinum* type F in phosphate buffer and crab meat. *Journal of Food Science*, **44**, 108–111.

Lynt, R.K., Soloman, H.M., Lilly, J.R. and Kautter, D.A. (1977) Thermal death time of *Clostridium botulinum* type E in meat of the blue crab. *Journal of Food Science*, **42** (4), 1022–1025.

Modi, N.K., Rose, S.A. and Tranter, H.S. (1990) The effects of irradiation and temperature on the immunological activity of *staphylococcal enterotoxin A*. *International Journal of Food Microbiology*, **11**, 85–92.

Notermans, S., Du Frene, J. and Keijberts, M.J.H. (1981) Vacuum-packaged cooked potatoes: Toxin production by *Clostridium botulinum* and shelf-life. *Journal of Food Protection*, **44** (8), 572–575.

Notermans, S., Dufrene, J. and Lund, B.M. (1990) Botulism risk of refrigerated, processed foods of extended durability. *Journal of Food Protection*, **53** (12), 1020–1024.

Ohlsson, T. (1994) Minimal processing-preservation methods of the future: an overview. *Trends in Food Science and Technology*, **5**, 341–344.

Ohye, D.R. and Scott, W.J. (1957) Studies on the physiology of *Clostridium botulinum* type E. *Australian Journal of Biological Science*, **10**, 85–94.

Ollinger-Snyder, P., El-Gazzar, F., Matthews, M.E., Marth, E.H. and Unklesbay, N. (1995) Thermal destruction of *Listeria monocytogenes* in ground pork prepared with and without soy hulls. *Journal of Food Protection*, **58** (5), 573–576.

Peck, M.W., Fairbairn, D.A. and Lund, B.M. (1993) Heat resistance of spores of non-proteolytic *Clostridium botulinum* estimated on medium containing lysozyme. *Letters in Applied Microbiology*, **16**, 126–131.

Petran, R.L., Sperber, W.H. and Davis, A.B. (1995) *Clostridium botulinum* toxin in romaine lettuce and shredded cabbage: effect of storage and packaging. *Journal of Food Protection*, **58** (6), 624–627.

Post, L.S., Lee, D.A., Solberg, M., Furang, D., Specchio, J. and Graham, C. (1985) Development of botulinal toxin and sensory deterioration during storage of vacuum and modified atmosphere packaged fish fillets. *Journal of Food Science*, **50**, 990–996.

Roberts, T.A. and Ingram, M. (1965) The resistance of spores of *Clostridium botulinum* type E to heat and radiation. *Journal of Applied Bacteriology*, **28** (1), 125–141.

Rose, S. (1986) Temperature observations on chilled foods from refrigerated retail displays. *Technical Memorandum No. 423*, Campden & Chorleywood Food Research Association, Chipping Campden, Glos., UK.

Sakaguchi, G. (1979) In *Foodborne Infections and Intoxications* (ed. H. Reiman), Academic Press.

Schellekens, M. (1996) New research issues in *sous vide* cooking. *Trends in Food Science and Technology*, **7**, 256–262.

Schmidt, C.F., Lechowich, R.V. and Folinazzo, J.F. (1961) Growth and toxin production by type E *Clostridium botulinum* below 40°F. *Journal of Food Science*, **26**, 626–630.

Scott, V.N. and Bernard, D.T. (1982) Heat resistance of spores of non-proteolytic type B *Clostridium botulinum*. *Journal of Food Protection*, **45** (10), 909–912.

Sheridan, J.J., Doherty, A., Allen, P., McDowell, D.A., Blair, I.S. and Harrington, D. (1995) Investigations on the growth of *Listeria monocytogenes* on lamb packaged under modified atmospheres. *Food Microbiology*, **12**, 259–266.

Siliker, J.H. and Wolfe, S.K. (1980) Microbiological safety considerations in controlled atmosphere storage of meats. *Food Technology*, **34**, 59.

Smith, J.P. (1990) A Hazard Analysis Critical Control Point approach to ensure the microbiological safety of *sous vide* processed meat pasta product. *Food Microbiology*, **7** (3), 177.

Snyder, O.P. (1995) The application of HACCP for MAP and *sous vide* products, in *Principles of Modified Atmosphere and* sous vide *Product Packaging* (eds J.M. Farber and K.L. Dodds), Technomic Publishing Company, Pennsylvania.

Solomon, H.M., Kautter, D.A. and Lynt, R.K. (1982) Effect of low temperature on growth of non-proteolytic *Clostridium botulinum* types B and F and proteolytic type G in crabmeat and broth. *Journal of Food Protection*, **45** (6), 516–518.

Solomon, H.M., Lynt, R.K., Lilly T., Jr. and Kautter, D.A. (1977) Effect of low temperature on growth of *Clostridium botulinum* spores in meat of the blue crab. *Journal of Food Protection*, **40**, 5–7.

Sörquist, S. (1994) Heat resistance of different serovars of *Listeria monocytogenes*. *Journal of Applied Bacteriology*, **76**, 383–386.

Stringer, S.C., Fairbairn, D.A. and Peck, M.W. (1997) Combining heat treatment and subsequent incubation temperature to prevent growth from spores of non-proteolytic *Clostridium botulinum*. *Journal of Applied Microbiology*, **82**, 128–136.

Taylor, L.Y., Cann, D.D. and Welch, B.J. (1990) Antibotulinal properties of nisin in fresh fish packaged in an atmosphere of carbon dioxide. *Journal of Food Protection*, **53** (11), 953–957.

Tui-Jyi Chai and Kuang T. Liang (1992) Thermal resistance of spores from five type E *Clostridium botulinum* strains in eastern oyster homogenates. *Journal of Food Protection*, **55** (1), 18–22.

Walker, S.J., Archer, P. and Banks, J.G. (1990) Growth of *Listeria monocytogenes* at refrigeration temperatures. *Journal of Applied Bacteriology*, **68** (2), 157.

Willcocx, F., Hendrickx, M. and Tobback, P. (1994a) A preliminary survey into the temperature conditions and residence time distribution of minimally processed MAP vegetables in Belgian retail display cabinets. *International Journal of Refrigeration*, **17** (7), 436–443.

Willcocx, F., Tobback, P. and Hendrickx, M. (1994b) Microbial safety assurance of minimally processed vegetables by implementation of the Hazard Analysis Critical Control Point (HACCP) System. *Acta Alimentaria*, **23** (2), 221–238.

Wimpfheimer, L., Altman, N.S. and Hotchkiss, J.M. (1990) Growth of *Listeria monocytogenes* Scott A, serotype 4 and competitive spoilage organisms in raw chicken packaged under modified atmospheres and in air. *International Journal of Food Microbiology*, **11**, 205–214.

Woodburn, M.J., Somers, E., Rodriguez, J. and Schantz, E.J. (1979) Heat inactivation rates of botulinum toxins A, B, E and F in some foods and buffers. *Journal of Food Science*, **44**, 1658–1661.

7 Shelf life and safety of minimally processed fruit and vegetables

ROBYN O'CONNOR-SHAW

7.1 Introduction

Shewfelt (1987) has defined minimal processing as 'the handling, preparation, packaging and distribution of agricultural commodities in a fresh-like state'. This definition includes processes such as trimming, slicing, shelling, use of chemical dips, reduction of pH and water activity (a_w), mild heat treatment, low level irradiation and modified atmosphere packaging (MAP) (Huxsoll and Bolin, 1989; Shewfelt, 1986). Consumers perceive minimally processed fruit and vegetables to be 'all natural' and 'healthy'. Thus, the use of chemical additives, other than possibly those with some nutritional benefit, or irradiation, is limited (Ronk et al., 1989) and will not be further discussed.

Two positive attributes of minimally processed fruit and vegetables are convenience and freshness, and thus these foods are becoming an important component of the food supply (Dougherty, 1990). A wide range of vegetables are retailed in a minimally processed form, with lettuce and chicory salad being the most popular, whereas the production of minimally processed fruit has not been developed to the same extent (Nguyen-the and Carlin, 1994). Examples of minimally processed fruit and vegetables include vegetables sticks, soup mixes, tossed salads, peeled and cored whole pineapple cylinders and microwaveable fresh vegetable trays (Shewfelt, 1987).

Unlike other forms of processing, minimal processing increases perishability rather than making products more stable (Rolle and Chism, 1987). Biological processes such as respiration, ripening and senescence, which continue in fruit and vegetables after harvest, are accelerated by minimal processing (Watada et al., 1990). Cutting hastens microbial spoilage through transfer of microorganisms from the nutritionally poor surface to the nutritionally rich tissues of fruit or vegetable where they may grow rapidly. Confinement of product in a package results in modification of the package atmosphere through tissue and microbial respiration. This impacts on the rates of physiological and microbiological deterioration.

The major challenge of future research is to produce minimally processed fruit and vegetables having premium quality and safety allowing an acceptable time for distribution (Shewfelt, 1987).

7.2 Physiological deterioration

Shewfelt (1987) cites factors such as growing conditions, cultural practices, cultivar and maturity at harvest, harvesting and handling methods which influence the quality of minimally processed fruit and vegetables, although little information is available on the influence of these factors on product stability during storage and handling.

Shewfelt (1986) defines quality as 'the composite of those characteristics that differentiate individual units of a product, and have significance in determining the degree of acceptability of that unit by the buyer'. Different characteristics are important in different fruit and vegetables. O'Connor-Shaw *et al.* (1994) developed lists of adjectives to allow them to describe sensory changes occurring in five minimally processed fruit – honeydew, kiwifruit, papaya, pineapple and cantaloupe – during storage. While flavour descriptors of sweetness and typical, were used in all cases, additional descriptors, 'bitterness and sharpness', 'other', 'acidity', and 'bitterness and sourness', for kiwifruit, papaya, pineapple, and cantaloupe, respectively, were necessary to fully describe fruit flavour.

Fruit and vegetables continue to respire after harvest. The net result of respiration is the production of CO_2, water vapour and heat through oxidation of carbohydrates, organic acids and lipids. Respiration supplies energy for biological processes including synthesis of typical pigments, odours and flavours, maintenance of membrane integrity, and synthesis of ethylene and enzymes involved in fruit ripening. Respiration rate reflects the plant's metabolic activity, and, in general, a higher respiration rate results in shorter shelf life (Labuza and Breene, 1989).

Ethylene has many effects on plant physiology with some of its most important effects including induction of rapid ripening, and over-ripening, in many fruit, and premature yellowing in many vegetables. Ethylene is normally produced by many kinds of ripening fruit (Zagory, 1995). Research is required to understand the mechanism of how quality attributes are regulated and affected by ethylene (Watada *et al.*, 1990).

Respiration rate is temperature dependent, with a two- to three-fold (or more) increase in respiration rate for each temperature increase of 10°C. For this reason, Zagory (1995) states that 'temperature control is the single most important factor in maintaining product quality'. Temperature should be maintained as close as possible to 0°C. Exceptions to this are those tropical and subtropical fruit and vegetables which suffer chilling injury when temperatures are reduced below about 10 to 12°C. Symptoms of chilling injury include failure to ripen properly, sunken epidermal lesions and increased susceptibility to decay organisms (Zagory, 1995). Products which suffer chilling injury should be stored at the lowest temperature at which no injury occurs (Shewfelt, 1986).

Relative humidity (Rh) is another important environmental factor in the shelf life of fruit and vegetables. Low Rh environments reduce spread of most decay organisms but increase transpiration and, thus, increase moisture loss. Moisture loss is associated with structural damage in fruit and wilting in leafy vegetables (Shewfelt, 1986). Optimum temperature and Rh conditions for storage of whole fruit and vegetables can be found in several sources (Shewfelt, 1987).

Senescence encompasses 'those processes that follow maturity and lead to death of tissue' (Shewfelt, 1986). These processes result in changes in appearance, flavour and texture of fruit or vegetable (Powrie and Skura, 1991). Damage to the cell membrane caused by cutting induces or enhances the degradative changes that occur during senescence (Shewfelt, 1986). The net result is shorter shelf life compared with whole fruit and vegetables because of increased rates of respiration, ethylene production, protein degradation, water loss and phenylalanine ammonialyase and polyphenol oxidase activity (Zagory, 1995). In some fruit and vegetables, an immediate effect of cutting is the production of undesirable brown pigments through a complex series of reactions involving polyphenol oxidases (King and Bolin, 1989). The effects of slicing on rates of respiration and ethylene production differ between climacteric and non-climacteric fruit and with the physiological age of climacteric fruit (Watada *et al.*, 1990).

Another factor, that of acid migration, comes into consideration when considering the shelf life of dressed salads. Brocklehurst *et al.* (1983) and Brocklehurst and Lund (1984) showed that addition of vegetables to mayonnaise caused an increase in acetic acid concentrations in vegetable tissue, and a decrease in pH and acidity of the mayonnaise, within six hours of mixing. Acetic acid partitioned predominantly in the water phase of the mayonnaise and rapidly equilibrated with water in vegetable tissues (Rose, 1985). This migration phenomenon limits shelf life in two ways. Firstly, it permits growth of microorganisms in mayonnaise, where they would otherwise have been inhibited because of the low pH (Smittle, 1977). Secondly, loss of water from plant tissue causes it to become translucent and modifies its texture (Campbell-Platt and Anderson, 1988).

In a novel paper, O'Connor-Shaw *et al.* (1994) described the sequence of sensory changes leading to spoilage in minimally processed cantaloupe, honeydew, kiwifruit, papaya and pineapple. The shelf life of fruit pieces related to the species at 4°C. Kiwifruit and papaya had the shortest shelf lives, 2 days, whereas honeydew melon had the longest shelf life, 14 days. The symptoms of spoilage differed between species. For example, brown discoloration was observed only in pineapples, while bitter flavours developed only in kiwifruit. Microbial growth did not appear to contribute to spoilage in diced kiwifruit, papaya and pineapple, but may have done so in the case of melons.

Processing procedures used to produce minimally processed fruit and vegetables affect quality. Bolin *et al.* (1977) reported that sharpness of cutting blade and slicing motion influenced the shelf life of shredded lettuce. A sharp blade exercising a slicing motion resulted in about twice the shelf life at 2°C compared with a sharp blade chopping or a dull blade slicing or chopping. Microscopic examination of a lettuce leaf being sliced showed exudation of cellular fluids. In contrast, when the lettuce leaf was torn into strips, no noticeable exudation occurred. Bolin and Huxsoll (1991) later confirmed that lettuce pieces (9 to 20 cm²) produced by manual tearing had a longer shelf life than pieces cut with a sharp knife. Lettuce shredded into 3 mm strips received a significantly higher organoleptic score compared with lettuce shredded into 1 mm strips after 11 days storage at 2°C (Bolin *et al.*, 1977). Cellular fluids released by slicing were enzymatically active (Bolin *et al.*, 1977). A significantly longer shelf life resulted if cellular fluids were removed with a water rinse and centrifugation (Bolin *et al.*, 1977; Bolin and Huxsoll, 1991). Sanguansri (1997) recently reviewed cutting techniques and their consequences on final product quality for minimally processed vegetables.

King and Bolin (1989) and Huxsoll and Bolin (1989) have reviewed methods used to control enzymic, both plant and microbial, degradation in minimally processed fruit and vegetables. These methods include use of mild heat treatments, sulphite solutions, ascorbic acid solutions (for inhibition of browning), calcium ions (for maintenance of texture) and high-pressure treatments. Optimising atmospheric storage conditions using modified or controlled atmosphere storage is an effective method for shelf life extension of minimally processed fruit and vegetables because it reduces rates of respiration and ethylene production.

7.3 Modified atmosphere packaging

Modified atmosphere packaging (MAP) and controlled atmosphere (CA) storage are two methods for extending the shelf life of perishable foods by altering the proportions of atmospheric gases surrounding the food. Controlled atmosphere storage relies on the continuous measurement of the composition of the storage atmosphere, and, when warranted, injection of appropriate gas(es) to maintain the desired gas composition. With MAP, no monitoring of package atmospheres is done. In the case of horticultural commodities, the composition of the atmosphere is a function of the initial gas composition, the product : headspace ratio, the gas permeability characteristics of the packaging film and plant tissue, and the respiration rates of the tissue (O'Connor *et al.*, 1992).

Labuza *et al.* (1992) state that 'oxygen levels in a package of respiring produce are affected by many factors: O_2 solubility in formulation, package permeability to CO_2 and O_2, O_2 consumption rate due to microbial growth, metabolic and chemical reactions of food, food temperature at gas flushing or evacuation and processing conditions. But, in essence, given any initial O_2 level, the package interior will always end up at some critical O_2 level in a few days if a semipermeable film is used.' Similarly, 'CO_2 levels will also reach some critical concentration based on rate of reactions at food surface, any metabolic production, and CO_2 permeation rate of the film' (Labuza et al., 1992).

Modified atmosphere packaging extends shelf life of fresh fruit and vegetables through suppression of respiration, ripening and subsequent senescence, due to the diversity of effects caused by low O_2, elevated CO_2, by inhibition of ethylene induced effects, and by reduction of moisture loss due to moisture barrier properties of the plastic film (Zagory, 1995).

Respiration rates of many fruit and vegetables, and indirectly the rates at which they ripen, age and decay are slowed when O_2 concentrations are decreased below about 10%. These reduced O_2 levels inhibit oxidation of phenolic compounds and thus the development of undesirable brown discoloration. Produce becomes more resistant to pathogen invasion due to delay in the onset of senescence. However, O_2 is required for normal metabolism to proceed. Concentrations below 1 to 2% can lead to anaerobic respiration and associated production of ethanol and acetaldehyde resulting in off-odours and off-flavours (Zagory, 1995).

Zagory (1995) writes that elevated CO_2 can also slow respiratory processes, although not as dramatically as low O_2. In some cases high CO_2 together with low O_2 have a greater effect than either gas alone. Another important effect of CO_2 is that at concentrations above 1 to 2% it reduces the sensitivity of plant tissues to ethylene. Like O_2, CO_2 also has an influence on phenolic metabolism. At concentrations >10%, CO_2 suppresses growth of a number of decay-producing fungi and bacteria. For example, 10 to 20% CO_2 suppresses growth of Botrytis cinerea on strawberries, and Monilinia fructicola on sweet cherries (El-Goorani and Sonmer, 1981).

Although the mode of action of higher than optimal CO_2 levels is not completely understood, these CO_2 concentrations are known to aggravate certain physiological or ethylene-induced disorders, such as a coreflush in apples or brownheart in apples and pears, and cause the development of off-flavours, such as in strawberries, apples, oranges and bananas (Brecht, 1980; Rolle and Chism, 1987). The product may become more susceptible to pathogens or ripen irregularly if stored under non-ideal atmospheres (Kader et al., 1989).

The amount of shelf life gained by the use of modified atmospheres is dependent upon variety and cultivar, physiological age and initial quality, degree of processing and storage temperature and Rh. In general, fruits and vegetables which undergo ripening after harvest are more responsive to MA than those that do not. The potential benefits of MAP for a given commodity can be predicted from knowledge of the primary causes of deterioration (Kader et al., 1989).

The CO_2, O_2 and ethylene concentrations within fruit or vegetable tissue determine its response to MA (Brecht, 1980; Kader *et al.*, 1989). Gas diffusion is dependent upon the surface area and respiration rate of the commodity, the composition of the enveloping atmosphere and the diffusion characteristics of the skin and tissue. As a plant organ enters senescence, cell sap fills the intracellular air spaces. This may significantly decrease the rate of gas diffusion within tissue.

Carbon dioxide concentrations of 1 to 20% and O_2 concentrations of 2 to 5% are used for CA storage of whole fruits; CO_2 concentrations of 1 to 20% and O_2 concentrations of 1 to 10% are used for the CA storage of whole vegetables. Nitrogen (N_2) is the balance in both cases (Kader, 1985; Meherivk, 1985; Richardson, 1985; Saltveit, 1985). Minimally processed fruit and vegetables tolerate lower O_2 and higher CO_2 concentrations than do their intact counterparts.

Little information on the ideal atmospheric storage conditions for sliced fruit and vegetables has been published. O'Connor-Shaw *et al.* (1996) prepared diced cantaloupe flesh that was microbiologically sterile in order to study the physiological deterioration of fruit when stored under a range of CAs at 4.5°C. Storage atmospheres were in continuous flow and contained from 0 to 26% CO_2 and 3.5 to 17% O_2. Acceptable product, as assessed by a highly trained taste panel, up to 28 days was obtained for three treatments: 6% CO_2 and 6% O_2, 9.5% CO_2 and 3.5% O_2, and 15% CO_2 and 6% O_2. Overall treatment with 0, 19.5% or 26% CO_2 (irrespective of O_2 concentration) caused significant deterioration in sensory characteristics. This compared with a shelf life of 4 days for cantaloupe pieces stored in air (O'Connor-Shaw *et al.*, 1994).

7.4 Microbial spoilage

7.4.1 Introduction

There is no clear distinction between field disease and postharvest spoilage of fruit and vegetables, as microbial attack may commence in the field, or produce may be contaminated at harvest or during subsequent handling (Lund, 1982).

Although the inner tissue of sound fruit and vegetables is often considered to be microbiologically sterile, several reports indicate that healthy fruit and vegetables may contain low levels of bacteria (Lund, 1992). In healthy tissue bacterial multiplication may be very limited, but if plant tissue is subjected to stress, e.g. chilling, bacterial growth may result (Lund, 1982). Several types of pathogenic fungi initiate an infection in sound developing fruit which then becomes quiescent until the resistance of the host decreases because of ripening and senescence, e.g. anthracnose of mangoes and pawpaw (Wills *et al.*, 1989). Whether in the field or after harvest, bacteria may enter tissue through natural openings, e.g. hydathodes, stomata or lenticels, or through damaged tissue (Lund, 1982). Immersion in water during commercial washing procedures can result in water entering air spaces of fruit or vegetable, e.g. through the stem scar in tomatoes and lenticels in potatoes, and to an increase in decay (Lund, 1992).

The most economically important bacterial spoilage of fruit and vegetables is probably that caused by soft rots. Bacteria also cause other types of defect which may allow soft rot bacteria to enter tissues and cause more extensive spoilage. Bacterial soft rot is characterised by collapse of plant tissue. The most important of the bacteria causing soft rots are *Erwinia carotovora* (subsp. *carotovora* and subsp. *atrosepticum*) and pectinolytic strains of *Pseudomonas fluorescens (P. marginalis)* (Lund, 1982, 1983). Lund (1982) states that 'a feature of soft rot bacteria is the lack of specificity with regard to tissue attacked, so that a wide range of fruit and vegetables can be affected'. The ability of bacteria to cause rotting of plant tissue results from their ability to enzymically degrade pectic substances of the middle lamella and the primary cell walls of plants (Lund, 1992). Lund (1983) provides details of the taxonomy, ecology, spoilage characteristics, and isolation procedures for soft rot bacteria.

Lund (1983) lists the factors which influence the development of post-harvest spoilage as 'extent of damage during harvesting, plant maturity, turgidity of plant tissue, cell wall structure and its vulnerability to bacterial enzymes, concentration in plant of bacterial nutrients and of antibiotic compounds, and ability of plant tissue to form barriers to bacterial infection'. Environmental conditions which maintain a fruit or vegetable at premium quality will most successfully protect it from microbial invasion (Lund, 1983). Brackett (1994) discusses the effects of these environmental factors on microbial ecology of minimally processed fruit and vegetables.

The cutting procedures involved in minimal processing transfer microorganisms from the surface of the whole fruit or vegetable to its interior. With certain constraints, such as tissue pH, content of organic acids, and the presence of antibacterial compounds – such as are produced by carrots, cabbage and onions – the nutrient rich plant tissue is most conducive to microbial growth (Lund, 1992). Abbey *et al.* (1988) showed that the major isolates from watermelon slices were the same as those detected on the rind. Contamination from equipment, personnel and processing environment may also introduce greater numbers and types of microorganism into internal tissues (Brackett, 1994). Parish and Higgins (1990) showed that the microbial counts of grapefruit sections removed from a commercial processing line increased by 0.5 $\log_{10}$ cfu/ml during an 18 h production run due to the build up of microorganisms on equipment surfaces over this period. With certain exceptions, e.g. cucumbers, peppers and tomatoes, the low pH of fruit, usually <4.5, favours fungal growth whereas the higher pH of vegetables, 4.5 to 7.0, allows bacterial growth (Lund, 1982). Little information is available on the microflora of minimally processed fruit (Nguyen-the and Carlin, 1994). However, microbial spoilage of leafy vegetables, carrots and dressed salads has been addressed and will be discussed in the following sections.

A hypochlorite solution, containing 50 to 100 ppm available chlorine, is widely used to remove organic material and reduce microbial contamination from the skin of fruit and vegetables, although the efficacy of chlorine is in dispute (Brackett, 1994; Nguyen-the and Carlin, 1994). Brackett (1992), for example, found virtually the same residual counts on brussel sprouts after dipping in a 200 mg/l chlorine solution, the maximum concentration approved by the United States Food and Drug Administration (FDA) for use with food (Cords, 1983), as after dipping in water. A serious disadvantage associated with the use of chlorine is the formation of carcinogenic trihalomethane compounds through its reaction with organic materials (Cords, 1983), and chlorine is not permitted to be used in direct contact with food in Belgium for this reason (Willocx *et al.*, 1994). Washing, if done improperly, can actually spread contaminants over produce, thus intensifying the potential for microbial spoilage (Brackett, 1992). Bacteria have different abilities to adhere to a leaf surface and this influences their resistance to removal by gentle washing. There is evidence that plant pathogens adhere most strongly to leaves of their host plant (Lund, 1992).

Heating the surface of whole fruit to a few degrees below the injury threshold has been shown to eradicate or delay development of incipient infection of pathogenic fungi. Hot water treatments involve heating at temperatures between 43 and 60°C for between 1 and 20 min (Eckert, 1975). O'Connor-Shaw *et al.* (1996) prepared sterile cantaloupe dice by immersing whole melons in boiling water for 3 min, then dicing aseptically.

Chemical preservatives, such as sodium benzoate, potassium sorbate, acetic acid and lactic acid, can be used to control microbial growth in plant tissue. Irradiation has also been proposed as a means of preventing decay and delaying senescence (Brackett, 1994; Nguyen-the and Carlin, 1994). However, the use of chemical additives and irradiation to prevent microbial spoilage is probably limited because consumers have an expectation that these products be 'all natural' (Ronk *et al.*, 1989). The hurdle concept involves the use of several food preservation techniques in combination to prevent microbial growth. This approach permits a less extreme use of individual treatments and less damage to product quality. The primary hurdle used in the preservation of minimally processed fruit and vegetables is refrigerated storage. Additional hurdles which could be used include low O_2 concentrations, through use of MAP, and the use of non-pathogenic lactic acid bacteria or their metabolic products (Grant and Patterson, 1995).

7.4.2 Lettuce

Total bacterial counts of freshly prepared lettuce, endive and chicory have been reported to range from 2.4 to 8 $\log_{10}$ cfu/g (Jacques and Morris, 1995; King et al., 1991; Marchetti et al., 1992; Morris and Lucotte, 1993; Nguyen-the and Prunier, 1989). King et al. (1991) and Morris and Lucotte (1993) showed that total counts of the outer leaves of these vegetables were at least an order of magnitude greater than counts of inner leaves. Morris and Lucotte (1993) also found that sowing date had a significant effect on total count of inner leaves, and explained this result by the observation that more intensive cropping at certain times of the year increased airborne contamination which then increased counts on inner leaves.

Jacques and Morris (1995) conducted experiments with endive to determine the effect of leaf age on total, fluorescent and pectinolytic counts. Similarly to King et al. (1991) and Morris and Lucotte (1993), they found population levels were lower, in this case about 100-fold, on inner leaves compared with outer leaves, at harvest. This difference persisted during 6°C, 7 day storage of the leaves. They concluded that leaf physiology had a significant influence on bacterial population densities, and to confirm this observation they inoculated endive leaves of two different age groups with a strain of P. fluorescens, T53 (P. marginalis). Strain T53 was shown to grow to higher numbers on older leaves. Leaf age was also shown to influence its susceptibility to decay, with younger leaves being less susceptible to marginal necrosis than older leaves. Thus, Jacques and Morris (1995) recommended that only the inner three-quarters of a head should be used for minimal processing.

Jacques and Morris (1995) demonstrated that P. fluorescens strain T53 produced soft rot in endive stored at 6°C. Soft rot on individual leaf pieces occurred when T53 populations were at least 6.7 $\log_{10}$ cfu/cm^2. Decay in 90% of leaf pieces occurred when T53 counts were ≥ 7.5 $\log_{10}$ cfu/cm^2. Atmospheric changes in packs resulting from vegetable respiration and microbial activity influenced spoilage production, with greatest decay being found in packs with lower CO_2 (<10%) and higher O_2 (>12%). The proportions of fluorescent and pectinolytic bacteria in total aerobic populations of leaves varied greatly. This had previously been found by Magnusson et al. (1990). Thus fluorescent and pectinolytic populations should be quantified independently due to their potential role in decay.

Nguyen-the and Prunier (1989) had earlier established that P. marginalis caused soft rot by experimentally producing brown discoloration and liquefaction in chicory leaves and observing greater deterioration in salads with higher numbers of this bacteria. Microbial counts were made on three batches of 20 salads which had been stored for 10 days at 10°C. Each set of 20 salads was a subset of 100 salads and included the 10 worst and best salads, determined by percentage of decayed leaves. In the best batch, where 5% of leaves were decayed, about 20% of pseudomonads caused soft rot. In the other two batches where about 20% of leaves had decayed, about 60% of pseudomonads caused soft rot. *Pseudomonas marginalis* was shown to be a weak pathogen: >8 $\log_{10}$ cfu/g were required to produce disease symptoms (in 4 days at 10°C); not all strains produced soft rot (in 7 days); and frequently only one or two of the three leaves used in pathogenicity testing developed soft rot. The threshold count for disease production established in this study was equivalent to that found by Jacques and Morris (1995).

Pseudomonads accounted for between 50 to 60% of isolates from freshly prepared and chill stored leafy vegetables, with *P. fluorescens (P. marginalis)* the dominant species; between 20 and 40% of isolates were enterobacteria (Magnusson *et al.*, 1990; Marchetti *et al.*, 1992; Nguyen-the and Prunier, 1989). Nguyen-the and Prunier (1989) found that the enterobacteria isolated in their study were mostly *Enterobacter agglomerans*. Magnusson *et al.* (1990) found that 25% of enterobacteria strains isolated from 14 day stored lettuce (1°C in air) were *Erw. carotovora*. Given the importance of *Erw. carotovora* as an agent of soft rot, it is surprising that it was not isolated more frequently in the above studies. Nguyen-the and Carlin (1994) explained this by noting that selective enrichment procedures are needed for its isolation and that removal of decayed leaves prior to processing may have reduced its numbers. Simon-Sarkadi *et al.* (1994) established a relationship between putrescine content of leafy vegetables stored at 5°C for 6 days and the Enterobacteriaceae count, which at this time was ≥90% of the total bacterial count, and suggested that putrescine concentration might be used to indicate spoilage.

7.4.3 Carrot

Carlin and Nguyen-the (1989) found that grated carrot, sampled at the end of a production line in a vegetable factory, had initial total counts of 6.7 $\log_{10}$ cfu/g. This count increased during the first week of storage (at 10°C), and then remained at a constant level of 8.7 $\log_{10}$ cfu/g. Growth of lactic acid bacteria was different. After 7 days, there was a high level of variability between packages, with counts in some bags remaining at the initial level (4 $\log_{10}$ cfu/g), while in other bags counts increased to 8 $\log_{10}$ cfu/g. Colonies were isolated from total count and lactic acid bacteria (MRS agar) plates at 0, 7 and 14 days and identified. Enterobacteria comprised 51% of isolates from total count plates, with *Ent. agglomerans* (32.7%) and *Ent. intermedium* (16.3%) being the dominant species. Another 36.5% of isolates from total count plates were identified as *Pseudomonas* spp., with *P. viridiflava* (18% of isolates) being the dominant species. All isolates from MRS agar plates were identified as *Leuconostoc mesenteroides*. Softening and development of off flavours in carrots were associated with atmospheres with excessively high levels of CO_2 (>30%) and excessively low O_2 (<1.5%), high numbers of lactic acid bacteria and yeasts, production of ethanol, acetic and lactic acid, which are characteristic signs of growth of *L. mesenteroides*. Marchetti *et al.* (1992) also reported that marked growth of lactic acid bacteria in carrots during 7 day storage at 5°C was accompanied by increases in CO_2, lactic acid, acetic acid and ethanol concentrations. Carlin and Nguyen-the (1989) found that the presence of pectinolytic pseudomonads on grated carrots was very variable, and no relationship between rate of deterioration and populations of these bacteria was found.

The work of Carlin and Nguyen-the (1989) poses two questions. Firstly, given that initial counts of *L. mesenteroides* did not vary greatly between packs, why was there great variation in these counts between spoiled and unspoiled packs? Secondly, was the high CO_2 and ethanol, low O_2 atmosphere in packs of spoiled carrots the result of anaerobic metabolism of carrots and/or microbial production?

7.4.4 Other vegetables

Albrecht *et al.* (1995) surveyed microbial counts of cut lettuce, tomatoes, broccoli and cauliflower, obtained from salad bars in delicatessens. Temperature of salad ingredients ranged from 5.1 to 18°C. Initial total counts of individual ingredients were between 5.3 and 6.6 $\log_{10}$ cfu/g, yeast and mould counts were between 6.3 and 8.4 $\log_{10}$ cfu/g, and coliform counts accounted for 44 to 50% of total count.

Manzano *et al.* (1995) found that the initial total count of pre-packaged soup vegetables was 8 $\log_{10}$ cfu/g. Between 17 and 51% of microorganisms were pectinolytic. Pseudomonads formed between 7 and 41% of total count. Coliforms, faecal coliforms, enterococci, lactic acid bacteria and yeast and moulds formed <10% of the population. Soup vegetables were packaged in air and in three MAs and stored for 7 days at 4°C. Total counts of stored samples were between 8.3 and 8.8 $\log_{10}$ cfu/g, pectinolytic counts comprised between 35 and 85% of total counts, and all other counts were <15% of total count.

Garg *et al.* (1993) found that the total counts of spinach, cauliflower heads and florets, and whole carrots and carrot sticks varied from 3 to 5 $\log_{10}$ cfu/g. In fresh spinach, coliform counts exceeded total counts, but in other vegetables formed <1% of total count. Lactobacilli counts comprised <1% of total counts in all fresh vegetables. After 7 day storage at 3.3°C, total counts ranged from 5.7 to 7.9 $\log_{10}$ cfu/g, and coliforms and lactobacilli formed <1% of total count.

Brocklehurst *et al.* (1987) studied the microflora of two salad varieties at their use-by-date after 7°C storage. Variety A consisted of lettuce, cabbage, carrot, onion and capsicum and variety B contained cabbage, cress, carrot, sweetcorn, celery and capsicum. Appearance and odour of vegetables in all packs was acceptable. Total counts of variety A were 8 $\log_{10}$ cfu/g, and consisted primarily of *P. fluorescens, P. putida* and *Ent. agglomerans*. Of the pseudomonads, ≤10% were pectinolytic. Total counts of variety B were about 9 $\log_{10}$ cfu/g, with lactic acid bacteria mostly dominating the microflora. Again pectinolytic bacteria, in this case *P. fluorescens, P. putida* and *Erw. carotovora*, formed ≤10% of the total count.

Ercolani (1997) found clostridial spores in numbers from <1 to >50 cfu/cm² on mature leaves of 19 species of horticultural plant under commercial cultivation in five localities in Italy. Clostridial isolates mainly belonged to six species, *C. pasteurianum, C. sporogenes, C. felsineum, C. butyricum, C. perfringens* and *C. roseum. Clostridium felsineum* and *C. roseum* are pectinolytic. When spore suspensions of *C. pasteurianum* and *C. perfringens* were inoculated onto the leaves of basil in the greenhouse, spore counts at first declined, then rose again significantly. A similar pattern was observed with spores of *C. sporogenes* inoculated onto tomato leaves. These results suggested that clostridial species may be able to multiply on leaf surfaces. Detection of *C. perfringens* on rocket-salad and basil is of concern because leaves of these plants are often eaten raw.

7.4.5 Dressed vegetable salads

After surveying the microflora of 330 chilled, mayonnaise-based vegetable salads produced in the United Kingdom, Rose (1984) concluded that lactobacilli and yeasts were the predominant microorganisms in these products. Brocklehurst *et al.* (1983) found that the yeasts, *Saccharomyces dairensis* and *S. exiguus*, dominated the microflora of coleslaw after 16 days storage at 10°C and were present at numbers of $\log_{10}7$ cfu/g. Brocklehurst and Lund (1984) confirmed the dominance of *S. exiguus* in mayonnaise-based salads at 10°C. Vegetable and potato salads spoiled because of gas production by *S. exiguus* and possibly lactobacilli, and due to visible surface colonies formed by *Pichia membranaefaciens* and *Geotrichum candidum*. Gas production by *S. exiguus* caused spoilage of Florida salad and coleslaw.

Microbial growth will not occur in mayonnaise formulated with low pH (2.7 to 3.4) and high levels of undissociated acetic acid (Smittle, 1977). However, Brocklehurst and Lund (1984) showed that the addition of cabbage or carrot to a mayonnaise allowed the growth of two spoilage yeasts, *S. dairiensis* and *S. exiguus,* as a result of acid migration between vegetables and dressing. The acid system of salads is further complicated by the production of substantial amounts of non-volatile lactic acid *in situ* by *Lactobacillus* spp. (Rose, 1985).

Rose (1985) demonstrated the interactive effect of temperature and acidity on microbial growth in salads, when she found that growth of spoilage microorganisms was largely prevented in high acid salads (pH 3.01) at 20°C, in intermediate acid salads (pH 3.70) at 10°C, and in low-acid salads (pH 4.13) at 4°C. Low temperature storage resulted in extended survival of *Clostridium perfringens*, *Escherichia coli*, *Salmonella* spp. and *Staphylococcus aureus* in salads, probably due to microbial metabolic inactivity and reduced acid mobility at low temperatures.

7.5 Effect of modified atmospheres on microorganisms

7.5.1 Introduction

There is a continuous spectrum of O_2 tolerance among microorganisms from the most sensitive strict anaerobe to the least sensitive strict aerobe (Morris, 1976). Adams and Moss (1995) define obligate aerobes as 'those organisms that are respiratory, generating most of their energy from oxidative phosphorylation using O_2 as the terminal electron acceptor in the process'. Morris (1976) defines an obligate anaerobe as an organism which is capable of generating energy without recourse to molecular O_2, and demonstrates a degree of adverse sensitivity to O_2 which renders it unable to grow in air at 1 atm.

Total exclusion of O_2 is not sufficient to ensure growth of most strict anaerobes. Anaerobes generally grow best under 'reducing conditions' where there is minimal drainage from the organism of reducing power which otherwise could more productively be used for energy yielding or biosynthetic reactions. Anaerobes differ in their ability to produce this excess reducing power, and those with the least capacity in this respect are unable to grow in media that have not been prepoised at a low redox potential (E_h) (Morris, 1976). The relationship between partial pressure of O_2 (pO_2) and E_h depends on the chemical nature of the culture medium or food, and both parameters should be measured in order to assess their effect on microbial growth (Lund *et al.*, 1984). Similarly, in order to isolate the microflora of an anaerobic environment, it is essential to match both E_h and O_2 concentrations of laboratory media with that of the *in vivo* environment (Hentges and Maier, 1972).

Oxygen is potentially toxic to microorganisms due to the formation of toxic reduction products, such as hydrogen peroxide, superoxide anion, hydroxyl radicle and singlet O_2. Aerobes possess mechanisms to dispose of these compounds, e.g. enzymes such as catalase, peroxidase and superoxide dismutase, but the majority of strict anaerobes are devoid of these enzymes (Morris, 1976). Lactic acid bacteria are examples of aerotolerant anaerobes. They are incapable of aerobic respiration but despite this can grow in air (Adams and Moss, 1995).

Farber (1991) has reviewed the effects of elevated levels of CO_2 on microorganisms. He says that although the specific way in which CO_2 exerts its bacteriostatic effect is unknown, the overall effect on microorganisms is an extension of lag phase of growth and a decrease in growth rate during logarithmic phase. The inhibitory effects of CO_2 on microorganisms are dependent upon: alteration of cell membrane function including effects on nutrient uptake and absorption; direct inhibition of enzymes or decreases in the rate of enzyme reactions; penetration of bacterial membranes, leading to intracellular pH changes; and direct changes to the physico-chemical properties of proteins.

Oxidative Gram-negative microorganisms and moulds are most sensitive to CO_2. Gram-positive bacteria, particularly lactic acid bacteria, tend to be most resistant. Growth inhibition is usually greater under aerobic conditions than anaerobic (Adams and Moss, 1995). For maximum antimicrobial effect, storage temperatures should be as low as possible because solubility of CO_2 decreases dramatically with increasing temperature. Interestingly, an increased CO_2 level seems to raise the temperature minimum for some pathogens (Farber, 1991).

In general, MAs used with horticultural commodities are usually not low enough in O_2 or high enough in CO_2 to prevent the growth of microorganisms (Labuza and Breene, 1989; Lund, 1982).

7.5.2 *Effect of modified atmospheres on spoilage microorganisms*

Brackett (1988, 1989 and 1990) investigated the effect of MA on the growth of microorganisms and the development of spoilage in tomatoes, broccoli stalks and bell peppers, respectively. Individual items were shrink wrapped in film, sealed in film pouches which were flushed with an atmosphere consisting of 10% CO_2, 5% O_2 and 85% N_2, and stored in cardboard cartons. Storage temperatures were 13°C for tomatoes and peppers and 1°C for broccoli. In all cases packaging inhibited the development of microbial spoilage. However, the effect of packaging on microbial growth was variable. For example, packaged tomatoes had higher total aerobic counts than unpackaged fruit; although total counts decreased in stored peppers, higher counts were detected in packaged peppers; total counts in stored broccoli were highest in unpackaged vegetables. These results show that the packaging treatments used 'somehow' retarded microbial spoilage without necessarily inhibiting microbial growth. Brackett (1988, 1990) suggested that packaging may have prevented fruit/vegetable dehydration thus making produce more resistant to microbial attack. Packaging appeared to favour the development of Gram-positive bacteria. At onset of microbial spoilage, about 50% of isolates from packaged tomatoes and 60% of isolates from shrink-wrapped broccoli, were identified as coryneform bacteria (Brackett, 1988, 1989). At a storage temperature of 5°C, Abdul-Raouf *et al.* (1993) reported that a MA of 3% O_2 and 97% N_2 delayed deterioration in the appearance of shredded lettuce but enabled psychrotrophs to increase to significantly higher numbers compared with lettuce packaged under ambient atmospheres.

In order to clarify the issue regarding the effect of MA on spoilage microflora, Hao and Brackett (1993) investigated the effect of six atmospheres (% $CO_2/O_2/N_2$: 0/5/95; 0/10/90; 5/10/85; 5/20/75; 10/5/85 and 10/20/70) and air on the growth of four vegetable spoilage bacteria: *Erw. carotovora* subsp. *carotovora, P. fluorescens*, an unidentified pectinolytic pseudomonad isolated from lesions occurring on bell pepper (pepper #15) and *Xanthomonas campestris*, in liquid artificial media. Atmospheres chosen were based on the relative tolerances of vegetables to CO_2 and O_2 (Brecht, 1980), and were maintained within 1% of their stated value for the length of the experiment. At 5°C, growth of *Erwinia, Xanthomonas* and pepper #15 was slightly but significantly reduced by some gas mixtures (% $CO_2/O_2/N_2$: 0/5/95, 0/10/90 and 10/5/85; 10/5/85; 0/5/95 and 10/5/85, respectively). None of the gas mixtures used had a significantly negative effect on growth of *P. fluorescens*. To test whether atmospheres inhibitory to bacterial growth also delayed bacterial spoilage, bell peppers were stored under one of these atmospheres, that containing 10% CO_2 and 5% O_2, at 25°C. This MA did not change the percentage of peppers which spoiled after inoculation with the four test bacteria. However, the colour of bell peppers stored in this MA did not change during 14 day storage, whereas most peppers stored in air started turning red after 7 days. From this study, as from the work by Brackett (1988, 1989, 1990), it would appear that extension of shelf life due to MAP was due to its effect on the physiological state of vegetables, rather than any inhibitory effect on spoilage microorganisms.

7.5.3 Effect of modified atmospheres on pathogenic microorganisms

Nguyen-the and Carlin (1994) list examples of foodborne infections linked to the consumption of raw fruit and vegetables. From this list it can be seen that, since 1980, outbreaks in Western countries have involved the microorganisms *Clostridium botulinum, Listeria monocytogenes, Salmonella* and *Shigella sonnei* and the products shredded cabbage in coleslaw, shredded lettuce, raw salad vegetables, bean sprouts, cantaloupe and watermelon. Various surveys have shown the presence of *Aeromonas hydrophila, Escherichia coli, L. monocytogenes, Salmonella, Staphylococcus aureus* and *Yersinia enterocolitica* in minimally processed vegetables (Lund, 1992; Nguyen-the and Carlin, 1994).

The transmission of *L. monocytogenes* by food was first convincingly demonstrated in an outbreak involving 41 cases that occurred in Canada in 1981, and was linked to the consumption of cabbage in coleslaw (Adams and Moss, 1995). Growing cabbage was thought to have been contaminated with *L. monocytogenes* from sheep manure, as manure came from a herd in which two sheep had listeriosis. Cabbage was subsequently cold stored for an extended period prior to processing permitting growth of *L. monocytogenes* (Schlech *et al.*, 1983).

Two large outbreaks of salmonellosis, the first of which involved >25 000 individuals, in North America in 1990 and 1991 were associated with the consumption of cantaloupe from salad bars (Madden, 1992). The causative microorganisms, *S. chester* and *S. poona*, were thought to have originated from the surface of melons and been introduced into fruit tissue by cutting, although this was not proved. However, in response to these outbreaks, the FDA instructed retailers to wash melons before cutting, use clean and sanitised utensils and surfaces when preparing cut melon, maintain melons at –7°C, and limit display of cut melon to <4 h if not refrigerated.

Pathogens which contaminate fruit and vegetables originate from soil, from manure, sewage sludge and polluted irrigation water, from animals and birds and from unhygienic practices after harvest. The risk of contamination of fruit or vegetables with a particular pathogen is dependent upon the ability of that pathogen to survive in the environment. For example, *L. monocytogenes* survived in soil to which sewage sludge had been added at initial population levels for at least eight weeks, and on decaying plant debris in soil for years (Brackett, 1992); *Salmonella* spp. survived on lettuce, cabbage and carrots stored at room temperature for about 28 days, and for much longer periods when vegetables were refrigerated (Madden, 1992).

The major concern associated with MAP of minimally processed fruit and vegetables is that the low O_2 concentrations used may inhibit aerobic spoilage microorganisms, which produce detectable defects in food, while permitting the growth of anaerobic pathogens which will not produce these indications of their growth and yet may pose a public health risk (Farber, 1991; Kader *et al.*, 1989). At temperatures ≤5°C, psychrotrophic pathogens of concern in these products are *A. hydrophila*, non-proteolytic strains of *C. botulinum*, *L. monocytogenes* and *Y. enterocolitica* (Hanlin *et al.*, 1995). *Aeromonas hydrophila*, *L. monocytogenes* and *Y. enterocolitica* are facultative anaerobes, and *C. botulinum* is an obligate anaerobe. Consequently these microorganisms should grow in the low O_2 levels encountered in MA packed minimally processed horticultural products. The minimum growth pH of *A. hydrophila* is 6.0, so this organism will not be of concern in high-acid fruit and vegetables.

Yersinia enterocolitica is a member of the Enterobacteriaceae, and is a Gram-negative facultatively anaerobic catalase positive rod. It can grow at a minimum temperature of –1°C and a minimum pH of between 4.1 and 5.1 depending upon acidulant and temperature (Adams and Moss, 1995). This bacterium has been frequently isolated from minimally processed vegetables, but in all but one case – an isolate from vegetables salad (Brocklehurst *et al.*, 1987) – strains were regarded as non-pathogenic (Nguyen-the and Carlin, 1994). Two other members of the Enterobacteriaceae, enterotoxigenic *E. coli* and *Salmonella*, also have low minimum growth temperatures,

4 and 5°C respectively (Adams and Moss, 1995), and have been associated with foodborne infections linked to consumption of raw vegetables and fruit (Nguyen-the and Carlin, 1994). No work on the effects of MA on the growth of *Y. enterocolitica*, enterotoxigenic *E. coli* and *Salmonella* in minimally processed fruit and vegetables at temperatures ≤5°C has been reported.

The following discussion focuses on the effect of MA storage on growth of *A. hydrophila, C. botulinum* and *L. monocytogenes* in horticultural products.

(a) Aeromonas hydrophila   Aeromonas hydrophila is a Gram-negative, catalase positive, rod which ferments glucose, grows at a minimum pH of 6.0 and a minimum temperature of –0.1°C. On the basis of mainly epidemiological evidence, *A. hydrophila* is suspected as a cause of gastroenteritis, and there is some controversy over its status as a human pathogen. Its principle reservoir is the aquatic environment, freshwater lakes and streams and wastewater systems. *Aeromonas hydrophila* has been isolated from salad vegetables (Adams and Moss, 1995).

Berrang *et al.* (1989a) examined the effects of CA storage on the growth of *A. hydrophila* on fresh asparagus spears, broccoli and cauliflower florets at 4°C. Two test strains were used. As well, the growth of *A. hydrophila* strains naturally occurring on these vegetables was followed. The atmospheres were different for each vegetable and had previously been shown to extend the shelf life of each vegetable. Atmospheres contained between 3 and 10% CO_2, between 11 and 18% O_2, with N_2 providing the balance. Results showed that CA storage did not significantly influence the growth of *A. hydrophila*, but did extend the length of time vegetables remained acceptable for consumption.

(b) Clostridium botulinum   Clostridium botulinum is a Gram-positive, spore-forming rod which is obligately anaerobic. Eight serologically distinct toxins are produced by the species although individual strains usually only produce one toxin type. Strains of *C. botulinum* display sufficient variety of physiological and biochemical attributes that they have been divided into four groups. Most cases of human botulism are caused by strains which produce toxin types A, B and E, and these strains are confined to Groups I and II. Group I contains proteolytic strains which produce toxin types A, B and F. Minimum growth temperature of these strains is 10 to 12°C; minimum growth pH is 4.7. Group II contains non-proteolytic strains which produce toxin types B, E and F and have a minimum growth temperature of 3 to 5°C and a minimum growth pH of 5.0 to 5.2 (Adams and Moss, 1995). Toxin production by *C. botulinum* usually occurs under the same conditions which permit vegetative cell growth. Requirements for spore germination are much less restrictive (Sperber, 1982).

Adams and Moss (1995) list four features common to outbreaks of botulism: (1) food is contaminated with spores or vegetative cells of *C. botulinum*; (2) food receives some treatment which restricts competitive microflora and normally *C. botulinum*; (3) food storage conditions are suitable for growth of *C. botulinum*; and (4) food is consumed cold or after a mild heat treatment insufficient to inactivate toxin.

Soil contamination is a major source of *C. botulinum* in food and one to which horticultural produce is prone (Adams and Moss, 1995). McClure *et al.* (1994) review surveys describing the distribution of *C. botulinum* spores in fruit and vegetables.

Hotchkiss *et al.* (1992) demonstrated that MA stored tomatoes supported toxin production when inoculated with a mixture of type A and proteolytic and non-proteolytic type B strains of *C. botulinum* at 23°C. However these tomatoes had been held for 2 to 9 days beyond the time they were first determined to be inedible. Several tomatoes had a pH of 4.4 to 4.5 at time of assay for toxin although authors speculated that, as pH measurements were made on homogenised tomatoes, they may not accurately reflect pH at point at which *C. botulinum* was growing. Oxygen concentrations in packages at time of testing were between 1.1 and 1.6%.

Carlin and Peck (1995) followed growth and toxin production by *C. botulinum* in 28 puréed vegetables, prepared under strictly anaerobic conditions, and incubated anaerobically at 30°C for 60 days. Eight vegetables supported growth and toxin production by non-proteolytic, B, E and F, strains of *C. botulinum*. This is the first report of toxin production by non-proteolytic strains of *C. botulinum* in a range of vegetables. Twenty vegetables supported growth and toxin production by proteolytic, A and B, *C. botulinum* strains. The ability of vegetables to support growth and toxin production appears related to their pH, with a threshold for non-proteolytic strains of 5.3, and for proteolytic strains of 4.6 (with some exceptions).

McClure *et al.* (1994) write that

> there is evidence from a substantial body of research that the minimum pH for growth of *C. botulinum* is in the range 4.6 to 4.8. There are, however, examples which have demonstrated that *C. botulinum* can grow and produce toxin at pH values below 4.6. These data have been obtained in laboratory media or under laboratory conditions where strict anaerobiosis prevails and there is a high concentration of protein present and various acidulants used. It has been suggested that the protein acts as a reducing agent, source of essential metabolites or as a buffer, slowing down the decrease in the internal pH of the cell. It has also been demonstrated that growth of organisms such as moulds and bacilli will increase the pH of the substrate, allowing growth and toxin production to occur.

Young-Perkins and Merson (1987) explained *C. botulinum* survival at low pH levels under optimal anaerobic conditions by saying that

> the ability of a microorganism to survive in an acidic medium may be related to the rate of proton migration into the cell relative to the proton-ejecting capacity

of the cell. In an O_2 free substrate, cell energy which would otherwise be expended in scavenging O_2 and lowering the redox potential of the immediate environment could be used to motivate expulsion of lethal hydrogen ions from the organism's interior.

Therefore in the presence of O_2 the minimum pH for growth would increase.

The ability of non-proteolytic strains of *C. botulinum* to grow at temperatures <10°C makes them of concern in refrigerated food, and it is important to know the conditions which permit/prevent their growth at low temperatures. Graham and Lund (1993) calculated doubling times of 42, 29 and 23 h at 4, 5 and 6°C respectively for vegetative cells of a non-proteolytic type B strain of *C. botulinum* in laboratory medium (pH 6.7). This compared to 22 min at 35°C. Initiation of growth at 4°C was very variable and this may have been because only a small proportion of bacteria in the inoculum had the ability to grow at this low temperature, and indeed the probability of growth (P) from a single vegetative cell at 4°C was estimated as 1 in 10^4. These results

all indicate that care must be taken in studies at extremes of growth, as unless inocula are very high and are conditioned to the conditions under test, no growth may be registered in a situation where growth is indeed possible but has a low probability.

Growth and toxin production by spores of non-proteolytic *C. botulinum* types B, E and F in a strictly anaerobic medium occurred at 3°C in 5 to 6 weeks and at 5°C in 2 to 3 weeks (Graham *et al.*, 1997).

Under conditions close to limits for growth, 5°C, pH 5.98 and 0.1% NaCl, Graham *et al.* (1996) predicted lag time and times to 2-fold and 1000-fold increases in numbers for spores of non-proteolytic *C. botulinum* in laboratory medium of 15, 16 and 24 days respectively. That is, the major component in time to a substantial increase in numbers is lag phase.

Lund *et al.* (1984) determined the effect of pO_2 and E_h of culture medium on the number of spores of *C. botulinum* type F required to produce growth. At 20°C, and pO_2 of ≤0.0045 atm (0.45%), corresponding to E_h of +217 mV, the probability of growth from single spores within 5 days was equal to that in strictly anaerobic conditions at E_h of −400 mV. At pO_2 values of between 0.0053 and 0.0084 atm, corresponding to E_h values of between +226 and +254 mV, between 10 and 1000 spores were required to produce growth, and at pO_2 values of between 0.012 and 0.016 atm, corresponding to E_h values of between +271 and +294 mV, numbers of spores required to produce growth were between 2×10^4 and $>2 \times 10^5$. In this experiment, all other parameters were near to optimum for growth and the degree of tolerance to O_2 may be considerably reduced when other conditions are less favourable.

Graham *et al.* (1996) developed a predictive model for the effect of temperature, pH and NaCl on growth from spores of non-proteolytic *C. botulinum*, mainly type B strains, suitable for use with fish, meat and poultry products. However, so far no model has been developed which takes into account the effects of O_2 and CO_2 concentrations in addition to temperature and pH.

(c) Listeria monocytogenes *Listeria monocytogenes* is a Gram-positive, catalase positive, facultatively anaerobic, cocci-rod, with a minimum growth temperature of 0°C and a minimum growth pH of between 4.4 and 5.6 depending upon acidulant. *Listeria* is ubiquitous, and food is the major source of infection. The transmission of *L. monocytogenes* by food was first established in an outbreak which was linked to the consumption of cabbage in coleslaw. Several surveys have revealed the presence of *L. monocytogenes* on various vegetables (Beuchat and Brackett, 1991).

Berrang *et al.* (1989b) studied the effect of CA storage on the growth of two strains of *L. monocytogenes* on fresh asparagus spears, broccoli and cauliflower florets stored at 4°C, using the procedures described for their work with *A. hydrophila* (Berrang *et al.,* 1989a). 'The most important finding' of this study was that although CA storage extended the length of time the vegetables remained acceptable for consumption by 7 days, it had no effect on the growth of *L. monocytogenes*. In the case of asparagus, after 21 day storage, *L. monocytogenes* populations were significantly higher under CA compared with populations stored under air for 14 days. Yet both storage times were endpoints with regard to consumer acceptability. However, with respect to broccoli and cauliflower the risk to public health would be no greater with CA storage compared to air storage as these vegetables did not support the growth of *L. monocytogenes*. Thus the potential safety hazard of a MA packed minimally processed fruit or vegetable should be assessed on a case by case basis.

Kallander *et al.* (1991) also found that MA inhibited spoilage but did not affect growth of *L. monocytogenes*. They investigated the effect of an anaerobic, 70% CO_2 atmosphere (30% N_2) on the growth of *L. monocytogenes* on shredded cabbage at 5°C. Counts of *L. monocytogenes* increased similarly, but gradually, in air and MA packaged cabbage for 13 days, when counts in normal atmospheres decreased sharply. This reduction coincided with decrease in cabbage pH and the onset of spoilage. These changes were not evident in MA packaged cabbage, in which counts remained constant.

Beuchat and Brackett (1990) found that *L. monocytogenes* grew similarly in whole and shredded lettuce leaves packaged in air and in 3% O_2 and 97% N_2 at 5°C. Significant increases in numbers occurred after 8 days of storage. Beuchat and Brackett (1991) showed that significant decreases in *L. monocytogenes* populations occurred in chopped cherry tomatoes packaged under air or a MA consisting of 3% O_2 and 97% N_2 at 10°C. Changes in population were not affected by packaging method.

7.6 Application of HACCP to minimally processed fruit and vegetables

HACCP (hazard analysis critical control points) is a tool for the systematic analysis of a food process for the identification of hazards and the planning for control of these hazards. The HACCP approach provides a highly disciplined foundation for systematic continuous process analysis to achieve products which are safe, meet customer expectations all the time and are produced with high productivity and effectiveness (Snyder, 1995). Willocx *et al.* (1994) describe the application of HACCP principles to the production of minimally processed endive. Safety and quality specifications for final product are given and process flow diagrams and hazard audit table are developed for pre-harvest, harvest and production operations.

Many MAP minimally processed fruit and vegetables fall under a class of foods designated as 'potentially hazardous', as they are low-acid (pH >4.6), high-moisture (a_w >0.85) products packaged in hermetically sealed packages with a modified atmosphere, have a shelf life <14 days, and require refrigeration for microbiological safety and maintenance of quality (Farber, 1995). Thus, there is a real need for regulations for these foods. Farber (1995) has reviewed current Canadian, US and European guidelines for MAP fruit and vegetables. He comments that

> it is recognised . . . that a major part of the success of extending the shelf life and quality of MAP foods, whilst still maintaining safety, is via a combination of the type of MAs, using proper refrigerated temperature, the use of additional hurdles, and the control of contamination and use of GMPs through the application of HACCP principles.

All guidelines recommend that a use-by-date for these products be displayed on package labels. Also included should be temperature time indicators, because temperature control is frequently lost once the product leaves the production facility, and yet is the primary preservation factor for these products (Hanlin *et al.*, 1995). These indicators should be adequately correlated with development of unacceptable numbers of pathogens or levels of toxin and development of spoilage (Labuza *et al.*, 1992).

References

Abbey, S.D., Heaton, E.K., Golden, D.A. and Beuchat, L.R. (1988) Microbiological and sensory quality changes in unwrapped and wrapped sliced watermelon. *Journal of Food Protection* **51**, 531–533, 537.

Abdul-Raouf, U.M., Beuchat, L.R. and Ammar, M.S. (1993) Survival and growth of *Escherichia coli* O157:H7 on salad vegetables. *Applied and Environmental Microbiology* **59**, 1999–2006.

Adams, M.R. and Moss, M.O. (1995) *Food Microbiology*, Royal Society of Chemistry, Cambridge, UK.

Albrecht, J.A., Hamouz, F.L., Sumner, S.S. and Melch, V. (1995) Microbial evaluation of vegetable ingredients in salad bars. *Journal of Food Protection* **58**, 683–685.

Berrang, M.E., Brackett, R.E. and Beuchat, L.R. (1989a) Growth of *Aeromonas hydrophila* on fresh vegetables stored under a controlled atmosphere. *Applied and Environmental Microbiology* 55, 2167–2171.

Berrang, M.E., Brackett, R.E. and Beuchat, L.R. (1989b) Growth of *Listeria monocytogenes* on fresh vegetables stored under a controlled atmosphere. *Journal of Food Protection* 52, 702–705.

Beuchat, L.R. and Brackett, R.E. (1990) Survival and growth of *Listeria monocytogenes* on lettuce as influenced by shredding, chlorine treatment, modified atmosphere packaging and temperature. *Journal of Food Science* 55, 755–758, 870.

Beuchat, L.R. and Brackett, R.E. (1991) Behaviour of *Listeria monocytogenes* inoculated into raw tomatoes and processed tomato products. *Applied and Environmental Microbiology* 57, 1367–1371.

Bolin, H.R., Stafford, A.E., King, J.R. and Huxsoll, C.C. (1977) Factors affecting the storage stability of shredded lettuce. *Journal of Food Science* 42, 1319–1321.

Bolin, H.R. and Huxsoll, C.C. (1991) Effect of preparation procedures and storage parameters on quality retention of salad-cut lettuce. *Journal of Food Science* 56, 60–62, 67.

Brackett, R.E. (1988) Changes in the microflora of packaged fresh tomatoes. *Journal of Food Quality* 11, 89–105.

Brackett, R.E. (1989) Changes in the microflora of packaged fresh broccoli. *Journal of Food Quality* 12, 169–181.

Brackett, R.E. (1990) Influence of modified atmosphere packaging on the microflora and quality of fresh bell peppers. *Journal of Food Protection* 53, 255–257.

Brackett, R.E. (1992) Shelf stability and safety of fresh produce as influenced by sanitation and disinfection. *Journal of Food Protection* 55, 808–814.

Brackett, R.E. (1994) Microbiological spoilage and pathogens in minimally processed refrigerated fruits and vegetables, in *Minimally Processed Refrigerated Fruits and Vegetables* (ed. R.C. Wiley), Chapman & Hall, London, UK, pp. 269–312.

Brecht, P.E. (1980) Use of controlled atmosphere to retard deterioration of produce. *Food Technology* 34 (3), 45–50.

Brocklehurst, T.F., White, C.A. and Dennis, C. (1983) The microflora of stored coleslaw and factors affecting the growth of spoilage yeasts in coleslaw. *Journal of Applied Bacteriology* 55, 57–63.

Brocklehurst, T.F. and Lund, B.M. (1984) Microbiological changes in mayonnaise-based salads during storage. *Food Microbiology* 1, 5–12.

Brocklehurst, T.F., Zaman-Wong, C.M. and Lund, B.M. (1987) A note on the microbiology of retail packs of prepared salad vegetables. *Journal of Applied Bacteriology* 63, 409–415.

Campbell-Platt, C. and Anderson, K.G. (1988) Pickles, sauces and salad products, in *Food Industries Manual*, 22nd edn (ed. M.D. Ranken), Blackie, Glasgow, UK, pp. 285–334.

Carlin, F. and Nguyen-the, C. (1989) Microbiological spoilage of fresh, ready-to-use grated carrots. *Sciences des Aliments* 9, 371–386.

Carlin, F. and Peck, M.W. (1995) Growth and toxin production by non-proteolytic and proteolytic *Clostridium botulinum* in cooked vegetables. *Letters in Applied Microbiology* 20, 152–156.

Cords, B.R. (1983) Sanitizers: halogens and surface active agents, in *Antimicrobials in Foods* (eds A.L. Branen and P.M. Davidson), Marcel Dekker, New York, NY, pp. 257–298.

Dougherty, R.H. (1990) Future prospects for processed fruit and vegetable products. *Food Technology* 44, 124, 126.

Eckert, J.W. (1975) Postharvest diseases of fresh fruits and vegetables – etiology and control, in *Postharvest Biology and Handling of Fruits and Vegetables* (eds N.F. Haard and D.K. Salunkhe), AVI Publishing Company, Westport, CT, pp. 81–117.

El-Goorani, M.A. and Sonmer, N.F. (1981) Effects of modified atmospheres on postharvest pathogens of fruits and vegetables. *Horticultural Reviews* 3, 412–461.

Ercolani, G.L. (1997) Occurrence and persistence of culturable clostridial spores on the leaves of horticultural plants. *Journal of Applied Microbiology* 82, 137–140.

Farber, J.M. (1991) Microbiological aspects of modified-atmosphere packaging technology – A review. *Journal of Food Protection* 54, 58–70.

Farber, J.M. (1995) Regulations and guidelines regarding the manufacture and sale of MAP and *sous vide* products, in *Principles of Modified Atmosphere and Sous vide Product*

Packaging (eds J.M. Farber and K.L. Dodds), Technomic Publishing Co., Lancaster, PA, pp. 425–458.

Garg, N., Churey, J.J and Splittstoesser, D.F. (1993) Microflora of fresh cut vegetables stored at refrigerated and abuse temperatures. *Journal of Food Science and Technology* **30**, 385–386.

Graham, A.F. and Lund, B.M. (1987) The combined effect of sub-optimal temperature and sub-optimal pH on growth and toxin formation from spores of *Clostridium botulinum*. *Journal of Applied Bacteriology* **63**, 387–393.

Graham, A. and Lund, B.M. (1993) The effect of temperature on the growth of non-proteolytic type B *Clostridium botulinum*. *Letters in Applied Microbiology* **16**, 158–160.

Graham, A.F., Mason, D.R. and Peck, M.W. (1996) Predictive model of the effect of temperature, pH and sodium chloride on growth from spores of non-proteolytic *Clostridium botulinum*. *International Journal of Food Microbiology* **31**, 69–85.

Graham, A.F., Mason, D.R., Maxwell, F.J. and Peck, M.W. (1997) Effect of pH and NaCl on growth from spores of non-proteolytic *Clostridium botulinum* at chill temperature. *Letters in Applied Microbiology* **24**, 95–100.

Grant, I.R. and Patterson, M.F. (1995) The potential use of additional hurdles to increase the microbiological safety of MAP and *sous vide* products, in *Principles of Modified Atmosphere and Sous vide Product Packaging* (eds J.M. Farber and K.L. Dodds), Technomic Publishing Co., Lancaster, PA, pp. 263–286.

Hanlin, J.H., Evancho, G.M. and Slade, P.J. (1995) Microbiological concerns associated with MAP and *sous vide* products, in *Principles of Modified Atmosphere and Sous Vide Product Packaging* (eds J.M. Farber and K.L. Dodds), Technomic Publishing Co., Lancaster, PA, pp. 69–104.

Hao, Y.-Y. and Brackett, R.E. (1993) Influence of modified atmospheres on growth of vegetable spoilage bacteria in media. *Journal of Food Protection* **56**, 223–228.

Hentges, D.J. and Maier, B.R. (1972) Theoretical basis for anaerobic methodology. *The American Journal of Clinical Nutrition* **25**, 1299–1305.

Hotchkiss, J.H., Banco, M.J., Busta, F.F, Genigeorgis, C.A., Kociba, R., Rheaume, L., Smoot, L.A., Schuman, J.D. and Sugiyama, H. (1992) The relationship between botulinal toxin production and spoilage of fresh tomatoes held at 13 and 23°C under passively modified and controlled atmospheres and air. *Journal of Food Protection* **55**, 522–527.

Huxsoll, C.C. and Bolin, H.R. (1989) Processing and distribution alternatives for minimally processed fruits and vegetables. *Food Technology* **43**, 124–128.

Jacques, M.A and Morris, C.E. (1995) Bacterial population dynamics and decay on leaves of different ages of ready-to-use broad-leaved endive. *International Journal of Food Science and Technology* **30**, 221–236.

Kader, A.A. (1985) A summary of CA requirements and recommendations for fruits other than pome fruits, in *Proceedings of the Fourth National Controlled Atmosphere Research Conference, 1985, Raleigh*, Raleigh Department of Horticultural Science, North Carolina State University, NC, pp. 445–470.

Kader, A.A., Zagory, D. and Kerbell, E.L. (1989) Modified atmosphere packaging of fruits and vegetables. *Critical Reviews in Food Science and Nutrition* **28**, 1–30.

Kallander, K.D., Hitchins, A.D., Lancette, G.A., Schmieg, J.A., Garcia, G.R., Solomon, H.M. and Sofos, J.N. (1991) Fate of *Listeria monocytogenes* in shredded cabbage stored at 5 and 25°C under a modified atmosphere. *Journal of Food Protection* **54**, 302–304.

King, A.D. and Bolin, H.R. (1989) Physiological and microbiological storage stability of minimally processed fruits and vegetables. *Food Technology* **43**, 132–135, 139.

King, A.D. Jr., Magnuson J.A., Török, T. and Goodman, N. (1991) Microbial flora and storage quality of partially processed lettuce. *Journal of Food Science* **56**(2), 459–461.

Labuza, T.P. and Breene, W.M. (1989) Applications of 'active packaging' for improvement of shelf life and nutritional quality of fresh and extended shelf life foods. *Journal of Food Processing and Preservation* **13**, 1–69.

Labuza, T.P., Fu, B. and Taoukis P.S. (1992) Prediction for shelf life and safety for minimally processed CAP/MAP chilled foods: a review. *Journal of Food Protection* **55**, 741–750.

Lund, B.M. (1982) The effect of bacteria on post-harvest quality of vegetables and fruits with particular reference to spoilage, in *Bacteria and Plants* (eds M.E. Rhodes-Roberts and F.A. Skinner), Society for Applied Bacteriology Symposium Series No. 10, Academic Press, London, UK, pp. 133–153.

Lund, B.M. (1983) Bacterial spoilage, in *Post-Harvest Physiology of Fruit and Vegetables* (ed. C. Dennis), Academic Press, London, UK, pp. 119–257.

Lund, B.M. (1992) Ecosystems in vegetable foods. *Journal of Applied Bacteriology Symposium Supplement* **73**, 115S-126S.

Lund, B.M., Knox, M.R. and Sims, A.P. (1984) The effect of oxygen and redox potential on growth of *Clostridium botulinum* type E from a spore inoculum. *Food Microbiology* **1**, 277–287.

Lund, B.M. and Peck, M.W. (1994) Heat resistance and recovery of spores of non-proteolytic *Clostridium botulinum* in relation to refrigerated, processed foods with an extended shelf life. *Journal of Applied Bacteriology Symposium Supplement* **76**, 115S-128S.

Madden, J.M. (1992) Microbial pathogens in fresh produce – the regulatory perspective. *Journal of Food Protection* **55**, 821–823.

Magnusson, J.A., King, A.D. Jr and Török, T. (1990) Microflora of partially processed lettuce. *Applied and Environmental Microbiology* **56**, 3851–3854.

Manzano, M., Citterio, B., Maifreni, M., Paganessi, M. and Comi, G. (1995) Microbial and sensory quality of vegetables for soup packaged in different atmospheres. *Journal of Science of Food and Agriculture* **67**, 521–529.

Marchetti, R., Casadei, M.A. and Guerzoni, M.E. (1992) Microbial population dynamics in ready-to-use vegetable salads. *Italian Journal of Food Science* **2**, 197–208.

McClure, P.J., Cole, M.B. and Smelt, J.P.P.M. (1994) Effects of water activity and pH on growth of *Clostridium botulinum*. *Journal of Applied Bacteriology Symposium Supplement* **76**, 105S–114S.

Meherivk, M. (1985) Controlled atmosphere storage conditions for some of the more commonly grown apple cultivars, in *Proceedings of the Fourth National Controlled Atmosphere Research Conference, 1985, Raleigh*, Raleigh Department of Horticultural Science, North Carolina State University, NC, pp. 395–421.

Morris, J.G. (1976) Oxygen and the obligate anaerobe. *Journal of Applied Bacteriology* **40**, 229–244.

Morris, C.E. and Lucotte, T. (1993) Dynamics and variability of bacterial population density on leaves of field-grown endive destined for ready-to-use processing. *International Journal of Food Science and Technology* **28**, 201–109.

Nguyen-the, C. and Carlin, F. (1994) The microbiology of minimally processed fresh fruit and vegetables. *Critical Reviews in Food Science and Nutrition* **34**, 371–401.

Nguyen-the, C. and Prunier, J.P. (1989) Involvement of pseudomonads in deterioration of 'ready-to-use' salads. *International Journal of Food Science and Technology* **24**, 47–58.

O'Connor, R.E., Skarshewski, P. and Thrower, S.J. (1992) Modified atmosphere packaging of fruits, vegetables, seafood and meat: state of the art. *ASEAN Food Journal* **7**, 127–136.

O'Connor-Shaw, R.E., Roberts, R., Ford, A.L. and Nottingham, S.M. (1994) Shelf life of minimally processed honeydew melon, kiwifruit, papaya, pineapple and cantaloupe. *Journal of Food Science* **59**, 1202–1206, 1215.

O'Connor-Shaw, R.E., Roberts, R., Ford, A.L. and Nottingham, S.M. (1996) Changes in sensory quality of sterile cantaloupe dice stored in controlled atmospheres. *Journal of Food Science* **61**, 847–851.

Parish, M.E. and Higgins, D.P. (1990) Investigation of the microbial ecology of commercial grapefruit sections. *Journal of Food Protection* **53**, 685–688.

Powrie, W.D. and Skura, B.J. (1991) Modified atmosphere packaging of fruits and vegetables, in *Modified Atmosphere Packaging of Foods* (eds B. Ooraikul and M.E. Stiles), Ellis Horwood, Sydney, NSW, pp. 169–245.

Richardson, D.G. (1985) CA recommendations for pears (including Asian pears), in *Proceedings of the Fourth National Controlled Atmosphere Research Conference, 1985, Raleigh*, Raleigh Department of Horticultural Science, North Carolina State University, NC, pp. 422–444.

Rolle, R.S. and Chism, G.W. (1987) Physiological consequences of minimally processed fruits and vegetables. *Journal of Food Quality* **10**, 157–177.

Ronk, R.J., Carson, K.L. and Thompson, P. (1989) Processing, packaging and regulation of minimally processed fruits and vegetables. *Food Technology* **43**, 136–139.

Rose, S.A. (1984) Studies on the microbiological status of pre-packed delicatessen salads collected from retail chill cabinets. *Technical Memorandum No. 37*, Campden Food Preservation Research Association, Chipping Campden, UK.

Rose, S. (1985) Microbiological studies on delicatessen salads. *Chilled Foods Symposium Papers, 7–9 May, Stratford-upon-Avon*, Campden Food Preservation Research Association, Chipping Campden, UK, pp. 131–144.

Saltveit, M.E. (1985) A summary of CA requirements and recommendations for vegetables, in *Proceedings of the Fourth National Controlled Atmosphere Research Conference, 1985, Raleigh*. Raleigh Department of Horticultural Science, North Carolina State University, NC, pp. 471–492.

Sanguansri, P. (1997) Cutting techniques for minimally processed vegetables. *Food Australia* **49**, 135–138.

Schlech, W.F., Lavigne, P.M., Bortolussi, R.A., Allen, A.C., Haldane, E.V., Wort, A.J., Hightower, A.W., Johnson, S.E., King, S.H., Nicholls, E.S. and Broome, C.V. (1983) Epidemic listeriosis – evidence for transmission by food. *The New England Journal of Medicine* **308**, 203–206.

Shewfelt, R.L. (1986) Postharvest treatment for extending the shelf life of fruits and vegetables. *Food Technology* **40**, 70–72, 74, 76–78, 80, 89.

Shewfelt, R.L. (1987) Quality of minimally processed fruits and vegetables. *Journal of Food Quality* **10**, 143–156.

Simon-Sarkadi, L., Holzapfel, W.H. and Halasz, A. (1994) Biogenic amine content and microbial contamination of leafy vegetables during storage at 5°C. *Journal of Food Biochemistry* **17**, 407–418.

Smittle, R.B. (1977) Microbiology of mayonnaise and salad dressing: a review. *Journal of Food Protection* **40**, 415–422.

Snyder, O.P. (1995) The applications of HACCP for MAP and *sous vide* products, in *Principles of Modified Atmosphere and Sous Vide Product Packaging* (eds J.M. Farber and K.L. Dodds), Technomic Publishing Co., Lancaster, PA, pp. 325–384.

Sperber, W.H. (1982) Requirements of *Clostridium botulinum* for growth and toxin production. *Food Technology* **36**(12), 89–94.

Watada, A.E., Abe, K. and Yamuchi, N. (1990) Physiological activities of partially processed fruits and vegetables. *Food Technology* **44**, 116–122.

Willocx, F., Tobback, P. and Hendrix, M. (1994) Microbial safety assurance of minimally processed vegetables by implementation of the Hazard Analysis Critical Control Point (HACCP) system. *Acta Alimentaria* **23**, 221–238.

Wills, R.B.H., McGlasson, W.B., Graham, D., Lee, T.H. and Hall, E.G. (1989) *Postharvest: An Introduction to the Physiology and Handling of Fruit and Vegetables*, 3rd edn, New South Wales University Press, Kensington, NSW.

Young-Perkins, K.E. and Merson, R.L. (1987) *Clostridium botulinum* spore germination, outgrowth and toxin production below pH 4.6; interactions between pH, total acidity and buffering capacity. *Journal of Food Science* **52**, 1084–1088.

Zagory, D. (1995) Principles and practice of modified atmosphere packaging of horticultural commodities, in *Principles of Modified Atmosphere and Sous Vide Product Packaging* (eds J.M. Farber and K.L. Dodds), Technomic Publishing Co., Lancaster, PA, pp. 175–206.

8 The sensory quality, microbiological safety and shelf life of packaged foods

IVOR CHURCH

8.1 Introduction

Cook–chill technology has been used in both catering and food manufacturing for a number of years, during which considerable effort has been devoted to the optimisation of sensory quality, microbiological safety and shelf life.

These factors are primarily dependent upon the product formulation and the process parameters applied. Regarding the latter, a number of bodies have made recommendations as to good practice, for example the (UK) Department of Health (DoH, 1989) for catering produced products, and the European Chilled Food Federation (ECFF, 1996) for manufactured products. These recommendations almost invariably aim to ensure microbiological safety by specifying a cooking treatment to produce a significant (i.e. 6 decimal log) reduction of *Listeria monocytogenes* (the most heat-resistant psychrotolerant pathogen) and restricting subsequent proliferation by rapid chilling and low temperature /short shelf life storage.

Generally however, cook–chill products become unacceptable sensorially prior to their becoming unsafe. This is due to a number of deteriorative changes, the most important of which (especially in poultry and pork meat), is the development of 'Warmed Over Flavour' (WOF) by the oxidation of lipids and the accumulation of thiobarbituric acid-reacting substances (TBARS). This effect, in conjunction with others, limits the shelf life of most cook–chill products to a few days at temperatures <5°C (e.g. Zacharias, 1980; Young *et al.*, 1989).

The relatively short shelf lives possible with cook–chill products have operational and financial disadvantages, and consequently shelf life extension is widely considered to be desirable. The packaging of cook–chill products in anaerobic environments can, in theory, enable this by inhibiting both oxidative reactions and the growth of aerobic bacteria. Several options exist in that the products may be packaged either prior to or post cooking, and in the latter, either vacuum or gas (i.e. carbon dioxide/nitrogen) environments may be applied. These will be discussed under the following major headings:

- vacuum packaging prior to cooking
- vacuum/gas packaging post cooking.

8.2 Vacuum packaging prior to cooking

This category includes products produced using three methods, namely AGS, (the cook-tank form of) Capkold and *sous vide*. AGS (McGuckian, 1969) and Capkold (Unklesbay *et al.*, 1977) were both designed for and restricted to catering use, whereas *sous vide*, although originally a catering technology, is now used by both caterers and food manufacturers.

8.2.1 Sous vide

As with cook–chill, the sensory quality, microbiological safety and shelf life of *sous vide* products are determined, to a great extent, by the process parameters applied (Church and Parsons, 1993). A number of national and international, statutory and non-statutory, recommendations exist as to Good Manufacturing Practice (GMP). As with cook–chill, these invariably focus upon microbiological safety. In this case however, the target organism is different as the vacuum packaging element introduces a potential risk from anaerobic pathogens, specifically all strains of type E and non-proteolytic strains of types B and F *Clostridium botulinum* (Betts and Gaze, 1995; Gould, 1996). In the spore form, this organism is significantly more heat resistant than *Listeria monocytogenes*, and consequently most of the GMP recommendations specify a treatment calculated to produce a significant (i.e. 6 decimal log) reduction in the number of *Clostridium botulinum* spores. Subsequent germination and toxin formation is addressed by the specification of rapid chilling and low temperature/short time storage.

Significant process parameter differences are evident however between the various GMP recommendations. Shelf life and storage temperatures vary, according – it would appear – to perceived need and practicality, respectively. Heat treatments generally vary due to the use of different raw data and methodologies to calculate the lethal rates; however the acceptance or otherwise of claims (e.g. Genigeorgis, 1993; Bovee *et al.*, 1996) that the treatments necessary to achieve a significant reduction of *Clostridium botulinum* spores causes unacceptable thermal damage also is a factor. Those recommendations that do take these claims into account use less severe heat treatments but generally compensate by other means. These involve avoiding products (e.g. aquatic) with a higher probability of contamination, the limited manipulation of shelf life and/or chilled storage temperature and the use of the 'hurdle concept' (Leistner, 1978). This involves the application of a number of individually sub-lethal preservation methods (e.g. pH, water activity and the presence of spices, natural acids, lysozyme and competing organisms) which together inhibit microbial proliferation and thus ensure product safety.

These differences are evident in the various GMP recommendations; for example, in the UK, the *Sous vide* Advisory Committee (SVAC, 1991) recommend (among others) a thermal core heat treatment of 90°C for 4.5 minutes for catering produced *sous vide* products with a shelf life of up to 8 days at 3°C, whereas the Advisory Committee on the Microbiological Safety of Food (ACMSF, 1992) state that the safety of *sous vide* products can be ensured by the following:

1. Storage at ≤3°C.
2. Shelf life of 10 days at 5°C or 5 days at 10°C.
3. Plus one of the following:
   - heat treatment (at thermal core) of 90°C for 10 min or equivalent
   - pH ≤ 5.0 throughout the food
   - salt concentration ≥ 3.5% throughout the food.
   - water activity ≤ 0.97 throughout the food
   - combination of heat and preservative factors that have been demonstrated to consistently prevent proliferation and toxin production by non-proteolytic *C. botulinum*.

Similar variation is also evident internationally.

Schellekens (1996) notes that, in the Netherlands, industrially produced products may be stored for up to 42 days at 5°C if the (thermal core of the) product attains a temperature of 90°C for 10 minutes. Products with a pH = 4.6 or a water activity <0.96 however require only a core heat treatment of 70°C for 2 minutes (i.e. 6 decimal log reduction in *Listeria monocytogenes*) for the same shelf life. A different set of standards applies to catering produced products in which the shelf life is, in institutions, limited to 4 days.

In France, the heat treatment applied determines the shelf life. A core temperature of 65°C with the application of a minimum Pasteurising Value (PV_{70}^{10}) of 100 enables products to have a shelf life of between 6 and 21 days, whereas a core temperature and PV_{70}^{10} of 70°C and 1000 respectively enables a shelf life of up to 42 days. Storage temperatures are specified variously between 0 and 8°C. SVAC (1991) claim that products processed according to their recommendations would have been subject to a PV_{70}^{10} of between 260 and 632 (Sheard and Church, 1992) and thus would qualify for a storage life of between 6 and 21 days under French legislation.

Rhodehamel (1992) notes that, in North America, the National Advisory Committee on Microbiological Criteria for Foods (NACMCF) recommend a minimum heat treatment sufficient to achieve a 4 decimal log reduction of *Listeria monocytogenes*.

It is evident from the previous discussion that little or no standardisation exists in terms of the process parameters applied to *sous vide* products. This is a potential cause of confusion in use and has, to a great extent, been reflected in the published work relating to the sensory and microbiological shelf life of *sous vide* products. These data will be discussed in the following sections.

8.2.2 Sensory quality and shelf life

It has been widely claimed that *sous vide* food is sensorially superior to that produced using cook–chill. This appears to be primarily based upon two factors: firstly, that the low oxygen tension inside the pack will inhibit both chemical oxidation and microbial (proteolytic or lipolytic) activity and, secondly, that the packaging will prevent evaporative losses of water and flavour volatiles during heat treatment. There is some evidence to support these claims (Creed *et al.*, 1993), however other data indicate that *sous vide* induced changes are not always desirable in terms of sensory acceptability (Church, 1996).

As mentioned previously, it has been claimed that the heat treatments necessary to address the *Clostridium botulinum* risk cause unacceptable thermal damage. Although some experimental evidence exists to support this claim in some, mainly vegetable, products (e.g. Varoquaux *et al.*, 1995; Church, 1996), a general effect remains unproven.

The available data regarding the sensory shelf life of *sous vide* products are somewhat inconclusive.

Creed *et al.* (1993) investigated *sous vide* chicken in red wine sauce. The samples were cooked to a core temperature of 75°C, cooled to a core temperature of 7°C within 50 minutes and stored at 3°C for up to 12 days. Reheating was carried out by placing the packs in boiling water until the contents reached 80°C. The assessors could discriminate between the freshly cooked control and *sous vide* samples after 0 (i.e. immediately post cooking) ($P < 0.001$), 5 ($P < 0.001$) and 12 ($P < 0.001$) days of chilled storage. It was claimed, however, that this discrimination was the result of a lightening of the sauce colour caused by the *sous vide* treatment. This was corroborated by the fact that, when freshly made sauce was added to a 12-day-old sample immediately prior to testing it, no significant difference was found between it and the freshly prepared conventionally cooked sample ($P > 0.05$). The most common reasons for correctly selecting the *sous vide* version as different were the paler colour and blander flavour of the chicken.

Turner and Larick (1996) investigated chicken breast treated with brine (either 0.5% sodium chloride or 0.5% sodium chloride plus 2% sodium lactate w/w), cooked to 77 or 94°C at either a 'slow' (i.e. a product heating rate of 0.8–0.9°C/min) or 'fast' (i.e. 1.5–2.2°C/min) rate prior to storage at 4°C for up to 28 days.

Samples heated to 94°C were less juicy and tender and, particularly when 'slow' cooked, exhibited greater lipid and phospholipid oxidation during chilled storage. This however was not detected (as WOF) by the sensory panel. WOF intensity increased and desirable flavour intensity decreased significantly ($P < 0.05$) between days 0 and 14. No significant change was evident, however, between days 14 and 28.

8.2.3 Microbiological safety and shelf life

As mentioned previously, the majority of studies undertaken on the microbiological safety of *sous vide* products concern *Clostridium botulinum*. The public health risk posed by *sous vide* is exacerbated by its anaerobic nature which inhibits both the chemical oxidation and aerobic spoilage organisms that, in cook–chill, usually provide sensory indication of spoilage prior to it becoming unfit for consumption. Thus a potential risk exists for *sous vide* products to be palatable yet microbiologically unsafe (Conner *et al.*, 1989). Although this has been demonstrated (e.g. Simpson *et al.*, 1995) there is little evidence to indicate a significant risk except in cases of significant product abuse resulting from either a lack of uniformity of heat treatment or temperature/time abuse during chilled storage.

The heating equipment used in manufacturing (i.e. retorts and autoclaves) produces a treatment of acceptable uniformity – this however is not the case with many of the atmospheric steamers, combination ovens and waterbaths used in catering (Schafheitle and Light, 1989; Sheard and Rodger, 1995; Hansen *et al.*, 1995). The magnitude of variation found in some cases (e.g. Sheard and Rodger (1995), found a 54 minute variation in heating samples to 75°C in a large combination oven) would have a significant effect upon the safety and shelf life of products processed according to, for example, the SVAC (1991) or ACMSF (1992) recommendations.

The effect of temperature/time abuse during chilled storage can be dramatic: whereas *Clostridium botulinum* type B took 80 days to produce toxin at 3.3°C, at 5.6°C the period was reduced to 18 days (Eklund *et al.*, 1967). Brown and Gaze (1990) found *C. botulinum* type E toxin in *sous vide* cod after 21 days at 8°C and Brown *et al.* (1991) found type E and B toxin in *sous vide* cod and chicken after 7 days at 15°C. Practical difficulties in maintaining chilled storage temperatures have been recognised, particularly in distribution and retail display (e.g. Bogh-Sorensen and Olsson, 1990).

Limited work specific to *sous vide* has been undertaken on the hurdle concept, however Meng and Genigeorgis (1994) found that sodium lactate significantly delayed toxigenesis of (non-proteolytic types B and E) *Clostridium botulinum* in (commercially produced and inoculated) beef, chicken and salmon, and that the effect was enhanced by 'lower temperatures'. The authors suggested that, in meat and poultry in which low levels of *Clostridium* spores are anticipated, the use of ≥2.4% lactate and a storage temperature ≤12°C would ensure inhibition of toxigenesis for periods well beyond the anticipated shelf life of 3–6 weeks. Salmon, it was noted, would, for reason of its higher probability of contamination, require more stringent measures.

Simpson *et al.* (1995) investigated the effect of pH and additional salt on (proteolytic types A and B) *Clostridium botulinum* toxigenesis in a pasta and meat sauce product processed in a waterbath at 75°C for 36 minutes and stored at 15°C for up to 42 days. Toxin was detected in samples of pH > 5.5 after 14–21 days and in products of pH 5.25 after 35 days. Toxin was not detected in any samples of pH < 5.25 within 42 days storage at 15°C. All products of pH 5.75 and 6 were visibly spoiled (i.e. pack swollen) prior to toxigenesis, however for products of pH 5.5 and 5.25 toxigenesis preceded spoilage. Additional salt (i.e. >1.5% w/w) prevented toxin production throughout the 42 day storage period at 15°C. Neither Meng and Genigeorgis (1994) nor Simpson *et al.* (1995) discussed the sensory implications, if any, of the hurdles used. Research is required as sensory detectability may be a limiting factor in practical use.

It is evident that the microbiological shelf life of *sous vide* products depends upon a number of variables including the microbiological quality of the raw materials, the product formulation and the process parameters applied. In the past, and particularly in catering use, the shelf life was apparently determined in an arbitrary manner; however increasingly, and particularly in food manufacturing, more sophisticated methods (e.g. predictive shelf life models) are being used. Such methods should be more reliable and thus necessitate less of a 'safety margin'. Gould (1996), however, suggests that this is not the case and that most models of this type result in the significant underestimation of shelf life. Further research is evidently required.

A number of investigations have examined both sensory and microbiological aspects of *sous vide* products.

Light *et al.* (1988) studied chicken à la king (i.e. chicken in a white sauce containing peas and other – unspecified – vegetables) and courgettes provençale (i.e. in tomato sauce), heat treated 'depending upon the product' and chill stored at 1–3°C for up to 21 days. Both products were microbiologically acceptable (in terms of Total Viable Count – TVC) after 21 days, however, the chicken and courgettes were unacceptable sensorially after 14 and 7 days respectively.

Schafheitle and Light (1989) studied chicken ballotine (i.e. chicken breast filled with a brunoise of vegetables), heat treated to a core temperature of 80°C and chill stored at 1–3°C for up to 21 days. The product was found to be acceptable both sensorially and microbiologically (in terms of TVC) at the end of the storage period.

Gittleson *et al.* (1992) assessed the sensory and microbiological deterioration of commercially produced salmon packed in two types of plastic packaging over 12 weeks of storage at 0–4°C. The heat treatment was not specified. The major sensory deterioration was related to increased dryness/flakiness of texture, protein precipitation (causing a white deposit on the flesh) and off-odour. These factors potentially limited shelf life to approximately 8 weeks. TBAR values were insignificant (i.e. below the levels generally considered to be indicative of rancidity) in all samples throughout the chilled storage period. The microbiological data were variable and inconclusive.

Simpson *et al.* (1994) undertook shelf life studies on a pasta and meat sauce product, apparently processed in a waterbath at 65°C (for either 71 or 105 minutes) or 75°C (for either 37 or 40 minutes), prior to chilled storage at either 5 or 15°C for up to 35 days. Physical, chemical and microbiological factors were evaluated. It was found that, at 5°C, a shelf life of 35 days was attainable irrespective of heat treatment, whereas at 15°C sample packs were swollen (as a result of carbon dioxide production) after 14–21 days, depending upon the severity of the heat treatment.

Hansen *et al.* (1995) investigated spiced and unspiced beef heated in waterbaths at either 59 or 62°C for 5 hours (minimum PV_{70}^{10} of 7.8 and 9.0 respectively), chilled to 3°C and stored at 2, 5 or 10°C for up to 35 days. Heat treatment had no significant effect upon the microbiological load, however samples treated at the higher temperature were significantly more tender (according to shear force measurements). At the lower storage temperatures (i.e. 2 and 5°C), the products were 'microbiologically stable' for at least 35 days, however at 10°C, package distortion due to 'gas-producing *Clostridia*' was observed in >25% of samples within 21 days. In the unspiced samples, a significant increase in off-flavours and odours was detected between days 1 and 23 storage at 2°C. This was not evident, however, in the spiced samples. The off-flavour development was not considered to be related to TBAR accumulation as the values remained insignificant throughout.

Bergslien (1996) investigated the sensory and microbiological quality of pretreated (i.e. sodium carbonate at concentrations between 2.5 and 10%) and un-pretreated *sous vide* salmon, subjected to one of three (core) heat treatments (i.e. 65°C for 10 minutes; 70°C for 12.5 minutes; 70°C for 20 minutes) prior to storage at 2°C for up to 7 weeks. Samples processed at 65°C for 10 minutes were of 'poor' bacterial quality after 1 week according to TVC; whereas those subjected to 70°C for 20 minutes were 'acceptable' after 7 weeks. Sodium carbonate produced an inhibitory effect upon bacterial growth in direct relation to concentration. No specific reference was made to *Clostridium botulinum* in the experimental work. The sensory element of the study focused upon appearance, off-odour and flavour. The appearance (in terms of colour and protein precipitation) was directly related to heat treatment, with both 60°C samples having higher scores (i.e. more colour and less protein precipitation) than those subjected to 70°C for 20 minutes. The 60°C samples did however exhibit significant deterioration (details not provided) within 2 weeks. Off-odour rendered all 60°C samples unacceptable within 4 weeks. In terms of flavour, the 60°C samples had the highest scores on the day of processing, however those subjected to 60°C for 10 minutes exhibited a rapid decrease over time and were unacceptable within 1 week of chilled storage. All of the above sensory losses were judged to have been caused by bacterial activity, and were significantly inhibited by pre-treatment with sodium carbonate. This would appear to offer the potential extension of both sensory and microbiological shelf life.

8.2.4 AGS and (the cook-tank form of) Capkold

The AGS system involved 'pasteurisation' (details unspecified) and chilled storage between 0 and –2°C for up to 60 days (Mason *et al.*, 1990). No objective data are available regarding its sensory quality or shelf life, however it is interesting to note that its originator claimed, at least a decade prior to *sous vide*, that packaging the food prior to cooking would improve sensory quality by preventing the evaporative losses of flavour volatiles and moisture during cooking (McGuckian, 1969).

No data are available regarding microbiological quality or shelf life of AGS products.

The cook-tank form of Capkold is generally used on meat joints or roasts. The process involves vacuum packaging the product, heat treatment (in a waterbath) to a core temperature of 85°C and chilled storage for 14–45 days at 2°C (Mason *et al.*, 1990). The majority of the work undertaken on Capkold (or similar 'cook-in-bag' systems) involves the comparison of Capkold and conventionally produced meat and meat products.

Jones *et al.* (1987) compared 'cook-in-bag' pork (cooked to a core temperature of 70°C prior to storage at 4°C for 14 or 28 days) with freshly cooked (to a core temperature of 70°C following frozen storage in PVC wrap for 14 or 28 days). Little significant difference in juiciness, tenderness, flavour intensity and off-flavour intensity was found between the two treatments stored for the same period of time, although the 'cook-in-bag' product did exhibit some decline in sensory quality over time (i.e. tenderness and flavour intensity). The 'cook-in-bag' samples had significantly lower TVC, which remained unchanged over 28 days of storage. The authors concluded that the product had a sensory and microbiological shelf life in excess of 28 days.

Stites *et al.* (1989) undertook similar work using beef. Samples that had been vacuum packaged, cooked to a core temperature of 65°C and stored at 4°C for 14 or 28 days were compared with those that had been freshly cooked following chilled storage in oxygen-permeable film for 4 or 28 days. The sensory and microbiological results obtained were similar to those of

Jones *et al.* (1987), causing the authors to conclude that beef heat treated to a core temperature of 70°C would have a shelf life in excess of 28 days at 4°C.

Smith and Alvarez (1988) drew similar conclusions with turkey roll (i.e. that a core temperature of 71°C would allow a shelf life significantly in excess of 30 days at 4°C).

Some data are available regarding products other than meat.

Notermans *et al.* (1981) demonstrated (proteolyic and non-proteolytic) *Clostridium botulinum* toxin formation prior to spoilage in potatoes that had been vacuum packaged, cooked (in a waterbath at 95°C for 40 minutes) and stored at 10°C for 3 weeks. No toxigenesis was found in samples stored at 4°C after 60 days. The authors concluded that vacuum-packaged potatoes were an 'ideal substrate' for *Clostridium botulinum* and that the consequent potential public health risk can be eliminated by storage at or below 4°C.

8.3 Vacuum/gas packaging post cooking

This category includes products produced using two catering production systems, namely Nacka (Bjorkman and Delphine, 1966) and (the kettle form of) Capkold (Unklesbay *et al.*, 1977), plus conventional cook–chill products that have been vacuum or gas packed post cooking. In this chapter, this latter category will be referred to as modified atmosphere packaging (MAP) cook–chill.

8.3.1 Nacka and (the kettle form of) Capkold

The Nacka process involved food being cooked in the usual manner, although an internal temperature of 80°C was attained whenever possible (i.e. unless it would cause unacceptable thermal damage to the product). It was then vacuum packed and placed in boiling water for between 3 and 10 minutes. This period was dependent upon the perceived microbiological risk: a product was deemed to present a high risk (and therefore qualify for a 10 minute immersion) if 80°C could not be attained in the previous cooking stage or if it could be expected to have high levels of contamination. The products were chill-stored at 4°C for 7–28 days (Mason *et al.*, 1990).

Little data are available regarding either sensory or microbiological quality.

Jakobsson and Bengtsson (1972) investigated the effect of different storage temperatures (3 and 8°C), headspaces (<5 – i.e. a vacuum – and 9 ml/pouch) and chilled storage time (1 day to 3 weeks) on cooked sliced roast beef that had been fried to a core temperature of 70–75°C and then 'pasteurised' to a core temperature of 80°C. Vacuum packaging was found to have a significant effect upon flavour over time; for example, vacuum-packaged and non-vacuum-packaged samples were comparable in terms of flavour after 2 weeks and 4 days respectively. Chilled storage temperature was found to have little effect upon sensory quality, there being no significant difference between products stored at 3 and 8°C for up to 21 days. Unfortunately, no microbiological data were provided.

Kossovitsas *et al.* (1973) investigated the sensory and microbiological quality of chicken à la king (i.e. chicken in white sauce containing carrots, onions, potatoes and peas), cod in cream sauce and broccoli in cream sauce. Products were cooked at core temperature of 80°C, vacuum packed, immersed in boiling water for 3 minutes (cf. Jakobsson and Bengtsson, 1972) and chill-stored at *ca.* 3°C for up to 45 days. All products were unacceptable sensorially after 30 days. The process was found to have 'effectively destroyed' both the *Salmonella* spp. and *Clostridium perfringens* with which it had been inoculated (i.e. none was found after 15, 30 or 45 days of storage).

The kettle form of Capkold (also known as a 'hot-fill' system) is used for liquid-based products and involves pasteurisation to (a liquid temperature of) 85°C, followed by negligible-headspace (i.e. not specifically vacuum) packaging, chilling and chilled storage at *ca.* 2°C for 14–45 days (Mason *et al.*, 1990).

Cremer *et al.* (1985), assessed the sensory and microbiological quality of chicken and noodles in sauce. Freshly prepared control samples were compared with pasteurised (88 ± 6.4°C), chilled (to 8.7 ± 2.9°C in 2 ± 0.6 h) and packaged equivalents that had been chill-stored (1.3 ± 0.9°C) for up to 28 days. In sensory terms, the freshly produced samples were generally found to be superior to chill stored products, the difference increasing with duration of chilled storage. General acceptability and mushiness of the chicken and noodles were significantly ($P < 0.05$) different from fresh after 4 weeks, while appearance and firmness of the noodles, colour, consistency of the sauce and overall moistness were different after 2 weeks. Flavour and chicken tenderness were unchanged over the storage period. The product was found to be 'safe' (according to TVC data) for over 24 days.

8.3.2 MAP cook–chill

A number of studies have investigated the shelf life implications of vacuum or gas packaging 'conventional' cook–chill products prior to chilled storage. These products have, almost invariably, been subjected to a heat treatment specifically designed to address the risk of psychrotolerant pathogens other than *Clostridium botulinum*.

Anaerobic systems have been shown to enable significant shelf life extension of a variety of cooked products (Church and Parsons, 1995). The systems applied usually involve packaging in either a vacuum or an atmosphere consisting of 20–40% carbon dioxide (CO_2)/60–80% nitrogen (N_2) v/v. Oxygen (O_2) is excluded to prevent both the growth of aerobic bacteria and WOF development. CO_2 is used as a bacterial and fungal growth inhibitor and N_2 as an inert 'filler' – to replace O_2, either as an alternative to vacuum packaging when the product is fragile, or to limit pack collapse as CO_2 is (necessarily) absorbed into the product. MAP packaging of this type is generally removed prior to reheating.

A number of studies have investigated either sensory or microbiological aspects.

Glew *et al.* (1980) compared the sensory quality of battered and deep-fried chicken stored in air and nitrogen atmospheres. The nitrogen-stored samples were significantly better in terms of flavour after 3 days at 0°C, causing the authors to conclude that cook–chill catering systems 'might be better from the flavour retention viewpoint if storage could take place in the absence of oxygen'.

Parsons *et al.* (1986) assessed the effect of temperature on the sensory shelf life of cook–chill products that had been vacuum packaged after cooking to 75°C. Results indicated that, in the case of lasagne, the product stored at 1°C remained 'highly acceptable' over the 16-day test period, whereas at 10°C the product had become unacceptable by day 12. No comparison with air packaged samples was provided.

Young *et al.* (1989) investigated the sensory quality of chicken drumsticks and à la king (i.e. in a white sauce with vegetables), packed either in air or under vacuum and stored at 0–3°C. Vacuum packaging was found to significantly extend the sensory shelf life, for example, vacuum-packaged samples were, after 11–14 days, comparable in terms of sensory quality to conventional samples stored for 4 days. The vacuum-packaged samples did however receive increased 'blandness of flavour' and 'blandness of odour' ratings. This effect, which was less pronounced in the chicken *à la* king due to the masking effects of the other (vegetable and sauce) components, was attributed to the fact that the assessors were used to, and therefore regarded as normal, the stronger flavours and aromas of partially oxidised chicken. The colour of the green vegetable component of the chicken à la king faded during storage, however this was less pronounced with vacuum than gas packaging.

Bertelsen and Juncher (1996) mention several investigations comparing the sensory quality of cook–chill products stored in air and in modified atmospheres (typically, 20% CO_2/80% N_2). In all cases, the modified atmospheres significantly reduced WOF development as measured by TBA levels and/or sensory evaluation.

Young *et al.* (1987) investigated the microbiological quality of cook–chill fried chicken drumsticks and chicken à la king (cf. Young *et al.*, 1989). Samples were vacuum and gas (100% CO_2 v/v and 70% CO_2/30% N_2 v/v) packaged and stored at 0 and 4°C for up to 21 days. All treatments inhibited growth (of naturally occurring *Pseudomonas* spp., *Clostridium* spp., *Lactobacillus* spp. and *Brochothrix thermosphacta*) to levels below 5 log_{10}/g for up to 17 days, the most effective treatment being gas packaging (100% CO_2 v/v) at 1°C. The results were taken to show that vacuum and gas packaging, in combination with chilled storage temperatures below 3°C, 'can result in a high degree of confidence in the microbiological safety of cooked chilled foods over 5 days'.

A number of studies have investigated both sensory and microbiological aspects. The majority of this work has been undertaken on meat and poultry and demonstrates the complex relationship between atmosphere, shelf life and storage temperature (e.g. Jantavat and Dawson, 1980; Hintlian and Hotchkiss, 1987). The data are somewhat contradictory but generally indicate that, whereas gas packaging is a better inhibitor of microbiological proliferation, vacuum packaging often retains sensory quality more effectively. In this respect the data of McDaniel *et al.* (1984) are typical: cook–chill beef was cooked to a core temperature of 60°C, packaged in both vacuum and gas (100% CO_2 v/v and 15% CO_2/30% O_2/55% N_2 v/v) atmospheres and stored at 4°C for up to 21 days. No difference in microbial (mesophiles and psychrotrophs) numbers was found after 7 days, however from day 14, the gas packaged (100% CO_2 v/v) sample had a significantly ($P < 0.05$) lower count than the vacuum-packaged sample. In terms of sensory quality, the vacuum-packaged samples exhibited little deterioration over 21 days, whereas both gas-treated samples (and the 100% CO_2 v/v sample in particular) showed significant (appearance, flavour and texture) losses by day 14. At day 7 however, the gas packaged samples were preferred to the vacuum-packaged samples.

Coulon and Louis (1989) found that a bouchée à la reine (a *vol au vent* filled with chicken in a bechamel sauce) had a shelf life, in air at 5°C, of approximately 5 days. In a 50% CO_2/50% N_2 v/v atmosphere, however, this could be doubled without unacceptable sensory losses or pack collapse, and a microbiological count acceptable under French regulation maintained for in excess of 20 days.

8.4 Conclusions

A number of technologies exist whereby anaerobic environments may be applied to cook–chill food products. At the present time, however, *sous vide* is the focus of most commercial and research attention. The reason for this is open to debate, but is probably due to the fact that it is perceived to have evolved from (and thus be superior to) the other vacuum-based systems, and is technically simpler than MAP cook–chill.

It is evident that the anaerobic packaging of cook–chill products, either before or after cooking, has a significant inhibitory effect upon both oxidative spoilage mechanisms and the proliferation of aerobic bacteria. This potentially enables shelf life extension. The factors that limit shelf life remain unelucidated, however concern that the potential exists – in specific abuse situations – for processed products to be palatable yet unfit for consumption has, in conjunction with the often fatal nature of Botulism, focused the research effort upon microbiological safety issues. Despite this, however, the available data are inconclusive – although it is evident that relatively minor product and/or process variations can have a significant effect upon microbiological shelf life.

A similar knowledge gap is evident regarding sensory shelf life, and thus further research is required relating to both aspects. Of particular importance in this respect are the effects of minor product formulation and/or process variation (including that arising from equipment function) and the microbiological efficacy and sensory implications of additional hurdles. In addition, the development of more accurate predictive shelf life models merits attention.

The existing knowledge gap is particularly evident with *sous vide* and is primarily the result of a significant variation in the process parameters applied and/or the experimental methodologies used in the published research. This reflects significant process parameter variation both internationally and between the major user classifications (i.e. caterers and food manufacturers). The problem is exacerbated further by the fact that the term '*sous vide*' is used to describe a number of closely related but distinct methods. Differentiation (perhaps in terms of product type, processing equipment and shelf life) and renaming of the various forms could both reduce confusion and, ultimately, form the basis of a standard specifying Good Manufacturing Practice. Such a standard is desirable, particularly in areas of use (such as catering) where less technical expertise is available and a tradition of less rigorous control exists. The undoubted practical problems associated in gaining international recognition for such a standard should not discourage the attempt.

References

ACMSF (1992) *Report on Vacuum Packing and Associated Processes*. London: HMSO.

Bergslien, H. (1996) *Sous vide* treatment of salmon (Salmon salar) In: *Proceedings of the Second European Symposium on Sous vide*. 10–12 April Leuven, Belgium, pp. 281–282.

Bertelsen, G. and Juncher, D. (1996) Oxidative stability and sensory quality of *sous vide* cooked products, in *Proceedings of the Second European Symposium on Sous vide*, 10–12 April, Leuven, Belgium, pp. 134–145.

Betts, G. and Gaze, J. (1995) Growth and heat resistance of psychrotrophic *Clostridium botu-linum* in relation to *sous vide* products. *Food Control*, **6** (1), 57–63.

Bjorkman, A. and Delphine, K. (1966) Swedens Nacka hospital food system centralises prep-aration and distribution. *Cornell Hotel and Restaurant Association Quarterly*, **7** (3), 84–87.

Bogh-Sorensen, L. and Olsson, P. (1990) The chill chain, in *Chilled Foods: The State of the Art* (ed. T. Gormley), London: Elsevier, pp. 245–268.

Bovee, E., Hartog, B. and Kant-Muermans, M. (1996) Microbiological safety assurance of *sous vide* cooked cod fillets, in *Proceedings of the Second European Symposium on Sous vide*, 10–12 April, Leuven, Belgium, pp. 293–296.

Brown, G. and Gaze, J. (1990) *Determination of the Growth potential of* Clostridium botulinum *Types E and Non-proteolytic B in Sous vide Products at Low Temperatures. Technical Mem-orandum No. 593*, Chipping Campden: Campden Food Preservation Research Association.

Brown, G., Gaze, J. and Gaskell, E. (1991) *Growth of* Clostridium botulinum *Non-proteolytic Type B and Type E in Sous vide Products Stored at 2-15C. Technical Memorandum No. 635*, Chipping Campden: Campden Food Preservation Research Association.

Church, I. and Parsons, A. (1993) Review: *sous vide* cook-chill technology. *International Journal of Food Science and Technology*, **28**, 563–574.

Church, I. and Parsons, A. (1995) Modified atmosphere packaging technology: a review. *Journal of the Science of Food and Agriculture*, **67**, 143–152.

Church, I. (1996) The sensory quality of chicken and potato products prepared using cook-chill and *sous vide* methods, in *Proceedings of the Second European Symposium on Sous vide*, 10–12 April, Leuven, Belgium, pp. 317–326.

Conner, D., Scott, V., Bernard, D. and Kautter, D. (1989) Potential *Clostridium botulinum* hazards associated with extended shelf life refrigerated foods: a review. *Journal of Food Safety*, **10**, 131–153.

Coulon, M. and Louis, P. (1989) Modified atmosphere packaging of precooked foods, in *Con-trolled/Modified Atmosphere/Vacuum Packaging of Foods* (ed. A.L. Brody), Trumbull (CT): Food and Nutrition Press, pp. 135–148.

Creed, P., Reeve, W. and Pierson, B. (1993) The sensory quality of *sous vide* foods. Paper pre-sented at 'Food Preservation 2000', October, US Army Natick Center, Natick, Mass, USA.

Cremer, M., Yum, T. and Banwart, G. (1985) Time, temperature, microbial and sensory quality assessment of chicken and noodles in a hospital foodservice system. *Journal of Food Science*, **50**, 891–896.

DoH (1989) *Chilled and Frozen. Guidelines on Cook–Chill and Cook–Freeze Catering Systems*, London: HMSO.

ECFF (1996) *Guidelines for the Hygienic Manufacture of Chilled Foods*, London: European Chilled Food Federation (UK Chilled Food Association)

Eklund, M., Wieler, D. and Poysky, F. (1967) Outgrowth and toxin production of non-proteo-lytic type B *Clostridium botulinum* at 3.3 to 5.6°C. *Journal of Bacteriology*, **93**, 1461–1462.

Genigeorgis, C. (1993) Additional hurdles for *sous vide* products, in *Proceedings of the FLAIR First European Sous vide Cooking Symposium*, 25–26 March, Heverlee, Belgium, pp. 57–58.

Gittleson, B., Saltmarch, M., Cocotas, P. and McProud, L. (1992) Quantification of the physi-cal, chemical and sensory modes of deterioration in *sous vide* processed salmon. *Journal of Foodservice Systems*, **6**, 209–232.

Glew, G., Berg, R. and Yam, Y. (1980) Some effects of the presence or absence of oxygen on stored precooked chilled chicken and pork meat, in *Advances in Catering Technology* (ed. G.Glew), London: Applied Science, pp. 417–442.

Gould, G. (1996) Conclusions of the ECFF Botulinum Working Party, in *Proceedings of the Second European Symposium on Sous vide*, 10–12 April, Leuven, Belgium, pp. 173–180.

Hansen, T., Knochel, S., Juncher, D. and Bertelsen, G. (1995) Storage characteristics of *sous vide* cooked roast beef. *International Journal of Food Science and Technology*, **30**, 365–378.

Hintlian, C. and Hotchkiss, J. (1987) Microbial and sensory evaluation of cooked roast beef packaged in a modified atmosphere. *Journal of Food Processing and Preservation*, **11**, 171–179.

Jakobsson, B. and Bengtsson, N. (1972) A quality comparison of frozen and refrigerated cooked sliced beef. *Journal of Food Science*, **37**, 230–233.

Jantavat, P. and Dawson, L. (1980) Effect of inert gas and vacuum packaging on storage stability of mechanically deboned poultry meats. *Poultry Science*, **59**, 1053–1058.

Jones, S., Carr, T. and McKeith, F. (1987) Palatability and storage characteristics of precooked pork roasts. *Journal of Food Science*, **52**, 279–281, 285.

Kossovitsas, C., Navab, M., Chang, C. and Livingston, G. (1973) A comparison of chilled holding versus frozen storage on quality and wholesomeness of some prepared foods. *Journal of Food Science*, **38**, 901–902.

Leistner, L. (1978) The microbiology of ready to serve food. *Fleischwirtschaft*, **58**, 2008–2011.

Light, N., Hudson, P., Williams, R., Barrett, J. and Schafheitle, J. (1988) A pilot study on the of *sous vide* vacuum cooking as a production system for high quality foods in catering. *International Journal of Hospitality Management*, **7**, 21–27.

Mason, L., Church, I., Ledward, D. and Parsons, A. (1990) Review: the sensory quality of foods produced by conventional and enhanced cook-chill methods. *International Journal of Food Science and Technology*, **25**, 247–259.

McDaniel, M., Marchello, J. and Tinsley, A. (1984) Effect of different packaging treatments on microbiological and sensory evaluation of precooked beef roasts. *Journal of Food Protection*, **47**, 23–26.

McGuckian, A. (1969) The AGS food system – chilled pasteurised food. *The Cornell Hotel and Restaurant Association Quarterly*, **10** (1), 87–92.

Meng, J. and Genigeorgis, C. (1994) Delaying toxigenesis of *Clostridium botulinum* by sodium lactate in *sous vide* products. *Letters in Applied Microbiology*, **19**, 20–23.

Notermans, S., Dufrenne, J. and Keijbets, M. (1981) Vacuum packed cooked potatoes: toxin production by *Clostridium botulinum* and shelf life. *Journal of Food Protection*, **44**, 572–575.

Parsons, A., Sheard, M. and Foxcroft, G. (1986) The extension of shelf life of cook/cool meat dishes by aseptic packaging and temperature control, and recipe development for microwave reheating, in *Recent Advances and Developments in the Refrigeration of Meat by Chilling*, Bristol: International Institute of Refrigeration, pp. 369–372.

Rhodehamel, E. (1992) FDA's concerns with *sous vide* processing. *Food Technology*, **46** (11), 73–76.

Schafheitle, J. and Light, N. (1989) Technical note: *sous vide* preparation and chilled storage of chicken ballontine. *International Journal of Food Science and Technology*, **24**, 199–205.

Schellekens, M. (1996) *Sous vide* cooking: state of the art, in *Proceedings of the Second European Symposium on Sous vide*, 10–12 April, Leuven, Belgium, pp. 19–28.

Sheard, M. and Church, I. (1992) *Sous vide Cook–Chill*, Leeds: Leeds Polytechnic.

Sheard, M. and Rodger, C. (1995) Optimum heat treatments for *sous vide* cook–chill products. *Food Control*, **6** (1), 53–56.

Simpson, M., Smith, J., Simpson, B., Ramaswamy, H. and Dodds, K. (1994) Storage studies on a *sous vide* spaghetti and meat sauce product. *Food Microbiology*, **11**, 5–14.

Simpson, M., Smith, J., Dodds, K., Ramaswamy, H., Blanchfield, B. and Simpson, B. (1995) Challenge studies with *Clostridium botulinum* in a *sous vide* spaghetti and meat-sauce product. *Journal of Food Protection*, **58**, 229–234.

Smith, D.M. and Alvarez, V.B. (1988) Stability of cook-in-bag turkey breast rolls during refrigerated storage. *Journal of Food Science*, **53**, 46–48, 61.

Stites, C., McKeith, F., Bechtel, P. and Carr, T. (1989) Palatability and storage characteristics of precooked beef roasts. *Journal of Food Science*, **54** (1), 3–6.

SVAC (1991) *Code of Practice for Sous vide Catering Systems*, Tetbury: SVAC.

Turner, B. and Larick, D. (1996) Palatability of *sous vide* processed chicken breast. *Poultry Science*, **75**, 1056–1063.

Unklesbay, N., Maxcy, R., Knickrehn, M., Stevenson, K., Cremer, M. and Matthews, M. (1977) *Foodservice Systems: Product Flow and Microbial Quality and Safety of Foods*, Columbia (MS): Missouri Agricultural Experimental Station.

Varoquaux, P., Offant, P. and Varoquaux, F. (1995) Firmness, seed wholeness and water uptake during the cooking of lentils (*Lens culinaris* cv. *anicia*) for *sous vide* and catering preparations. *International Journal of Food Science and Technology*, **30**, 215–220.

Young, H., Youngs, A. and Light, N. (1987) The effects of packaging on the growth of naturally occurring microflora in cooked, chilled foods used in the catering industry. *Food Microbiology*, **4**, 317–327.

Young, H., MacFie, H and Light, N. (1989) Effect of packaging and storage on the sensory quality of cooked chicken menu items served from chilled vending machines. *Journal of the Science of Food and Agriculture*, **48**, 323–338.

Zacharias, R. (1980) Chilled meals: sensory quality, in *Advances in Catering Technology* (ed. G. Glew), London: Applied Science, pp. 409–416.

9 Microbiological safety considerations when using hurdle technology with refrigerated processed foods of extended durability

LEON G.M. GORRIS and MICHAEL W. PECK

9.1 Introduction to REPFEDs

Despite our awareness of food safety and the many preservation techniques that may be used to ensure the safe production of food, the reported incidence of food poisoning continues to increase worldwide. Concomitantly, there is a strong desire among consumers for more fresh-like, lightly processed and preserved foods than those currently available. Only mild preservation techniques should be used to ensure the safety and quality of these minimally processed foods. With some types of these convenience food products, like ready-to-eat salads, freshness and healthiness are the paramount selling features. These foods rely on a very short shelf life (up to 5–7 days at 1–8°C) for safety and quality. Products with a longer shelf life developed recently are the so-called refrigerated processed foods of extended durability (REPFEDs), such as *sous vide* and cook–chill foods. Sales of REPFEDs are increasing at a tremendous rate in many countries in Europe. REPFEDs meet the consumer's requirement for food requiring minimal preparation time, and being a qualitatively superior product compared to foods made by more traditional methods, e.g. canning and freezing.

REPFEDs are composed of normal ingredients such as meat, fish, potatoes, rice, pastas and vegetables. They are processed at a lower temperature (maximum generally within the range 65–95°C) and for a longer period of time than, for example, canned foods. The lower temperatures (below 70°C) are used to cook delicate products such as fish, while the higher temperatures are employed with vegetables. The heating regimes are intended to minimise the processing impact on sensory and organoleptic aspects of product quality. After heating, products are cooled rapidly (generally down to storage temperature in 1–2 hours), and stored at refrigeration temperatures (1–8°C). The shelf life of the product is dependent upon the antimicrobial impact of the heat treatment applied and the storage temperature prevailing.

Typically, the shelf life of REPFEDs at refrigeration temperatures is up to 42 days, and preparation is in one of the following three ways:

1. Ingredients (which may include both raw and cooked components) are packed in a heat-stable pouch. With preparations referred to as '*sous vide*' products, a vacuum is applied (10–120 mbar) and the pouch sealed, whereupon the packaged product is cooked. In other cases, a modified atmosphere may be applied before sealing and heating the package.
2. Ingredients are cooked individually and subsequently packaged.
3. Ingredients are individually cooked, then packaged, and then heated again.

9.2 Microbiological safety considerations with REPFEDs

The mild preservation technologies utilised to produce REPFEDs reduce the initial level of contamination but do not give a sterile product. The mild heat treatment applied should eliminate cells of vegetative bacteria, but not bacterial spores. Essentially, surviving spore-formers that germinate will be able to proliferate without competition from the vegetative bacteria previously present. The surviving microorganisms find themselves in a new ecological niche for growth and, if pathogenic, may pose a serious health hazard to consumers (Schellekens, 1996; Peck, 1997). Since REPFEDs are mostly packed under vacuum or in an anaerobic atmosphere, growth of aerobic microorganisms is restricted, while growth of anaerobic bacteria is favoured. Storage of REPFEDs at refrigeration temperatures will restrict growth to those microorganisms capable of growth at low temperatures, and while these microorganisms may grow only slowly at such temperatures, the rather long storage times used with most REPFEDs may allow slowly growing microorganisms to reach hazardous levels. The ecological niche found in REPFEDs therefore favours colonisation by microorganisms that produce heat-resistant spores and grow in the absence of oxygen at refrigeration temperatures. Non-proteolytic *Clostridium botulinum* can multiply and form toxin at temperatures as low as 3.0–3.3°C (Schmidt *et al.*, 1961; Eklund *et al.*, 1967a, b; Graham *et al.*, 1997) and is the principal microbiological safety concern in REPFEDs. The absence of competition from other microorganisms and the extended product shelf life might give additional time for toxin production. Non-proteolytic *C. botulinum* has been associated with outbreaks of botulism following consumption of 'dried' fish, vacuum-packed salmon and also 'fermented' marine products by the northern native populations of Europe, Canada and the USA. Table 9.1 gives examples of time to toxin formation in foods inoculated with a low concentration of spores of non-proteolytic *C. botulinum*.

Spores of *C. botulinum* are ubiquitous in nature and no raw food material can be guaranteed free of spores. Under anaerobic conditions, the nutrients found in foods will allow growth of *C. botulinum* and countermeasures should involve either treatments that eliminate spores or storage conditions

Table 9.1 Time to toxin formation in several foods inoculated with low numbers of spores of non-proteolytic *Cl. botulinum* (based on data from Post *et al.*, 1985; Garcia *et al.*, 1987; Baker *et al.*, 1990; Meng and Genigeorgis, 1993; Carlin and Peck, 1996a)

Food	Inoculum level (spores/g product)	Time (d) to toxin formation at specified temperature						
		4°C	5°C	8°C	10°C	12°C	16°C	25/30°C
Cod	10^2	18	–[a]	8	–	6	–	1
Salmon	10^2	21	–	6	–	3	–	1
Red snapper	10^2	NT[b]	–	12	–	3	3	1
Turkey	10^2	–	--	8	–	5	2	1
Cauliflower	10^3	–	21	15	13	–	4	2
Mushroom	10^3	–	20	10	10	–	3	1

[a]Test not performed at the specified temperature.
[b]No toxin formation found within 60 days, however toxin was detectable within 21 days using an inoculum of 10^3 spores/g.

that prevent growth and toxin production. Proteolytic *C. botulinum* produces spores of a greater heat resistance but is unable to multiply at temperatures below 10°C. When storage below this temperature can be assured, these pathogens will not present a hazard in REPFEDs.

Pathogens other than non-proteolytic *C. botulinum* may pose a hazard to the safety of REPFEDs under certain conditions. Proteolytic *C. botulinum* and *C. perfringens* are of concern when the storage temperature is not adequately controlled and exceeds 10°C for a prolonged time. Several isolates of *Bacillus cereus* have been found able to grow slowly just under 10°C but, contrary to the situation with *C. botulinum*, high numbers of cells of *B. cereus* are needed to pose a genuine safety hazard.

Although non-spore-forming pathogens should be eliminated by the mild heat treatments applied during processing of the REPFEDs, recontamination after processing is of concern for REPFEDs of the type B mentioned above. These products are individually cooked and then packaged. The handling necessary for this may lead to recontamination of the treated product with pathogens such as *Listeria monocytogenes*, *Escherichia coli* O157:H7, *Yersinia enterocolitica* and *Salmonella* spp. Some of these pathogens grow at refrigeration temperatures (e.g. *L. monocytogenes, Y. enterocolitica*), while others survive for extended periods (e.g. *E. coli* O157:H7, *Salmonella* spp.). Again, the absence of a competitive microflora might allow the recontaminating pathogens a better chance to proliferate as compared to the situation with the original raw material.

9.3 The concept of hurdle technology

In hurdle technology, combinations of different preservation factors ('hurdles') are used to achieve multi-target, mild preservation effects

(Leistner, 1992, 1995; Leistner and Gorris, 1995). In combination these hurdles control growth of spoilage and pathogenic microorganisms, despite each hurdle not being sufficient on its own to achieve the same effect. The selection of hurdles needs to be carefully designed and take into account the quality attributes of a food product and the processing technologies that this product can be subjected to. In practice, the large variability in foods and microorganisms relevant to food preservation make this a demanding task. The sequence, type and intensity of hurdles need to be optimised according to the most important variable(s). When only a few specific microorganisms need to be controlled and there is little variation to be expected, less different hurdles or hurdles of lower intensity may achieve the desired effect. On the other hand, where high numbers of many different target microorganisms occur, extended sets of hurdles need to be designed to obtain reliable control.

Preservative hurdles may act at specific or a specific cellular sites, such as the cell membrane, DNA, or specific enzymes. The most effective pathogen control is achieved by targeting several different cellular sites simultaneously. This 'multiple-target' approach is often more effective than 'single-targeting' and enables the use of hurdles of lower intensity, thereby having a less detrimental effect on product quality. There is also the possibility that different hurdles in a food will not just have an additive effect, but could act synergistically. The chances of an organism acquiring some degree of tolerance or resistance to the treatment may also be much lower with a 'multiple-target' approach than with a 'single-target' approach.

When using hurdle technology to aid the safe development of REPFEDS, it is important to consider metabolic exhaustion of microorganisms. The importance of metabolic exhaustion in relation to food processing, especially the phenomenon of autosterilisation of food, was firstly described by Leistner and Karan-Djurdjić (1970). The autosterilization of foods was observed in experiments with mildly heated (95°C core temperature) liver sausage adjusted to different water activities, inoculated with *Clostridium sporogenes,* and stored at 37°C. It was found that spores of *C. sporogenes* survived the heat treatment but gradually vanished from the product during storage. Test with shelf-stable meats in the same laboratory reproduced this autosterilisation at ambient storage temperature with spores of *Bacillus* and *Clostridium* species as well as with salmonellae (Leistner, 1995). Supportive evidence was additionally reported by Alzamora *et al.* (1993) and Tapia de Daza *et al.* (1995), who found that viable counts of various groups of bacteria, yeasts and moulds that survived the mild heat treatment decreased quite fast in high-moisture fruit products during storage in the presence of additional hurdles (pH, a_w, sorbate, sulfite). In some circumstances in REPFEDs, microorganisms that survive the mild heat treatment might demonstrate a similar phenomenon.

9.4. Hurdle technology to ensure the safe production of REPFEDs

9.4.1. Optimisation of the combination of mild heating and subsequent refrigerated storage

Before designing new, often more complex hurdle technology preservation protocols for REPFEDs, the heat treatment and subsequent storage temperature chosen should be optimised to provide the desired degree of protection against growth and toxin production by non-proteolytic *C. botulinum*. This is generally taken to be a process that provides a 6 decimal reduction with respect to non-proteolytic *C. botulinum* (ACMSF, 1992; Peck *et al.*, 1995; ECFF, 1996). Studies in a model food investigated the effect of heat treatment at 65–95°C combined with storage at 5–25°C on time to growth from an inoculum of 10^6 spores of a mixture of strains of non-proteolytic *C. botulinum* (Table 9.2). It was found that heat treatments of 70°C for 2545 min, 75°C for 464 min, 80°C for 70 min, 85°C for 23 min and 90°C for 10 min each prevented outgrowth and toxin formation within

Table 9.2 Combined effect of lysozyme, heat treatment and subsequent storage temperature on time to visible growth from 10^6 spores of non-proteolytic *C. botulinum* types B, E and F (based on data from: Graham *et al.*, 1996; Peck and Fernandez, 1995, 1997; Peck *et al.*, 1995)

Heat treatment (set temperature and holding time)	Added lysozyme (units/ml)	Time (d) to growth at specified storage temperature					
		5°C	6°C	8°C	12°C	16°C	25°C
None	0	13	–[a]	7	4	2	1
70°C/105 min	0	14	–	9	6	2	1
70°C/1000 min	0	57	–	21	8	5	2
70°C/2545 min	0	NG[b]	–	50	22	8	3
75°C/464 min	0	NG	–	48	38	23	8
75°C/734 min	0	NG	–	NG	18	15	5
80°C/70 min	0	NG	–	44	19	8	5
80°C/184 min	0	NG	–	NG	37	21	11
85°C/23 min	0	NG	–	NG	30	38	15
90°C/10 min	0	NG	–	NG	NG	NG	NG
95°C/19 min	0	NG	–	NG	NG	NG	–
None	625	–	7	4	2	–	1
65°C/364 min	625	–	11	4	2	–	1
70°C/8 min	625	–	8	6	4	–	1
75°C/27 min	625	–	13	9	5	–	1
80°C/23 min	625	–	40	23	12	–	3
85°C/19 min	625	–	53	53	42	–	6
90°C/20 min	625	–	–	NG	51	29	20
95°C/15 min	625	–	NG	NG	NG	–	32

[a]Not tested.
[b]No growth in 60 days.

Tests (including heat treatment and subsequent storage) were conducted in a model food system (meat slurry). When present, lysozyme was added at 625 units (10 µg)/ml prior to heating.

a period of 42 days at 8°C. A total of 160 combinations of heat treatment and incubation temperature were tested in this study, and a predictive model was developed that described the effect of heat treatment and storage temperature on time to growth (Fernandez and Peck, 1997). This model can be used to identify the optimal combination process taking into account the heat treatment that can be applied to a food (without adversely affecting textural properties), the intended refrigeration temperature (this may vary from country to country) and the shelf life.

Among the non-spore-forming bacteria, *L. monocytogenes* is relatively heat resistant and the thermal destruction of its cells has been studied frequently. For example, in ground pork, Ollinger-Snyder *et al.* (1995) measured D-values at 50, 55 and 60°C of 109, 9.80 and 1.14 min, respectively. Thus, if an acceptable level of safety with respect to *L. monocytogenes* is deemed to be achieved by a 4 decimal process, this study shows that ground pork must be heated to an internal temperature of 60°C for at least 4.6 min (Ollinger-Snyder *et al.*, 1995).

A number of new or improved preservation technologies, such as modified atmosphere packaging, use of organic acids or biopreservatives and ultra high pressure, are being developed or already coming onto the market that, in combination with mild heating and refrigeration, might contribute to the continued safe development of REPFEDs. Some recent research has focused on identifying combinations of mild heating and subsequent refrigerated storage that can be used to ensure product safety, and also on combination processes that include additional preservative factors. Examples are now given for *L. monocytogenes* (recontaminant) and non-proteolytic *C. botulinum* (primary contaminant).

9.4.2 Single and combined processes with respect to L. monocytogenes

Current knowledge on the growth limits, in terms of single and combined processes, of pathogens relevant to REPFEDs is elaborate, though mostly pragmatically and not systematically acquired. Growth of *L. monocytogenes* in food has been reported at −1.5°C (Hudson *et al.*, 1994). In the same study, growth of two other psychrotrophic pathogens (*Aeromonas hydrophila* and *Yersinia enterocolitica*) was also reported at this low temperature. The limiting pH for growth of different strains of *L. monocytogenes* has been reported as pH 4.2–4.3 (George *et al.*, 1996; Ryser and Marth, 1991). The sensitivity of this pathogen to different weak acids varies (e.g. Sorrells *et al.*, 1989; Ahamad and Marth, 1989; George *et al.*, 1996). George *et al.* (1996) have developed predictive models that describe the effect of lactic acid, pH and temperature, and acetic acid, pH and temperature on the growth of *L. monocytogenes*. When considered in terms weight of acid per volume or molarity of acid, acetic acid was more inhibitory than lactic acid (Sorrells *et al.*, 1989; Young and Foegeding, 1990; George *et al.*, 1996), while when considered in terms of undissociated acid (as weight per volume) lactic acid was more inhibitory than acetic acid (Young and Foegeding, 1993; George *et al.*, 1996). The effect of other weak acids (e.g. citric acid, sorbic acid, tartaric acid) on the growth and survival of *L. monocytogenes* has also been examined (e.g. Ahamad and Marth, 1990; El Shenawy and Marth, 1991). El Shenawy and Marth (1991) studied the combined effect of potassium sorbate with a weak acid in a culture broth at 13°C. Using 0.3% sorbate, acidification to pH 5.0 with acetic, tartaric, lactic or citric acid in all cases suppressed growth from 10^3 cfu/ml. *L. monocytogenes* grew at pH 5.6 and pH 6.0, regardless of organic acid, except at pH 6.0 when acetic acid was used. Ahamad and Marth (1990) evaluated the effect of 0.3 and 0.5% acetic, citric and lactic acid on this pathogen at 13 and 35°C. Acetic acid caused the greatest inactivation, while citric acid caused the greatest degree of injury followed in order by lactic and acetic acid. Incubation at 13°C did not add to the acid injury effect, although both injured and uninjured organisms remained viable about nine times longer at 13°C than at 35°C. Acid-injured cells failed to grow with 6% added NaCl, which is somewhat lower than the 10% NaCl (equiv. $a_w = 0.92$) generally accepted as the limit for growth (Ryser and Marth, 1991; Hefnawy and Marth, 1993).

The effect of modified atmospheres on growth of *L. monocytogenes* has been studied in laboratory media and in foods (e.g. Hudson *et al.*, 1994; Garcia de Fernando *et al.*, 1995; Sheridan *et al.*, 1995; Fernandez *et al.*, 1997). A predictive model has been developed that described the effect of mixtures of carbon dioxide and nitrogen, pH, temperature and NaCl concentration on growth of this pathogen (Fernandez *et al.*, 1997). Predictions from the model compared well with the reported growth of *L. monocytogenes* in many foods, and also with two previous predictive models (Bennik *et al.*, 1995; Farber *et al.*, 1996). In their study on the growth of *L. monocytogenes* on sliced roast beef at low temperatures, Hudson *et al.* (1994) investigated the possibility of controlling pathogen growth by vacuum packaging or by applying a carbon dioxide saturated headspace. At –1.5°C, vacuum packaging did not prevent growth, but packing under 100% carbon dioxide was successful. At 3°C, neither treatment had the desired effect.

9.4.3 Single and combined processes with respect to non-proteolytic C. botulinum

The effect of single preservative factors (e.g. pH, salt concentration, temperature) on the growth of non-proteolytic *C. botulinum* has been studied. In general, growth and toxin production do not occur below pH 5.0, at a salt level > 5% or $a_w < 0.97$. The minimum temperatures at which growth and toxin production have been reported is 3.0–3.3°C (Schmidt *et al.*, 1961; Eklund *et al.*, 1967a, b; Graham *et al.*, 1997). The combined effect of pH, temperature and NaCl concentration on growth from spores of non-proteolytic *C. botulinum* at 3–10°C has been described (Graham *et al.*, 1997), and predictive models have been developed that describe the effect of combinations of environmental factors on growth from spores of non-proteolytic *C. botulinum* (e.g. Baker and Genigeorgis, 1992; Whiting and Call, 1993; Graham *et al.*, 1996a; Whiting and Oriente, 1997).

Hurdle technology considers together the effect of intrinsic and extrinsic (process) preservative factors, and identifies combinations of these factors that are necessary for the safe production of foods. One such detailed study by Graham *et al.* (1996b) examined the combined effects of pH (5.6 to 6.5), NaCl concentration (0.6 to 4.3% in the aqueous phase), lysozyme addition (0 or 25 µg/ml), heat treatment (up to 95°C for 19 min) and storage temperature (5, 8, 12 or 16°C) on time to growth from 10^6 spores of non-proteolytic *C. botulinum* in a model food system. Conditions of pH, NaCl concentration, heat treatment and storage temperature that did not individually prevent growth and toxin production, could be combined to prevent growth and toxin production. For example, following a heat treatment at 85°C for 18 min growth was observed at pH 6.5 and 0.5% NaCl at 12°C in 24 days. The time before growth was observed could be extended to over 42 days by one of: increasing the heat treatment, lowering the pH, increasing the NaCl concentration, or storage at 8°C (Graham *et al.*, 1996b).

The effect of combinations of other preservative factors, such as weak acid, modified atmospheres and bacteriocins, on growth of non-proteolytic *C. botulinum* has also been described. Addition of lactate has been reported to delay time to toxin production in beef, chicken and salmon (Meng and Genigeorgis, 1994; Gibbs, 1996). Addition of 1.8% sodium lactate increased time to toxin production in chicken breast at 8°C and 12°C, from 16 to 60 days and from 12 to >40 days, respectively (Meng and Genigeorgis, 1994).

Modifying the headspace in packages containing REPFEDs is a common practice, involving either the application of a vacuum (as with *sous vide* products) or by using a modified gas atmosphere. Stringer and Peck (unpublished results) tested the effect of a range of different gas atmospheres (mixtures of carbon dioxide, nitrogen, hydrogen, oxygen, argon, helium, and nitrous oxide) on growth from spores of non-proteolytic *C. botulinum*. With the exception of atmospheres containing oxygen, none of the other mixtures tested significantly reduced the number of unheated or heated spores leading to turbidity. Following a heat treatment at 80°C for 10 min, none of these gases reduced the number of spores resulting in growth within 6 weeks at 10°C by a factor of 10^6. No growth was observed when the laboratory medium was prepared under 21% oxygen. However, in related studies with a meat slurry prepared under different gas mixtures, growth was observed at a similar time in meat slurry prepared under air and an anaerobic gas mixture (Peck and Fairbairn, unpublished results). This latter observation is important as it is sometimes considered that traces of oxygen above a food are sufficient to prevent growth and toxin production by non-proteolytic

C. botulinum. Thus, even under aerobic gas atmospheres, foods themselves might contain sufficient reducing components to lower the redox potential enabling growth and toxin production by non-proteolytic *C. botulinum.* Recent published measurements of the redox potential of deli foods (Snyder, 1996), emphasises this point.

Some foods may have intrinsic hurdles (preservative factors) that contribute to their safety with respect to psychrotrophic pathogens. Examples of such foods appear to be pasteurised crab meat and minimally processed oysters (Peck, 1997). Although spores of non-proteolytic *C. botulinum* may be present in such foods and the heat treatment applied is minimal, they have never been implicated in botulism outbreaks.

9.4.4 Natural preservatives as additional hurdles

The use of artificial preservatives is currently under considerable debate. More natural preservatives are sought and some may be of interest for use with REPFEDs. Bacteriocins, for instance, produced by lactic acid bacteria can have an inhibitory activity against non-proteolytic *C. botulinum* and *L. monocytogenes* (Bennik *et al.*, 1997). Both young and old cultures of *L. monocytogenes* were sensitive to the bacteriocin PA-1 (Figure 9.1). Despite being proteinaceous, some bacteriocins are very heat stable and active over

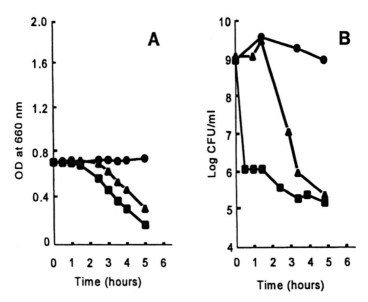

Figure 9.1 Effect of addition of cell free extract of *Pediococcus parvulus*, containing the bacteriocin pediocin PA-1, on growth of *Listeria monocytogenes* in brain heart infusion broth at 30°C. The following additions were made to cultures in the log phase (panel A) and stationary phase (panel B) of growth: (●) no pediocin; (▲) pediocin at 10 BU/ml and (■) pediocin at 100 BU/ml (BU = bacteriocin units). (Data are unpublished from Bennik and Gorris.)

a wide range of pH and could be added to REPFEDs prior to heat pro-cessing (Abee *et al.*, 1995). However, while the lactic acid bacteria as a group are generally recognised as safe (GRAS) and, thus, by definition food-grade, the bacteriocins they produce do not have that explicit status. Only in the case of nisin, a bacteriocin produced by *Lactococcus lactis* subsp. *lactis*, have a number of food applications been approved by regulatory authorities. The direct addition of lactic acid bacteria to REPFEDs as pro-tective cultures prior to heating is unlikely to be successful unless drastic protective measures are taken to maximise their survival.

9.5 Potential problems when using hurdle technology with REPFEDs

9.5.1 Homeostasis and stress reactions

A possible limitation to the successful use of hurdle technology with REPFEDs could be the different physiological response of microorganisms to stress conditions, in particular homeostasis and stress reactions. Such reactions may enable microorganisms to minimise the effect of specific con-straints, and there may also be non-specific effects. As already discussed by Knøchel and Gould (1995), such mechanisms may reduce the effectiveness of hurdle technology for food preservation. Nevertheless, the simultaneous use of stress factors with different cellular targets is still a valuable concept in the hurdle technology approach, since to counter multiple stresses will involve the expenditure of a lot of energy on the part of the target micro-organism.

Homeostasis is the tendency of microorganisms to balance their internal physicochemical conditions to the external environment. They do so by acti-vating specific or non-specific stress reactions. Microorganisms operate a range of different homeostasis mechanisms and stress reactions, including:

- pH-Homeostasis: manipulating the intracellular pH to stay between certain narrow limits even though the pH in the environment changes outside these limits.
- Osmohomeostasis: balancing the internal osmotic pressure to changes in water activity (a_w) due to wetting/drying or to changes in salt or sugar con-centrations outside the cell.
- DNA-repair: responding to damages in DNA due to UV-radiation.
- Heat-shock protein synthesis: *de novo* synthesis of proteins that increase cell heat resistance in response to non-lethal heating.
- Pressure-shock protein synthesis: production of proteins that enable the cell to withstand ultra high pressure stress.

Homeostasis and stress reactions enable microorganisms to keep important physiological systems operating, in balance and unperturbed even when the environment around them is greatly perturbed (Gould, 1988, 1995). When the homeostasis of a microorganism is disturbed (e.g. by a preservative factor), it will not multiply but remain in the lag-phase or even die before homeostasis is re-established. Individually, mild preservation technologies (e.g. refrigeration, mild heating, modified atmosphere packaging, biopreservation) that may be used with REPFEDs might not overcome most of the existing homeostasis mechanisms and stress reactions. Thus, they should be utilised in combination following the multi-target approach concept of hurdle technology.

(a) Stress adaptation counteracting preservation efficacy The formation of spores by some microorganisms is a well-known example of how microorganisms adapt to environmental stress. Non-spore-forming bacteria also react or adapt to environmental stress conditions in a way that helps them to overcome or minimise the impact of the stress. Stress adaptation involves processes on different cellular levels: membrane-transport systems, receptor functioning, signal transduction, control of gene expression, bioenergetics, reserve material status, etc. The membrane seems to be a very prominent site for stress adaptation (Russell *et al.*, 1995). Detailed studies at the physiological level on stress adaptation processes are in progess. The results of this research may give new leads that are essential for improving mild preservation technology.

In the paragraphs below, a brief account will be given of the mechanisms involved in adaptation to osmotic, pH and heat stress. When the a_w, pH or temperature that a microorganism is exposed to in a food is close to that experienced previously, the adaptation phase (lag phase) can be proportionally short, while if conditions are different the adaptation phase (lag phase) can be proportionally long. The doubling time is independent of previous growth conditions, and is influenced by the existing environmental conditions only. This is illustrated by data from a study by George, Defernez and Peck (unpublished results) presented in Table 9.3. Cells of *L. monocytogenes* strain F6861 were grown to late exponential phase in TSYGB prepared under different conditions, and growth curves were then constructed in TSYGB at pH 5.5, 0.5% lactic acid, 5% NaCl at 12°C. The duration of the lag phase and the doubling time were derived from the growth curves using the equation of Baranyi (Baranyi and Roberts, 1994). The doubling time varied from 7.6 to 9.7 hours and was not related to previous growth conditions, whilst the lag time varied from 15 to 71 hours and was influenced by previous growth conditions (Table 9.3). Cells grown at a higher pH, or in the absence of lactic acid, or at a lower NaCl concentration all had a longer lag phase than cells grown under the conditions used in the construction of the growth curves. A similar response was observed with cells of *L. monocytogenes* previously grown at 30°C. Pre-exposure to preservative factors such as heat, ethanol, metal ions, acid, oxidative compounds

Table 9.3 Effect of preconditioning of the inoculum on the lag time and doubling time of
L. monocytogenes when grown at pH 5.5, 0.5% (w/v) lactic acid, 5.0% NaCl and 12°C

pH	Lactic acid (% w/v)	Lag time (h) and doubling time (h) following preconditioning at indicated NaCl concentration (% w/v)		
		0.5	2.5	5.0
7.0	0	71/9.2[a]	52/8.5	54/9.2
7.0	0.5	41/9.0	–[b]	48/8.8
6.0	0	31/9.6	–	44/7.6
6.0	0.5	30/8.7	–	23/8.1
5.5	0	31/9.7	25/8.9	17/9.1
5.5	0.5	25/9.6	–	15/8.7

[a]Fitted lag time (h) and doubling time (h).
[b]Conditions not tested.

Cells of *L. monocytogenes* strain F6861 grown to late exponential phase in the indicated conditions of pH, lactic acid and NaCl concentration at 12°C, were used as an inoculum for growth curves at pH 5.5, 0.5% (w/v) lactic acid, 5.0% (w/v) NaCl and 12°C (George, Defernez and Peck, unpublished results).

or alkalinity, have all increased the number of microorganisms that survive subsequent exposure to severe stress of the same kind. Thus, stress adaptation depends on the history of the microorganism and the degree of change (stress) it experiences in its environment.

(b) Adaptation to osmotic stress Recent research has examined the physiological basis of osmohomeostasis with a number of pathogenic bacteria (Csonka, 1989; Booth *et al.*, 1994). Some microorganisms, including important food-poisoning ones like *L. monocytogenes* and *S. aureus* are able to grow at quite low a_w by responding at two levels: (i) activation of constitutive transport systems to accumulate certain osmoprotective compounds (compatible solutes) in the cell, and (ii) expression of osmotically induced genes. Transport mechanisms exist in all foodborne pathogens that allow the level of compatible solutes in the cytoplasm to be balanced to the degree of osmotic stress. Examples of compatible solutes are: potassium, betaine, glutamate, proline, carnitine and trehalose. Depending on the microorganism and the occurrence of other environmental stress factors, as little as 1 µM of the solute can have a protective effect against osmotic stress (Koo and Booth, 1994). This is a result of a very efficient scavenging transport system. In addition to physiological adaptation, genetic changes may occur at different levels. Some genes appear to respond specifically to signals related to the osmohomeostasis mechanism (such as changes in the activity of two-component regulatory systems and changes in DNA topology). Other genes seem to be regulated by the actual growth rate of a microorganism in such a way that they are increasingly expressed as the

growth rate is reduced. The latter is a non-specific countermeasure that can be effective against many different stress factors and one of the most powerful survival tools of microorganisms under stress. However, these physiological and genetic responses require a high energy input from the microorganism. In foods, microorganisms are likely to move between starvation and feast conditions. In some circumstances this may not give them the opportunity to accumulate sufficient energy reserves in order to survive multiple stress conditions.

In the case of *L. monocytogenes*, growth of the pathogen at low water activity is stimulated by the presence of proline and betaine in the growth medium (Beumer *et al.*, 1994). Both compounds occur naturally in many vegetable foods. In addition, carnitine, a common meat component, also acted as an osmoprotectant (Verheul *et al.*, 1995a). The most effective osmoprotectants, betaine and carnitine, when included at 1 mM, considerably reduced the lag time and increased the growth rate (Figure 9.2). Studies of the transport of carnitine by *L. monocytogenes* were undertaken using L-[^{14}C]carnitine. It was found that L-carnitine transport was ATP-dependent and that the compound was not metabolised further. Kinetic analysis of carnitine transport in glucose-energised cells revealed the presence of a

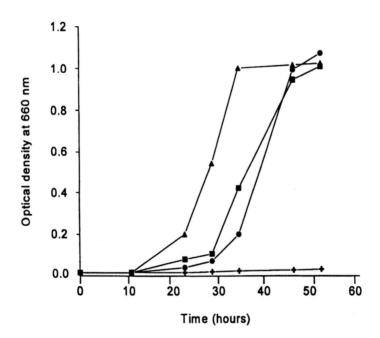

Figure 9.2 Culativation of *Listeria monocytogenes* at 37°C in (▲) a minimal medium (MM) (+) MM supplemented with 3% NaCl, (●) MM supplemented with 3% NaCl and 1 mM betaine (trimethylglycine) and (■) MM supplemented with 3% NaCl and 1 mM DL-carnitine(-hydroxy-*N*-trimethyl aminobutyrate). (Adapted from data in Beumer *et al.*, 1994.)

high-affinity uptake system, which enables the bacterium to effectively scavenge L-carnitine from the environment. Thus, the pathogen can utilise this osmoprotectant even when it is available in the food matrix in trace amounts only.

(c) Adaptation to pH stress   One of the most important homeostasis mechanisms is related to the intracellular pH of microorganisms. Tightly regulated as it is, it is essential for continued growth and viability by maintaining the functionality of key cell components. Food preservatives such as acetic, lactic, propionic and benzoic acids all affect the pH-homeostasis mechanism. Probably also carbon dioxide, used as an antimicrobial in modified atmosphere packaging, interferes with pH-homeostasis. Membrane associated transport systems for hydrogen ions are continuously operational to balance the internal pH to that of the environment. As the external pH is reduced, hydrogen ions are actively imported into the cytoplasm, and *vice versa*.

Adaptation to low pH stress is a phenomenon that has been frequently observed for a number of foodborne organisms. Leyer and Johnson (1992) showed that cells of *S. typhimurium,* adapted to acid by exposure to hydrochloric acid at pH 5.8 for one to two doublings, had increased resistance to inactivation by organic acids commonly present in cheese, including lactic, propionic, and acetic acids. Acid-adapted cells also showed enhanced survival over a period of two months in cheddar, Swiss, and mozzarella cheeses kept at 5°C as compared to non-adapted cells (Figure 9.3). From the same laboratory, acid adaptation of *E. coli* O157:H7 was reported to increase survival of the pathogen in acidic foods (Leyer *et al.*, 1995). Cells adapted to lactic acid (pH 5.0) survived better than non-adapted cells during a sausage fermentation, and showed enhanced survival in shredded dry salami (pH 5.0) and apple cider (pH 3.4).

Acid adaptation has been observed with *L. monocytogenes*. Kroll and Patchett (1992) reported that exposure of the pathogen to pH 5.0 significantly increased the survival of cells at low pH (3.0–3.5). Gahan *et al.* (1996) investigated the acid tolerance response of *L. monocytogenes* in a variety of acidic foods and demonstrated that acid-adapted cells exhibited enhanced survival in acidified milk products, including cottage cheese (pH 4.71), natural yoghurt (pH 3.9) and full-fat cheddar cheese (pH 5.16). Also, they showed improved survival in low-pH foods containing acids other than lactic acid, e.g. orange juice (pH 3.76) and salad dressing (pH 3.0). However, in unripened mozzarella cheese (pH 5.6), a commercial cottage cheese (pH 5.15) and low-fat cheddar cheese (pH 5.25), acid adaptation did not seem to enhance survival.

Acid adaptation is an important phenomenon, and to ensure the safe production of food, more detailed information is required. Although it may allow a microorganism to survive acid stress, it may not be sufficient if

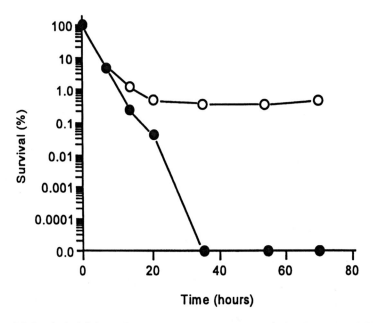

Figure 9.3 Survival of *Salmonella typhimurium* inoculated at 10^4 cfu/g in commercially produced mozzarella cheese (pH ± 5.3) during aerobic storage at 5°C. (●) Non-acid adapted cells (pre-grown at pH 7.6) (○) acid-adapted cells (pre-grown at pH 5.8). (Adapted from data in Leyer and Johnson, 1992.)

multiple hurdles are used. Of particular interest may be the additional hurdle of components that increase membrane permeability.

(d) Adaptation to heat stress Exposure to a non-lethal heat treatment triggers reactions in vegetative microorganisms that lead to an increase in the heat resistance of cells. With mildly preserved foods such as REPFEDs, heating is frequently slow because it often involves large portion sizes. The slow rising temperature when heat treatments are applied to REPFEDs may therefore induce heat shock in cells during the heating process. Linton *et al.* (1992) reported that the measured heat resistance (at 55°C) of *L. monocytogenes* Scott A cells previously exposed to 48°C for 10 min was significantly higher than that of non-exposed cells: 2.1-fold higher when recovered on TYSE ($D_{55°C}$ 18.7 and 8.9 min, respectively) and 1.4-fold higher when recovered on McBride Listeria agar ($D_{55°C}$ 9.6 and 6.7 min, respectively). A similar increase in the measured heat resistance of other important foodborne pathogens (e.g. *E. coli* O157:H7, *S. typhimurium, S. enteritidis*) has also been reported following non-lethal heat shock. The physiological processes involved in heat stress adaptation have been studied in some detail for many years. Mild heat treatments were shown to trigger the *de novo* synthesis of so-called heat shock proteins via the activation and

expression of certain heat shock related genes. The function and regulation of these heat shock proteins has been reviewed recently (Gross, 1996).

(e) Non-specific stress adaptation reactions   In many cases stress reactions have a non-specific effect, which means that exposing bacteria to certain sublethal stresses may significantly improve their response to other, apparently different types of stress that they are exposed to at a later stage. Examples of such aspecific or 'cross'-tolerance reactions are manifold.

Acid-adaptation (pH 5.4), for instance, has been found to increase, marginally, the resistance of *L. monocytogenes* Scott A towards the bacteriocin nisin (Okereke and Thompson, 1996). When challenged at pH 3.3 at 35°C, the acid-adapted cells (pre-grown at pH 5.4) were 150–7500-fold more resistant to acid stress at pH 3.3 than unadapted cells (pre-grown at pH 7.0). In the presence of 1.5 µg nisin/ml, 47% of the acid-adapted cells survived compared to 41% of the unadapted cells. With *S. typhimurium*, acid-adapted cells (pre-grown at pH 5.8) had an increased tolerance towards heat, salt, an activated lactoperoxidase system, and the surface-active agents crystal violet and polymyxin B (Leyer and Johnson, 1993).

Compatible solutes have also been implicated in conveying cross-tolerance. Accumulation of the compatible solute trehalose has been linked to increased thermo-tolerance of certain yeasts, as a result of a redistribution of water in the yeast cell by the trehalose (Hottiger *et al.*, 1994). The accumulation of betaine or carnitine by *L. monocytogenes* not only enables the organism to grow at reduced a_w levels or increased NaCl concentrations but also may increase the growth rate of the pathogen at refrigeration temperature (Figure 9.4; Verheul and Abee, unpublished). A possible link between osmotolerance and cryotolerance has been discussed before by Ko *et al.* (1994).

That cross-tolerance can result in a cell of considerably increased resistance is shown by studies on *E. coli* cells that were adapted to high or low pH levels and which subsequently exhibited a reduced sensitivity to normally lethal alkalinity, UV light and thermal stress (Goodson and Rowbury, 1991; Raja *et al.*, 1991; Rowbury *et al.*, 1996). In most cases the actual mechanisms underlying the occurrence of cross-tolerance remain to be elucidated at the cellular/molecular level. Further study of this is important, as an unexpected increase of tolerance or resistance in pathogens might invalidate apparently safe mild (hurdle technology) preservation protocols.

9.5.2 Booster compounds in foods

Part of understanding the potential of foodborne pathogens to grow and survive in foods requires knowledge of how the microorganisms interact with the food. Factors such as carbohydrates, proteins and minerals in the food may be essential for the growth and survival of pathogens. However,

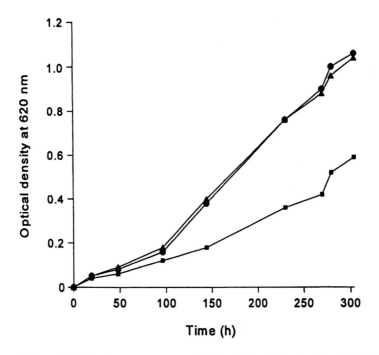

Figure 9.4 Growth of *Listeria monocytogenes* at 7°C in (■) a minimal medium (●) MM supplemented with 1 mM betaine and (▲) MM supplemented with 3% NaCl and 1 mM carnitine. (Data are unpublished from Verheul and Abee.)

non-pathogenic microorganisms that are often competitors of pathogens may also derive benefit from these components. Several types of food matrix components have been found recently to affect the growth and survival potential of *L. monocytogenes* and non-proteolytic *C. botulinum*. Knowledge about such 'booster' compounds in foods is important. When relying on mild preservation technologies, small variations from the unknown or unexpected may make the difference between growth or no growth of a pathogen. One group of booster compounds, the osmoprotectants, has already been mentioned.

(a) Essential amino acids   *L. monocytogenes* is relatively fastidious and requires a carbohydrate source, an iron source, vitamins and amino acids for growth. Additionally, the pathogen needs leucine and methionine. Since this pathogen does not exhibit any extracellular peptide or protein hydrolysis activity, it is generally accepted that it only can utilise the specific essential amino acids. However, it was recently demonstrated that *L. monocytogenes* is equipped with a constitutive, proton motive force-dependent carrier protein for di- and tri-peptides, which can provide the pathogen with the essential amino acids in a more complex form rather efficiently

(Amezaga *et al.*, 1995; Verheul *et al.*, 1995b). In fact, oligopeptides composed of up to eight amino acid residues have been shown to be a possible source of leucine and methionine. These amino acids are liberated by intracellular hydrolysis after uptake (Verheul *et al.*, 1997).

In many foods, essential amino acids and suitable small oligopeptides may not be available to pathogenic bacteria. Considering that mildly preserved foods are generally not sterile, other types of microorganisms may be present that produce the required proteases. Indeed, when *L. monocytogenes* was co-cultured with *Bacillus cereus* in milk, in which casein was present as a non-accessible source of nitrogen for *L. monocytogenes*, it was observed that growth of *B. cereus* sustained optimal growth of *L. monocytogenes*, indicating that the pathogen can benefit from the proteolytic activity of *B. cereus* towards casein for its source of amino acids (Gorris *et al.*, 1996).

(b) Lytic enzymes   Heating inactivates the germination system of spores of non-proteolytic *C. botulinum*, and foods may contain lytic enzymes that will germinate apparently heat-inactivated spores of non-proteolytic *C. botulinum*, thereby increasing measured spore heat resistance. For example, addition of hen egg-white lysozyme to the nutrient medium used to enumerate survivors after heating increased the measured heat resistance of spores of non-proteolytic *C. botulinum* (Peck *et al.*, 1992). In the absence of lysozyme, heating spores in phosphate buffer (pH 7.0) at 85°C showed a 5 decimal kill in less than 1 min. With 10 µg lysozyme/ml in the recovery medium, heating spores at 85°C for 5 min and for 120 min gave only a 2.6 and 4.1 decimal kill, respectively (Peck *et al.*, 1992). Hen egg-white lysozyme present in the recovery medium did not increase the measured heat resistance of the entire spore population, but only that of a sub-population of spores (0.1–20%) that possess coats that are naturally permeable to lysozyme (Peck *et al.*, 1993). Only with these spores can lysozyme diffuse from the recovery medium into heat-damaged spores and induce germination by hydrolysing peptidoglycan in the spore cortex (Gould, 1989). The coats of a majority of the spore population are impermeable to lysozyme, and hence lysozyme has no effect on the recovery of these spores. This mixture of lysozyme permeable and impermeable spores, each with a different measured thermal resistances, leads to biphasic thermal death curves when spores are enumerated on media containing lysozyme.

Hen egg-white lysozyme is relatively heat resistant (Lund and Peck, 1994) and when added to a model food system (meat slurry) prior to heating it increased the heat treatment required to prevent growth from 10^6 spores of non-proteolytic *C. botulinum* (Table 9.2). The heat treatments necessary without lysozyme to prevent subsequent growth at 25°C in 60 days were: 70°C for >2545 min, 75°C for 1793 min, 80°C for >363 min, 85°C for 36 min or 90°C for 10 min. In the presence of lysozyme, following heat treatments at 90°C for 20 min and 95°C for 15 min, growth was observed at 25°C after 20 and 32 days, respectively (Table 9.2).

Lytic enzyme activity has been reported in different plant tissues and is thought to result mainly from chitinases. These enzymes can be expressed constitutively or induced in response to wounding, microbial attack or exposure to chemicals. The influence of vegetable extracts on growth from heated spores of non-proteolytic *C. botulinum* was recently investigated (Stringer and Peck, 1996). It was found that the number of spores capable of colony formation decreased rapidly with increased periods of heat treatment at 85°C when subsequent incubation was on a nutrient medium (Figure 9.5). Heat treatment at 85°C for 2 min reduced the number of spores leading to colony formation by a factor of $>10^4$ when plated on a nutrient medium. However, heat treatment at 85°C for 10 min did not reduce the number of colonies formed by a factor of 10^4 when spores were subsequently incubated on recovery media containing extracts from turnip, rutabaga (swede), flat bean or cabbage juice (Figure 9.5). In these cases, biphasic thermal death curves comparable to those obtained in the presence of hen egg-white lysozyme were obtained. Extracts from zucchini (courgette), bean sprout juice did not have a protective effect. The mechanism

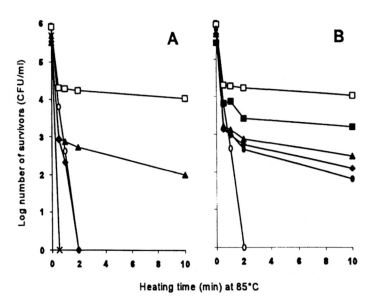

Figure 9.5 Viable counts of spores of non-proteolytic *Clostridium botulinum*, heated in peptone yeast extract broth at 85°C for up to 10 min and subsequently plated on (O) agar medium (containing peptone yeast extract, starch and glucose) supplemented with (□) hen egg-white lysozyme (10 µg/ml) or 50% of one of the following vegetable juices: panel A (▲) potato (♦) mungbean sprouts (●) carrot and (×) zucchini; panel B: (■) turnip (▲) rutabaga (♦) flat bean and (●) cabbage. (Adapted from data in Stringer and Peck, 1996.)

by which these vegetable juices increase measured thermal resistance may be similar to that of hen egg-white lysozyme.

Lytic enzymes other than hen egg-white lysozyme or those from vegetable extracts have also been reported to increase the measured heat resistance of spores of non-proteolytic *C. botulinum* (Peck *et al.*, 1993; Lund and Peck, 1994; Peck and Stringer, 1996; Stringer and Peck, 1996). These include: other type c lysozymes (e.g. from turkey egg, quail egg, duck egg, human milk), other enzymes (papain), fruit extracts and horse blood. Lysozyme-like activity has also been detected in many raw foods. In most cases the activity is higher than that required to increase the measured heat resistance of spores of non-proteolytic *C. botulinum* (Lund and Peck, 1994). Enzymes capable of permeating heat-damaged spores of non-proteolytic *C. botulinum* and initiating germination are also produced by other microorganisms, especially spore-formers. Spores of these organisms that have survived mild heating may lead to growth and, as such, facilitate growth from spores of non-proteolytic *C. botulinum*.

Suggestions have been made to use lysozyme and other lytic enzymes in foods as a preservative factor (Nielsen, 1991; Gould, 1992; Fuglsang *et al.*, 1995). In view of the potential protective effect of lytic enzymes towards non-proteolytic *C. botulinum*, it is strongly advised not to follow this suggestion for the production of REPFEDs.

9.6 Interactions between microorganisms

In general terms, REPFEDs are not sterile but contain a low level of microbial contamination. The organisms present are in most cases, spore-forming bacteria that have survived the heat treatment applied. With REPFEDs that are combined following heating, post-heating contaminants may also be present. Pathogens surviving heating or contaminating REPFEDs at packaging may interact in a number of different ways with other microorganisms present. Antagonistic or competitive interactions may be important. For example, microorganisms may compete for sources of carbon, nitrogen, minerals, vitamins, etc., while the production of acetic or lactic acid or other antimicrobial agents into the food matrix may be detrimental to the growth of competitive microorganisms. A bacteriocin-producing lactic acid bacterium that suppresses non-proteolytic *C. botulinum* or *L. monocytogenes* (Bennik *et al.*, 1997) may be of use with REPFEDs that are packed after heating.

Examples of metabiotic associations, where one organism makes conditions favourable for a second, have already been described. For example, the release of spore-coat diffusive germination enzymes from spore-formers helping heat-damaged spores of non-proteolytic *C. botulinum* to germinate, and the release of extracellular proteases that convert complex proteins in the food matrix to amino acids or oligopeptides required by *L. monocytogenes*. One metabiotic association that has been recently studied in some detail, is the interaction between non-proteolytic *C. botulinum* and psychrotrophic *Bacillus* species (Carlin and Peck, 1996b). The heat resistance of *Bacillus* spores is generally higher than that of spores of non-proteolytic *C. botulinum*, thus, *Bacillus* spores are likely to be present in mildly heated REPFEDs with a frequency that is at least equal to that of spores of *C. botulinum*. Using a culture broth, the study demonstrated that growth of, and toxin production by, non-proteolytic *C. botulinum* in aerobic cultures were promoted by the presence of psychrotrophic *Bacillus* strains. The metabiotic effect was attributed to oxygen consumption by *Bacillus* cells, which depleted the oxygen and reduced the redox potential to levels which permitted growth and toxin production by *C. botulinum*. That non-proteolytic *C. botulinum* grew and produced toxin following extensive growth of the *Bacillus* strains might not have been expected. The existence of this metabiotic interaction in REPFEDs remains to be ascertained, but all relevant conditions are probably met in that many foodborne *Bacillus* species are relatively psychrotrophic and can grow well on foods packed under an aerobic atmosphere or low-oxygen atmosphere. This study further underlines that the use of oxygen as a safety hurdle in REPFEDs may not be advisable.

9.7 Legislation

A number of relevant guidelines and codes of practice have been drawn up (MARF, 1974, 1988; ACMSF, 1992, 1995; ECFF, 1996; Betts, 1996) with respect to the safe production of REPFEDs, most are targeted at preventing growth and toxin production by non-proteolytic *C. botulinum*. In the UK, the Advisory Committee on the Microbiological Safety of Food (ACMSF) concluded that safety with respect to non-proteolytic *C. botulinum* could be ensured by one of the following (ACMSF, 1992, 1995):

a: storage at <3.3°C – this was the lowest reported growth temperature for non-proteolytic *C. botulinum* when the recommendations were made;
b: storage at ≤5°C and a shelf life of ≤10 days – this time/temperature combination was deemed sufficient to prevent toxin production if a small number of spores were present;
c: storage at 5–10°C and a shelf life of ≤5 days – this time/temperature combination was deemed sufficient to prevent toxin production if a small number of spores were present;
d: storage at chill temperature combined with one of the following processes to restrict pathogen survival or control pathogen growth:
   d.1: heat treatment of 90°C for 10 min or equivalent lethality, e.g. 70°C for 1675 min, 75°C for 464 min, 80°C for 129 min, 85°C for 36 min

(ACMSF, 1992); the European Chilled Food Federation recommended 80°C for 270 min, 85°C for 52 min (ECFF, 1996);

d.2: pH ≤5.0 throughout the food;

d.3: salt concentration ≥3.5% throughout the food;

d.4: a_w, ≤0.97 throughout the food;

d.5: combinations of heat treatment and other preservative factors which can be shown consistently to prevent growth and toxin production by *C. botulinum*.

It is recommended (ACMSF, 1992; ECFF, 1996) that the heat treatments or combination processes utilised reduce the number of viable spores of non-proteolytic *C. botulinum* by a factor of 10^6 (a 6 decimal process).

Although the recommendations from the ACMSF (1992, 1995) are an important step towards harmonising food manufacturing practice with governmental concern, several refinements may be merited. For example, the combinations of chill temperature and heat treatment (as in d.1) were included in the ACMSF recommendations on the basis of data on the recovery of spores at optimum and not a chilled temperature. Also, while the recommendations in d.2, d.3 and d.4 are of value, most REPFEDs have a pH > 5, a high water activity, and a salt concentration < 3.5%. In effect, for most REPFEDs, the principal factors controlling microbiological safety and quality are likely to be the heat process, the storage temperature (the set value as well as the deviation from this) and the maximum shelf life.

The minimum temperature at which growth and toxin production occurs is often reported as 3.3°C, and this temperature was therefore also included in the ACMSF recommendations. However, in a recent publication growth and toxin production are described at 3.0–3.2°C. Growth/toxin production have now been reported at 3.0°C in 49 days, 3.1°C in 42 days, 3.2°C in 35 days (Graham *et al.*, 1997), 3.3°C in 31 days (Schmidt *et al.*, 1961), and 4.0°C in 20 days (Solomon *et al.*, 1982). Between 2.1 and 2.5°C, growth has not been observed for up to 90 days (Schmidt *et al.*, 1961, Eklund *et al.*, 1967a, b; Solomon *et al.*, 1982; Graham *et al.*, 1997). Growth tests do not appear to have been carried out at 2.6–2.9°C, and for the time being it might be prudent to assume that growth and toxin production occur in this range until it is demonstrated otherwise. Maintaining a temperature of ≤2.5°C might be possible in some circumstances (e.g. institutions, catering establishments), but there is doubt as to whether temperatures in this range can always be maintained throughout the distribution chain, particularly in products intended for domestic use. Current regulations require chilled foods to be held at 0–3°C in Spain, 0–4°C in France, ≤7°C in Belgium and ≤8°C in the UK (Schellekens, 1996).

In France, regulations for the production of *sous vide* foods seem to be targeted at ensuring the overall microbiological (and sensory) quality of the product and not at achieving specific reductions in numbers of spores of non-proteolytic *C. botulinum* (MARF, 1974, 1988). These recommendations combine heating with refrigerated storage. For products with a shelf life of up to 21 days, a heat treatment equivalent to 70°C for 100 min is recommended. To take account of possible risks, the manufacturer is required to check compliance with microbiological criteria after storage at 4°C for 14 days followed by 8°C for 7 days. This recommendation leaves only a small margin of safety with regard to an overall 6 decimal process for non-proteolytic *C. botulinum* in foods in the absence of lysozyme (Table 9.2). For products with a shelf life of up to 42 days, the recommended heat treatment is pasteurisation equivalent to heating at 70°C for 1000 min. In this case, compliance is checked after storage at 3°C for 28 days followed by 8°C for 14 days. Although this treatment appears to provide a slightly greater margin of safety than that described above, the margin of safety is still small if foods without lysozyme are maintained at 8°C rather than a lower temperature (Table 9.2). Both combination processes might be even less effective if the food contains lysozyme or a similar enzyme.

9.8 Conclusions

Consumers' confidence in the safety of a food product is an important factor for successful sales of any food. Product quality is becoming more and more important but, in the longer term, it would be unwise to value quality over safety. After all, food manufacturers do have a large range of mild preservative factors which can be used to assure the safety of minimally processed, high-quality foods. With REPFEDs, microbial safety and stability are determined by the proper choice of a heat treatment, storage temperature and shelf life for specific food preparations. When necessary, additional hurdles for microbial growth can be built into food preparations to maintain an appropriate safety margin and to extend the shelf life further. Hurdle technology is a valuable tool in minimal processing. However, its use relies heavily on detailed knowledge of the effects of preservative factors individually and in combination as well as on factors or processes that interfere with these effects. Much research in this field has been done, but further research is required. Food legislation authorities do not yet have all the information necessary to define conditions that could be applied to ensure the continued safe production of REPFEDs with respect to non-proteolytic *C. botulinum* and other relevant psychrotrophic pathogens. It is important that this research continues, and a key feature is likely to be close communication and collaboration between food manufacturers and legislative bodies.

Acknowledgements

This chapter was compiled in the framework of the shared cost project 'Improvement of the safety and quality of refrigerated ready-to-eat foods using novel mild preservation techniques' (EC Contract AIR1-CT92-0125) and the concerted action 'Harmonisation of safety criteria for minimally processed foods' (EC Contract FAIR-CT96-1020). Annette Verheul and Tjakko Abee are greatfully acknowledged for providing their unpublished data. Author Peck gratefully acknowledges the Competitive Strategy Grant of the BBSRC for funding his work.

References

Abee, T., Krockel, L. and Hill, C. (1995) Bacteriocins: modes of action and potentials in food preservation and control of food poisoning. *Int. J. Food Microbiol.* **28**, 169–185.

ACMSF, Advisory Committee on the Microbiological Safety of Food (1992) *Report on Vacuum Packaging and Associated Processes*, Her Majesty's Stationery Office, London, UK.

Advisory Committee on the Microbiological Safety of Food (1995) *Annual Report 1995*, Her Majesty's Stationery Office, London, UK.

Ahamad, N. and Marth, E.H. (1989) Behavior of *Listeria monocytogenes* at 7°C, 13°C, 21°C and 35°C in tryptose broth acidified with acetic, citric, or lactic acid. *J. Food Prot.* **52**, 688–695.

Ahamad, N. and Marth, E.H. (1990) Acid-injury of *Listeria monocytogenes*. *J. Food Prot.* **53**, 26–29.

Alzamora, S.M., Tapia, M.S., Argaiz, A. and Welti, J. (1993) Application of combined methods technology in minimally processed fruits. *Food Res. Int.* **26**, 125–130.

Amezaga, M.-R., Davidson, I., McLaggan, D., Verheul, A., Abee, T. and Booth, I.R. (1995) The role of peptide metabolism in the growth of *Listeria monocytogenes* ATCC 23074 at high osmolarity. *Microbiol.* **141**, 41–49.

Baker, D.A. and Genigeorgis, C. (1992) Predictive modeling. in *Clostridium botulinum: Ecology and Control in Foods* (eds A.H.W. Hauschild and K.L. Dodds), Marcel Dekker, New York, pp. 343–406.

Baker, D.A., Genigeorgis, C., Glover, J. and Razavilar, V. (1990) Growth and toxigenesis of *Clostridium botulinum* type E in fishes packed under modified atmospheres. *Int. J. Food Microbiol.* **10**, 269–290.

Baranyi, J. and Roberts, T.A. (1994) A dynamic approach to predicting bacterial growth in food. *Int. J. Food Microbiol.* **23**, 277–294.

Bennik, M.H.J., Smid, E.J., Rombouts, F.M. and Gorris, L.G.M. (1995) Growth of psychrotrophic foodborne pathogens in a solid surface model system under the influence of carbon dioxide and oxygen. *Food Microbiol.* **12**, 509–519.

Bennik, M.H.J., Smid, E.J. and Gorris, L.G.M. (1997) Vegetable-associated *Pediococcus parvulus* produces pediocin PA-1. *Appl. Environ. Microbiol.* **63**, 2074–2076.

Betts, G.D. (ed.) (1996) *Code of Practice for the Manufacture of Vacuum and Modified Atmosphere Packed Chilled Foods with Particular Regard to the Risk of Botulism*, Campden and Chorleywood Food Research Association, Chipping Campden, UK.

Beumer, R.R., te Giffel, M.C., Cox, L.J., Rombouts, F.M. and Abee, T. (1994) Effect of exogenous proline, betaine, and carnitine on growth of *Listeria monocytogenes* in a minimal medium. *Appl. Environ. Microbiol.* **60**, 1359–1363.

Booth, I.R., Pourkomailian, B., McLaggan, D. and Koo, S.P. (1994) Mechanisms controlling compatible solute accumulation: a consideration of the genetics and physiology of bacterial osmoregulation. *J. Food Engin.* **22**, 381–397.

Carlin, F. and Peck, M.W. (1996a) Growth of, and toxin production by, non-proteolytic

Clostridium botulinum in cooked vegetables at refrigeration temperatures. *Appl. Environ. Microbiol.* **62**, 3069–3072.

Carlin, F. and Peck, M.W. (1996b) Metabiotic association between non-proteolytic *Clostridium botulinum* type B and foodborne *Bacillus* species. *Sciences des Aliments* **16**, 545–551.

Csonka, L.N. (1989) Physiological and genetic responses of bacteria to osmotic stress. *Microbiol. Rev.* **53**, 121–147.

ECFF, European Chilled Food Federation (1996) *Guidelines for the Hygienic Manufacture of Chilled Foods*, The European Chilled Food Federation, Helsinki, Finland.

Eklund, M.W., Poysky, F.T. and Weiler, D.I. (1967b) Characteristics of *Clostridium botulinum* type F isolated from the Pacific coast of the United States. *Appl. Microbiol.* **15**, 1316–1323.

Eklund, M.W., Weiler, D.I. and Poysky, F.T. (1967a) Outgrowth and toxin production of *non-proteolytic* type B *Clostridium botulinum* at 3.3 to 5.6°C. *J. Bacteriol.* **93**, 1461–1462.

El Shenawy, M. and Marth, E.H. (1991). Organic acids enhance the antilisterial activity of potassium sorbate. *J. Food Prot.* **54**, 593–597.

Farber, J.M., Cai, Y. and Ross, W.H. (1996) Predictive modeling of the growth of *Listeria monocytogenes* in CO_2 environments. *Int. J. Food Microbiol.* **32**, 133–144.

Fernandez, P.S. and Peck, M.W. (1997) A predictive model that describes the effect of prolonged heating at 70–80°C and incubation at refrigeration temperatures on growth and toxigenesis by non-proteolytic *Clostridium botulinum*. *J. Food Prot.* **60**, 1064–1071.

Fernandez, P.S., George, S.M., Sills, C.C. & Peck, M.W. (1997) Predictive model of the effect of CO_2, pH, temperature and NaCl on the growth of *Listeria monocytogenes*. *Int. J. Food Microbiol.* **37**, 37–45.

Fuglsang, C.C., Johansen, C., Christgau, S. and Adler-Nissen, J. (1995) Antimicrobial enzymes: applications and future potential in the food industry. *Trends Food Sci. Technol.* **6**, 390–396.

Gahan, C.G.M., O'Driscoll, B. and Hill, C. (1996) Acid adaptation of *Listeria monocytogenes* can enhance survival in acidic foods and during milk fermentation. *Appl. Environ. Microbiol.* **62**, 3128–3132.

Garcia, G., Genigeorgis, C. and Lindroth, S. (1987) Risk of growth and toxin production by *Clostridium botulinum* non-proteolytic types B, E, and F in salmon fillets stored under modified atmospheres at low and abused temperatures. *J. Food Protec.* **50**, 330–336.

García de Fernando, G.D., Nychas, G.J.E., Peck, M.W. and Ordóñez, J.A. (1995) Growth/survival of psychrotrophic pathogens on meat packaged under modified atmospheres. *Int. J. Food Microbiol.* **28**, 221–231.

George, S.M., Richardson, L.C.C. and Peck, M.W. (1996) Predictive models of the effect of temperature, pH and acetic and lactic acids on the growth of *Listeria monocytogenes*. *Int. J. Food Microbiol.* **32**, 73–90.

Gibbs, P.A. (1996) The microbiological safety and quality of foods processed by the *sous vide* system as a method of commercial catering, in *Proc. Second European Symposium on* Sous vide, 10–12 April 1996, Katholieke Universiteit Leuven, Leuven, pp.199–208.

Goodson, M. and Rowbury, R.J. (1991) RecA-independent resistance to irradiation with UV light in acid-habituated *Escherichia coli*. *J. Appl. Bacteriol* **70**, 177–180.

Gorris, L.G.M., Abee, T. and Peck, M.W. (1996) Food matrix influences on pathogen growth. *Inter. J. Food Technol., Marketing, Packag. Anal. (ZFL)* **47** (1/2), 62–65.

Gould, G.W. (1988) Interference with homeostasis – food, in *Homeostatic Mechanisms in Micro-organisms* (eds R. Whittenbury, G.W. Gould, J.G. Banks and R.G. Board), Bath University Press, Bath, pp. 220–228.

Gould, G.W. (1989) Heat-induced injury and inactivation, in *Mechanisms of Action of Food Preservation Procedures* (ed. G.W. Gould), Elsevier, London, pp. 11–42.

Gould, G.W. (1992) Ecosystem approach to food preservation. *J. Appl. Bacteriol.* **73**, 58s–68s.

Gould, G.W. (1995). Homeostatic mechanisms during food preservation by combined methods, in *Food Preservation by Moisture Control, Fundamentals and Applications* (eds G.V. Barbosa-Cánovas and J. Welti-Chanes), Technomic Publishing Co., Lancaster, pp. 397–410.

Graham, A.F., Mason, D.R. and Peck M.W. (1996a) A predictive model of the effect of temperature, pH and sodium chloride on growth from spores of non-proteolytic *Clostridium botulinum*. *Int. J. Food Microbiol.* **31**, 69–85.

Graham, A.F., Mason, D.R. and Peck M.W. (1996b) Inhibitory effect of combinations of heat treatment, pH and sodium chloride on growth from spores of non-proteolytic *Clostridium botulinum* at refrigeration temperatures. *Appl. Environ. Microbiol.* **62**, 2664–2668.

Graham, A.F., Mason, D.R., Maxwell, F.J. and Peck, M.W. (1997) Effect of pH and NaCl on growth from spores of non-proteolytic *Clostridium botulinum* at chill temperatures. *Lett. Appl. Microbiol.* **24**, 95–100.

Gross, C.A. (1996) Function and regulation of heat shock proteins, in *Escherichia coli and Salmonella (Cellular and Molecular Biology)*, 2nd edn, Vol. 1 (eds F.C. Neidhardt *et al.*), ASM Press, Washington DC, USA, pp 1382–1399.

Hefnawy, Y.A. and Marth, E.H. (1993) Survival and growth of *Listeria monocytogenes* in broth supplemented with sodium chloride and held at 4 and 13°C. *Lebens. Wissen. Technol.* **26**, 388–392.

Hottiger, T., De Virgilio, C., Hall, M.N., Boller, T. and Wiemken, A. (1994) The role of trehalose synthesis for the acquisition of thermotolerance in yeast: II. Physiological concentrations of trehalose increase the thermal stability of proteins *in vitro*. *Eur. J. Biochem.* **219**, 187–193.

Hudson, J.A. (1993) Effect of pre-incubation temperature on the lag time of *Aeromonas hydrophila*. *Lett. Appl. Microbiol.* **16**, 274–276.

Hudson, J.A., Mott, S.J. and Penney, N. (1994) Growth of *Listeria monocytogenes*, *Aeromonas hydrophila*, and *Yersinia enterocolitica* on vacuum and saturated carbon dioxide-controlled atmosphere packaged sliced roast beef. *J. Food Prot.* **57**, 204–208.

Knøchel, S. and Gould, G. (1995) Preservation microbiology and safety: Quo vadis? *Trends Food Sci. Technol.* **6**, 127–131.

Ko, R., Tombras Smith, L. and Smith, G.M. (1994) Glycine betaine confers osmotolerance and cryotolerance on *Listeria monocytogenes*. *J. Bacteriol.* **176**, 426–431.

Koo, S.P. and Booth, I.R. (1994) Quantitative analysis of growth stimulation by glycine betaine in *Salmonella typhimurium*. *Microbiol.* **140**, 617–621.

Kroll, R.G. and Patchett, R.A. (1992) Induced acid tolerance in *Listeria monocytogenes*. *Lett. Appl. Microbiol.* **14**, 224–227.

Leistner, L. (1992) Food preservation by combined methods. *Food Res. Int.* **25**, 151–158.

Leistner, L. (1995) Principles and applications of hurdle technology, in *New Methods of Food Preservation* (ed. G.W. Gould), Blackie Academic & Professional, London, pp. 1–21.

Leistner, L. and Karan-Djurdjić, S. (1970) Beeinflussung der Stabilität von Fleischkonserven durch Steuerung der Wasseraktivität. *Fleischwirtschaft* **50**, 1547–1549.

Leistner, L. and Gorris, L.G.M. (1995) Food preservation by hurdle technology. *Trends Food Sci. Technol.* **6**, 41–46.

Leyer, G.J and Johnson, E.A. (1992) Acid adaptation promotes survival of *Salmonella* spp. in cheese. *Appl. Environ. Microbiol.* **58**, 2075–2080.

Leyer, G.J. and Johnson, E.A. (1993) Acid adaptation induces cross-protection against environmental stresses in *Salmonella typhimurium*. *Appl. Environ. Microbiol.* **59**, 1842–1847.

Leyer, G.J. Wang, L.L. and Johnson, E.A. (1995) Acid adaptation of *Escherichia coli* O157:H7 increases survival in acidic foods. *Appl. Environ. Microbiol.* **61**, 3752–3755.

Linton, R.H., Webster, J.B., Pierson, M.D., Bishop, J.R. and Hackney, C.R. (1992) The effect of sublethal heat shock and growth atmosphere on the heat resistance of *Listeria monocytogenes* Scott A. *J. Food Prot.* **55**, 84–87.

Lund, B.M. and Peck, M.W. (1994) Heat-resistance and recovery of non-proteolytic *Clostridium botulinum* in relation to refrigerated, processed foods with an extended shelf life. *J. Appl. Bacteriol.* **76**, 115s–128s.

MARF, Ministère de l'Agriculture de la Republique Française (1974) Regulations for the hygienic conditions concerning preparation, preservation, distribution and sale of *ready-to-eat* meals. *Journal Official de la Republique Française*, 26 June 1974.

MARF, Ministère de l'Agriculture de la Republique Française (1988) Prolongation of the life span of pre-cooked food, modification of procedures enabling authorisation to be obtained. Memorandum DGAL/SVHA/N88/no. 8106, 31 May 1988, Paris.

Meng, J. and Genigeorgis, C.A. (1993) Modeling lag phase of non-proteolytic *Clostridium botulinum* toxigenesis in cooked turkey and chicken breast as affected by temperature, sodium lactate, sodium chloride and spore inoculum. *Int. J. Food Microbiol.* **19**, 109–122.

Meng, J. and Genigeorgis, C.A. (1994) Delaying toxigenesis of *Clostridium botulinum* by sodium lactate in 'sous-vide' products. *Lett. Appl. Microbiol.* **19**, 20–23.

Nielsen, H.K. (1991) Novel bacteriolytic enzymes and cyclodextrin glycosyl transferase for the food industry. *Food Technol.* **45** (1), 102–104.

Okereke, A. and Thompson, S.S. (1996) Induced acid-tolerance response confers limited nisin resistance on *Listeria monocytogenes* Scott A. *J. Food Prot.* **59**, 1003–1006.

Ollinger-Snyder, P., El-Gazzar, F., Matthews, M.E., Marth, E.H. and Unklasbay, N. (1995) Thermal destruction of *Listeria monocytogenes* in ground pork prepared with and without soy hulls. *J. Food Prot.* **58**, 573–576.

Peck, M.W. (1997) *Clostridium botulinum* and the safety of refrigerated processed foods of extended durability. *Trends Food Sci. Technol.* **8**, 186–192.

Peck, M.W. and Fernandez, P.S. (1995) Effect of lysozyme concentration, heating at 90°C, and then incubation at chilled temperatures on growth from spores of non-proteolytic *Clostridium botulinum*. *Lett. Appl. Microbiol.* **21**, 50–54.

Peck, M.W. and Stringer, S.C. (1996) *Clostridium botulinum*: mild preservation techniques. in *Proc. Second European Symposium on Sous vide*, 10–12 April 1996, Katholieke Universiteit Leuven, Leuven, pp. 181–197.

Peck, M.W., Fairbairn, D.A. and Lund, B.M. (1992) The effect of recovery medium on the estimated heat-inactivation of spores of non-proteolytic *Clostridium botulinum*. *Lett. Appl. Microbiol.* **15**, 146–151.

Peck, M.W., Fairbairn, D.A. and Lund, B.M. (1993) Heat-resistance of spores of non-proteolytic *Clostridium botulinum* estimated on medium containing lysozyme. *Lett. Appl. Microbiol.* **16**, 126–131.

Peck, M.W., Lund B.M., Fairbairn, D.A., Kaspersson, A.S. and Undeland, P.C. (1995) Effect of heat treatment on survival of, and growth from spores of non-proteolytic *Clostridium botulinum* at refrigeration temperatures. *Appl. Environ. Microbiol.* **61**, 1780–1785.

Post, L.S., Lee, D.A., Solberg, M., Furgang, D., Specchio, J. and Graham, C. (1985) Development of botulinal toxin and sensory deterioration during storage of vacuum and modified atmosphere packaged fish fillets. *J. Food Sci.* **50**, 990–996.

Raja, N., Goodson, M., Smith, D.G. and Rowbury, R.J. (1991) Decreased DNA damage by acid and increased repair of acid-damaged DNA in acid-habituated *Escherichia coli*. *J. Appl. Bacteriol.* **70**, 507–511.

Rowbury, R.J., Lazim, Z., and Goodson, M. (1996) Regulatory aspects of alkali tolerance induction in *Escherichia coli*. *Lett. Appl. Microbiol.* **22**, 429–432.

Russell, N.J., Evans, R.I., ter Steeg, P.F, Hellemons, J., Verheul, A. and Abee, T. (1995) Membranes as a target for stress adaptation. *Int. J. Food Microbiol.* **28**, 255–261.

Ryser, E.T. and Marth, E.H. (1991) *Listeria, Listeriosis, and Food Safety*, Marcel Dekker, New York.

Schellekens, M. (1996) New research issues in sous-vide cooking. *Trends Food Sci. Technol.* **7**, 256–262.

Schmidt, C.F., Lechowich, R.V. and Folinazzo, J.F. (1961) Growth and toxin production by type E *Clostridium botulinum* below 40°F. *J. Food Sci.* **26**, 626–630.

Sheridan, J.J., Doherty, A., Allen, P., McDowell, D.A., Blair, I.S. and Harrington, D. (1995) Investigations on the growth of *Listeria monocytogenes* on lamb packaged under modified atmospheres. *Food Microbiol.* **12**, 259–266.

Snyder, O.P. (1996) Redox potential in deli foods: botulism risk. *Dairy Food Environ. Sanitation* **16**, 546–548.

Solomon, H.M., Kautter, D.A. and Lynt, R.K. (1982) Effect of low temperature on growth of non-proteolytic *Clostridium botulinum* types B and F and proteolytic type G in crabmeat and broth. *J. Food Prot.* **45**, 516–518.

Sorrells, K.M., Enigl, D.C. and Hatfield, J.R. (1989) Effect of pH, acidulant, time and temperature on the growth and survival of *Listeria monocytogenes*. *J. Food Prot.* **52**, 571–573.

Stringer, S.C. and Peck, M.W. (1996) Vegetable juice aids the recovery of heated spores of non-proteolytic *Clostridium botulinum*. *Lett. Appl. Microbiol.* **23**, 407–411.

Tapia de Daza, M. S., Argaiz, A., López-Malo, A. and Díaz, R.V. (1995) Microbial stability assessment in high and intermediate moisture foods: special emphasis on fruit products, in *Food Preservation by Moisture Control, Fundamentals and Applications* (eds G.V. Barbosa-Cánovas and J. Welti-Chanes), Technomic Publishing Co., Lancaster, pp. 575–601.

Verheul, A., Beumer, R.R., Rombouts, F.M. and Abee, T. (1995a) An ATP-dependent carnitine transporter in *Listeria monocytogenes* Scott A is involved in osmoprotection. *J. Bacteriol.* **177**, 3205–3212.

Verheul, A., Hagting, A., Amezaga, M.-R., Booth, I.R. Rombouts, F.M. and Abee, T. (1995b) A di- and tripeptide transport system can supply *Listeria monocytogenes* Scott A with amino acids essential for growth. *Appl. Environ. Microbiol.* **61**, 226–233.

Verheul, A., Rombouts, F.M. and Abee, T. (1998) Utilization of oligopeptides by *Listeria monocytogenes* Scott A. *Appl. Environ. Microbiol.* **64**, 1059–1065.

Whiting, R.C. and Call, J.E. (1993) Time of growth model for proteolytic *Clostridium botulinum.* *Food Microbiol.* **10**, 295–301.

Whiting, R.C. and Oriente, J.C. (1997) Time to turbidity model for non proteolytic type B *Clostridium botulinum. Int. J. Food Microbiol.* **36**, 49–60.

Young, K.M. and Foegeding, P.M. (1993) Acetic, lactic and citric acids and pH inhibition of *Listeria monocytogenes* Scott A and the effects on intracellular pH. *J. Appl. Bacteriol.* **74**, 515–520.

10 Hazards associated with non-proteolytic *Clostridium botulinum* and other spore-formers in extended-life refrigerated foods

VIJAY K. JUNEJA

Note: Mention of brand or firm names does not constitute an endorsement by the US Department of Agriculture over others of a similar nature not mentioned.

10.1 Introduction

There is currently a trend in the marketplace towards the desire for fresh-tasting, high-quality, low salt, preservative free, convenience meals which require minimal preparation time. This demand by consumers has resulted in increased production of minimally processed, ready-to-eat, extended shelf life refrigerated foods in North America and Europe. Refrigerated foods such as fresh and cured meat and produce have been produced and successfully marketed for decades. However, the products of concern in regard to hazards associated with spore-formers are those designed for refrigerated storage that are marketed with extended shelf life expectations of 3–6 weeks (Meng and Genigeorgis, 1994). Food products of this type have been referred to as 'New Generation Refrigerated Food' (NFPA, 1988). According to Mossel and Thomas (1988), such foods are known as 'Refrigerated Processed Foods of Extended Durability (REPFED)'. The National Advisory Committee on Microbiological Criteria for Foods (NACMCF, 1990) defined these products as 'Refrigerated foods containing cooked, uncured meat or poultry products that are packaged for extended refrigerated shelf life and that are ready-to-eat or prepared with little or no additional heat treatment'. The REPFED contain little or no preservatives, receive minimal thermal processing, and may be packaged under vacuum or modified atmosphere to extend shelf life. According to NACMCF (1990), these products are characterized as: (1) pre-cooked, packaged products intended to receive little or no additional heat treatment prior to consumption, (2) extended refrigerated shelf life, (3) not typically prepared using a conventional preservation system, e.g. low water activity or acid pH, etc., and (4) marketed under vacuum, modified atmosphere packaging (MAP), and/or in sealed containers. These REPFED include *sous vide* products (Livingston, 1985) and also other products with pasteurization and

preservation combinations that ensure extended shelf life under refrigerated storage (Brown and Gould, 1992). Examples of REPFED include
canned, pasteurized crab meat; canned, pasteurized ham; and vacuum-
packed, hot-smoked fish products (Lund and Notermans, 1993). Other
examples may include sauces, soups, entrées, salads, including pasta,
seafood and meat salads, fresh pasta and complete meals (Conner *et al.*,
1989). Commercially available complete meals include ravioli, moussaka,
tortellini, cannelloni with meat, cannelloni with vegetables, lasagna, lasagna
bolognese, tagliatelle with pork, tagliatelle with chicken, tagliatelle with
mushrooms, and chilli con carne (Notermans *et al.*, 1990).

Rosset and Poumeyrol (1986) indicated that these foods are commonly
prepared by one of three systems: (1) foods that are packaged then cooked,
e.g. *sous vide* foods; (2) foods that are cooked, then packaged; (3) foods that
are cooked, then packaged, then heated again. Based on pre-processing and
post-processing microbial risks, NACMCF (1990) divided nine process
types into three categories (Table 10.1).

This chapter will deal with the potential threat of foodborne illness
through the consumption of REPFED contaminated with spore-forming
pathogens because of their ability to survive the heat treatment given to
these products and their subsequent germination, outgrowth and multiplication during cooling, storage and distribution of these products. The psychrotrophic pathogen of paramount concern in these products,
non-proteolytic *Clostridium botulinum* types B and E, will be discussed in
depth. Other spore-formers, such as *Clostridium perfringens* and *Bacillus
cereus*, though have relatively high heat resistance, are generally not considered to pose a threat in adequately processed and stored REPFED, will
be discussed briefly.

10.2 Justification for concern

The safety and preservation of these refrigerated foods are being ques-
tioned and therefore warrant a critical evaluation of the technology. The
spore-formers, particularly non-proteolytic *C. botulinum*, pose a potential
public health hazard in REPFED due to the methods of preparation, distribution and storage of these products. Unlike proteolytic *C. botulinum*
strains which digest proteins in foods and produce foul odors which may
warn consumers of a potential botulism hazard, non-proteolytic *C. botulinum* strains may not sufficiently alter the odor and appearance to warn
consumers of possible danger. The concerns regarding REPFED are justi-
fied for a variety of reasons: The aim of food processors in such foods is to
destroy the vegetative cells of spoilage and pathogenic bacteria or to reduce
the microbial load. The thermal process is not designed to destroy bacterial
spores or to result in commercial sterility; thus, the spores survive and

Table 10.1 Chilled food processes based on pre- and post-processing microbial risks

Category	Process type	Process	Examples of food
Products 'Assembled and Cooked'	1	Raw ingredients – Pre-cook (optional) – Formulate – Vacuum-package – Pasteurize – Chill – Distribute	'*Sous vide*'
	2	Raw ingredients – Formulate – Vacuum-package – Cook – Chill – Distribute	Rolls and roasts
	3	Raw ingredients – Formulate – Cook – Pump – Package hot – Chill – Distribute	Stews, sauces, soups, fruit pie fillings, pudding
Products 'Cooked and Assembled'	4	Raw ingredients – Cook-chill – Package – Distribute	Roast or fried chicken, other roasts, uncured sausages
	5	Raw ingredients – Formulate – Cook – Chill – Slice or dice – Package – Distribute	Uncured luncheon meat, Diced meat
	6	Raw ingredients – Cook – Chill – Assemble – Package – Distribute	Meat and pasta, meat and sauces, dinners, sandwiches, pizza
	7	Raw ingredients – Formulate – Cook (optional) – Fill into dough – Cook (optional) – Chill – Package – Distribute	Meat pies, quiches, patties, pâtés
Products 'Assembled with Cooked and Raw Ingredients'	8	Raw ingredients – Chill – Formulate – Recook and chill (optional) – Package – Distribute	Chef salad, chicken salad, sandwiches, pizza with raw ingredients
	9	Raw ingredients – Formulate – Cook – Fill while hot – Seal Hold (optional) – Chill – Distribute	Uncured jellied meats, gelled fruits

Adapted from NACMCF (1990).

persist in the final product. This increasing trend towards the use of low heat treatments in an attempt to retain the product quality provides an ideal substrate for germination and growth of non-proteolytic *C. botulinum* spores. More significantly though, since the spoilage microflora which play a significant role causing deterioration and spoilage are inactivated, the foods may become toxic while remaining organoleptically acceptable.

These products are generally formulated with little or no preservatives in view of the consumer demands for not only fresh or home-made appeal but also healthier foods, e.g. lower salt levels in foods. These days, it is not uncommon to find products such as ham with 2% or lower levels of salt whereas these products were traditionally made with levels as high as 3% or even 5% salt. Attempts to lower salt levels may increase botulinal risks. Some of these products would require refrigeration at temperatures as low as 3.3°C or below from production to consumption to prevent spoilage and assure microbiological safety with regard to prevention of non-proteolytic *C. botulinum* outgrowth and toxin production. Thus, increased and, in some cases, excessive reliance on refrigeration as a sole safety factor against growth of non-proteolytic *C. botulinum* involves substantial risk, especially as there is every likelihood that these products will be maintained at temperatures higher than 3.3°C. Sufficient evidence exists to document that temperature abuse is a common occurrence at both the retail and consumer levels. According to The National Food Processors Association (NFPA, 1988), manufacturers should assume that temperature abuse will occur at some point during the distribution of a refrigerated food product. While it is relatively easy to control temperature in the institutional foodservice settings (e.g. restaurants, etc.), temperature abuse is likely to occur when shipped to remote locations, marketed by retailers or handled by consumers. Surveys of retail food stores and consumer refrigeration units have revealed that holding temperatures of >10°C are common (Daniels, 1991; Bryan *et al.*, 1978; Wyatt and Guy, 1980; Hutton *et al.*, 1991; Anon., 1989; van Grade and Woodburn,, 1987). Thus, it is unrealistic to rely on refrigeration for the safety of REPFEDs. Furthermore, since many of these products are marketed in packages that consumers view as traditionally containing shelf-stable foods, there is a high risk of consumer temperature abuse and mishandling.

An additional risk factor associated with these products comes from the fact that they are often packaged with little or no headspace oxygen to extend the shelf life from a nutritional and sensory point of view, e.g. vacuum packaging or MAP. While vacuum packaging inhibits normal spoilage microorganisms, it provides a favorable environment for anaerobic pathogens such as *C. botulinum*. Mild heat treatment in combination with vacuum packaging may, in fact, select for *C. botulinum* and increase the potential for botulism. Similarly, a major food safety concern for MAP products is the possible growth of anaerobic psychrotrophic pathogens which can increase to hazardous levels while the food remains edible. The concern regarding MAP is based on the stimulatory effect of CO_2 on spore germination, the inhibition of normal and aerobic spoilage microbiota of meat which are indicators of spoilage and the potential for temperature abuse. The extended shelf life of MAP foods may allow pathogens to grow to sufficiently high infective dose levels.

Another factor to consider is that, while it is generally known that ready-to-eat products are more at risk than the products that will be cooked or reheated prior to consumption, recent evidence clearly indicates that heating REPFEDs before consumption does not completely destroy toxin and thus cannot be considered as a safety factor to reduce the botulism risk (Notermans *et al.*, 1990). These authors tested the effect of microwave heating as recommended by the manufacturer (700 watts for 5 min) on inactivation of botulinum toxin and observed partial inactivation; however, toxin was completely inactivated after 10 min heating. In this study, a negative impact on product quality after 10 min heating was observed.

Finally, an increased demand for extended shelf life chilled foods may mean that production of such products moves beyond the relatively well-controlled foodservice operations and larger retail operations to smaller, less experienced manufacturers.

10.3 Spore-forming pathogens of concern

10.3.1 Clostridium botulinum

Clostridium botulinum is a Gram-positive, spore-forming, anaerobic, rod-shaped, catalase-negative, soil organism that produces a potent neurotoxic protein known as botulinum neurotoxin. This toxin causes a neuroparalytic disease known as botulism. The organism is the most hazardous spore-forming foodborne pathogen. Presence of spores in foods is not a public health hazard unless they can germinate, outgrow and multiply into toxin-producing vegetative cells. However, it is worth emphasizing that presence of even <1 spore/g is enough to justify strict prophylactic measures to be taken because consumption of even less than 1 gram of the foods in which growth has occurred can be sufficient to produce symptoms of human botulism. Botulinum neurotoxin is a relatively heat-sensitive toxin which can cause various symptoms of paralysis, the most severe of which are respiratory impairment that can result in death.

The species of *C. botulinum* with an outstanding characteristic of elaborating the neurotoxin have been divided into seven types, A through G, depending on the serological specificity of the neurotoxin produced. Types A, B, E and F produce human botulism whereas C and D have been implicated with avian botulism. Based on serological and cultural characteristics, the different types of *C. botulinum* have been classified into four subgroups (Smith and Sugiyama, 1988), which correspond to grouping by DNA homology (Lee and Riemann, 1970) and by somatic antigens (Lynt *et al.*, 1971).

- Group I: type A and proteolytic strains of types B and F.
- Group II: type E and non-proteolytic strains of types B and F.
- Group III: types C and D.
- Group IV: type G.

Those groups that are of primary importance to food safety are group I strains (*C. botulinum* types A, B and F) and non-proteolytic, group II (*C.*

botulinum types B, E and F) strains. While proteolytic type A and B strains of *C. botulinum* are more tolerant to environmental stresses, produce highly heat-resistant spores (Lynt *et al.*, 1982) and have a minimal growth temperature of 10°C (Smelt and Haas, 1978), non-proteolytic *C. botulinum* strains are less tolerant to stresses, form considerably less heat-resistant spores (Lynt *et al.*, 1982; Scott and Bernard, 1977; Solomon *et al.*, 1982). Recently, Graham *et al.* (1997) reported growth and toxin production by non-proteolytic *C. botulinum* strains at 3°C; spores of these strains that survive the thermal process would pose a botulism hazard even under proper refrigeration temperatures, if a secondary barrier is not present. Conner *et al.* (1989) suggested that additional hurdles or barriers must be built into a particular REPFED product. Thus, without additional barriers, heat processing must be sufficient to destroy the non-proteolytic *C. botulinum* spores if the food is to be safe.

(a) Occurrence   *C. botulinum* is widely distributed all over the world and is ubiquitous in the soil. Spores of this organism may find their way into processed food through raw materials or by post-processing contamination of food. It is not possible to be certain that any food will not contain spores of *C. botulinum* other than the foods that have been aseptically packed or have received sporicidal heat treatment. Narayan (1967) examined tissues such as mesenteric lymph nodes, liver, spleen and kidney of clinically healthy cattle and pigs to determine the level of contamination. Out of 203 cultures, 17 were positive, of which two were non-proteolytic. In a survey conducted by Gibbs *et al.* (1994), the presence of non-proteolytic strains of *C. botulinum* were not demonstrated in over 500 samples of refrigerated packaged foods in the UK. Surveys in the United States also indicated low incidence of *C. botulinum*. Taclindo *et al.* (1967) examined ready-to-eat mainly vacuum-packaged, refrigerated foods such as luncheon meats, sausages, cheese, shredded vegetable mix, unshucked oysters, smoked salmon, smoked herring, barbecued cod, kippers, anchovy paste and a variety of oriental foods. Out of 113 samples examined, one sample of luncheon meat out of 73 was positive for type B and one out of 4 samples of unshucked oysters was positive for type E spores. Insalata *et al.* (1969) examined 100 samples of each of four categories of commercially available convenience foods. The categories examined were 'boil-in-the-bag' foods, vacuum-packed foods, pasteurized foods and dehydrated and freeze-dried foods. Of the 400 samples analyzed, only one sample of vacuum-packed frankfurters was found to be positive. These studies clearly indicate low incidence of non-proteolytic strains of *C. botulinum* spores in meat and meat products but a few positive samples stress the severe need for adoption of measures to prevent possible risk to consumers.

(b) Heat resistance   The thermal resistance of bacterial spores (or vegetative cells) is generally described by two parameters, D and *z*. The D-value,

at a particular temperature, is the time in minutes during which the number of a specific bacterial (cells or spores) population is reduced by 90% or one log_{10}. The z-value is the temperature change required to bring about a 10-fold change (increase or decrease) in the D-value. While D-value reflects the resistance of an organism to a specific temperature, z-value provides information on the relative resistance of the organism to different destructive temperatures. From the knowledge of these two parameters, it is possible to calculate the microbial inactivation during heating.

The spores of *C. botulinum* types A and B, the most heat resistant, are targeted for destruction to ensure microbiological safety of low-acid foods. The canning industry adopted a D-value at 121°C of 0.2 min and a 12 log reduction as a standard for designing a required thermal process for an adequate degree of protection against *C. botulinum*.

The thermal process for REPFED is not designed for the destruction of proteolytic strains of *C. botulinum* because these strains do not grow at or below 10°C. However, quantitative data on the thermal characteristics of non-proteolytic *C. botulinum* are definitely needed to completely assess the importance of these organisms in the preservation of minimally processed refrigerated foods. The heat resistance of *C. botulinum* type E spores has been thoroughly investigated in aqueous media as well as in a variety of foods (Tables 10.2 and 10.3). The heat resistance of non-proteolytic *C. botulinum* type B spores has been investigated in buffer by a number of researchers (Table 10.4), but there are only a few reported studies in foodstuffs (Juneja *et al.*, 1995a, b; Juneja and Eblen, 1995). While information on the heat resistance of non-proteolytic *C. botulinum* in aqueous media is very useful, different food substrates confer varying degrees of protection to spores against heat (Simunovic *et al.*, 1985; Juneja *et al.*, 1995a).

Scott and Bernard (1982) investigated the heat resistance of *C. botulinum* type E strains: Minnesota, Whitefish and Saratoga in the same pH 7.0 phosphate buffer and reported D-values at 82.2°C of 0.52, 0.4 and 0.37 min, respectively; the D-values for non-proteolytic type B strains in phosphate buffer at 82.2°C ranged from 1.49 to 32.30 min depending upon the strain. However, in their studies, the higher heat-resistance values were not obtained consistently. Interestingly, the heat resistance of the non-proteolytic type B spores increased or decreased after storage for several months. The authors suggested that there may be significant variations among: strains, cultures of the same strain, and the reported D-values by different investigators within the same strain. In general, non-proteolytic type B strains were more resistant than type E strains. These conclusions were in agreement with a study by Notermans *et al.* (1990). In this study, the D-values at 80°C for type E strains were 1.7 min and the D-values for the type B strains were approximately 5.2 min.

Lynt *et al.* (1977) investigated the thermal death time of *C. botulinum* type E spores heated in crab meat. Sterilized blue crab meat was inoculated with

Table 10.2 Heat resistance of *Clostridium botulinum* type E spores in an aqueous medium

Number of strains	Heating menstruum	Heating temperature	D-values (min)	z-value (°C)	Reference
9	Sterile water	80	<0.33–1.25	7.4–10.8	Roberts and Ingram (1965)
2	Phosphate buffer pH7	80	0.4–3.3	14.0–17.0	Ohye and Scott (1957)
4	Phosphate buffer pH7	77	0.77–1.95	–	Ito et al. (1967)
1	Phosphate buffer pH7	80	2.3	15.0	Schmidt (1964)
1	Phosphate buffer pH7	79.4	1.3	–	Alderton et al. (1974)
1	Phosphate buffer pH7*	91	13.5	–	Alderton et al. (1974)
1	Phosphate buffer pH7*	93	3.8	–	Alderton et al. (1974)
1	Water	79.4	0.6	–	Alderton et al. (1974)
1	Water*	91	5	–	Alderton et al. (1974)
3	Phosphate buffer pH7	82.2	0.37–0.52	8.7	Scott and Bernard (1982)
1	Phosphate buffer pH7	80	1.4	7.2	Bohrer et al. (1973)
1	Phosphate buffer pH7	76.7	4.1	7.2	Bohrer et al. (1973)
3	Phosphate buffer pH7	80	0.39–3.91		Juneja et al. (1995a)
3	Phosphate buffer pH7*	80	1.03–4.35		Juneja et al. (1995a)

*Lysozyme in the recovery medium.

Table 10.3 Heat resistance of *Clostridium botulinum* type E spores in a variety of foods

Number of strains	Heating menstruum	Heating temperature	D-value (min)	z-value (°C)	Reference
5	Blue crab	74 82.2	6.8–13.0 0.49–0.74	6.5 6.5	Lynt et al. (1977)
1	Oyster homogenate	50 80	7943 0.78	7.5 7.5	Bucknavage et al. (1990)
5	Mehnaden surimi	73.9 82.2	8.66 1.22	9.78 9.78	Rhodehamel et al. (1991)
1	Clam liquor	82.2	0.2	–	Licciardello (1983)
5	White fish chubs	80	1.6–4.3	7.6	Crisley et al. (1968)
5	Oyster homogenate	73.9 82.2	2.0–8.96 0.07–0.43	4.2–7.1 4.2–7.1	Tui-Jyi and Kuang (1982)
1	Crab meat	80	4.4	6.3	Bohrer et al. (1973)
1	Evaporated milk	76.7 80	6.4 1.9	6.3 6.3	Bohrer et al. (1973)
1	Tuna in oil	76.7 80	40.9 10.5	6.1 6.1	Bohrer et al. (1973)
1	Corn in brine	76.7 80	11.2 3.2	6.0 6.0	Bohrer et al. (1973)
1	Sardines in tomato sauce	76.7 80	20 4.5	6.3 6.3	Bohrer et al. (1973)
1	Turkey*	70 85	51.89 1.18		Juneja et al. (1995a)

*Lysozyme in the recovery medium.

Table 10.4 Heat resistance of non-proteolytic *Clostridium botulinum* type B spores in buffer

Strains	Heating menstruum	Heating temp.	D-value (min)	z-value (°C)	Reference
2129B	Phosphate buffer pH 7	82.2	32.3	9.7	Scott and Bernard (1982)
17B	Phosphate buffer pH 7	82.2	16.7	6.5	Scott and Bernard (1982)
ATCC 17844	Phosphate buffer pH 7	82.2	4.17	16.5	Scott and Bernard (1982)
CBW 25	Phosphate buffer pH 7	82.2	1.49	8.3	Scott and Bernard (1982)
CBW 25	Phosphate buffer pH 7*	82.2	73.6	–	Scott and Bernard (1985)
ATCC 17844	Phosphate buffer pH 7*	82.2	24.5	–	Scott and Bernard (1985)
17B	Phosphate buffer pH 7*	82.2	28.2	–	Scott and Bernard (1985)
3	Phosphate buffer pH 7	80	0.60–1.90		Juneja *et al.* (1995a)
	Phosphate buffer pH 7*	80	3.22–4.31		Juneja *et al.* (1995a)

*Lysozyme in the recovery medium.

spores and then 1 g samples were heated in thermal death time tubes at 73.9–85°C. The Beluga strain was the most heat resistant of the five strains tested. The D-values at 82.2°C were 0.74, 0.51, 0.63, 0.62 and 0.49 min for Beluga, Alaska, Crab G21, Crab 25V-1 and Crab 25V-2 strains, respectively. The D-value at 80°C for spores of five strains of *C. botulinum* type E in ground white fish chubs were 4.3, 2.1, 1.8, 1.6 and 1.6 min for Alaska, Beluga, 8E, Iwanai and Tenno strains, respectively (Crisley *et al.*, 1968).

Gaze and Brown (1990) assessed the heat resistance over the temperature range 70 to 92°C of non-proteolytic *C. botulinum* type B and E spores heated in homogenates of cod (pH 6.8) and carrot (pH 5.7; Table 10.5). It was calculated that a heat treatment at 90°C for 7 min or equivalent time/temperature combination would be sufficient to achieve a 6D process for the most heat-resistant non-proteolytic *C. botulinum* spores. This is a much more severe process than that recommended for safe production of *sous vide* foods in the USA.

The estimated D-value depends upon the method used to calculate the values. The studies mentioned above used the fraction negative method to estimate the D-values. In this method, spores are recovered in an optimal medium after heat exposure for various times and survival is determined by observing growth or no growth after incubation. In such cases, the D-values reported may be influenced by a few heat-resistant spores since the shape of the curve relating the heating time to the number of survivors is not known. In another method, regression lines are fitted to the experimental

Table 10.5 Heat resistance of non-proteolytic *Clostridium botulinum* type B and E spores heated in cod and carrot homogenates over the temperature range 75 to 90°C

Type/strain	Substrate	D-value (min)			z-value (°C)
		75°C	85°C	90°C	
Type B/ATCC 25765	Cod	53.9	4.0	1.1	8.64
Type B/ATCC 25765	Carrot	19.39	1.57	0.43	9.76
Type E/ATCC 9564	Cod	58.5	4.75	0.79	8.26
Type E/ATCC 9564	Carrot	18.05	0.73	0.48	9.84

Adapted from Gaze and Brown (1990).

data points that contribute to shouldering or tailing by logistic survival equations (Whiting, 1993); this approach is an ideal way for estimating the heat resistance for non-linear heat inactivation data. In a study reported by Juneja *et al.* (1995a), heated non-proteolytic type B and type E *C. botulinum* spores exhibited lag periods or shoulders, where the population remained constant, followed by a linear decline. Also, a subpopulation of more persistent spores was observed that declined at a slower rate. D-values at 80°C for type E spores in buffer ranged from 1.03 min for strain Whitefish to 4.51 min for strain Saratoga. D-value for the most heat-resistant non-proteolytic type B strain KAP B5 in buffer was 4.31 min at 80°C. The z-values in buffer for all strains were very similar, ranging from 8.35 to 10.08°C. Turkey slurry offered protection to the spores with concomitant increase in heat resistance. D-values in turkey slurry ranged from 51.89 min at 70°C to 1.18 min at 85°C for type E strain Alaska ($z = 9.90$°C) and from 32.53 min at 75°C to 0.80 min at 90°C for non-proteolytic type B strain KAP B5 ($z = 9.43$°C).

Determination of spore survival after exposure to heat is characterized by the ability of the injured spores, which are more nutritionally demanding than unheated spores, to grow and form colonies on the recovery media. Moreover, sporulation media and the temperature used for spore preparation, heating menstruum, recovery conditions including the composition and pH of the medium, the presence of inhibitors, and temperature and time of incubation, affect the calculated spore heat resistance (Foegeding and Busta, 1981; Russell, 1982). Additionally, sufficient evidence exists to document that the addition of lysozyme to the recovery medium resuscitates and increases the recovery of the heat-injured spores, thereby increasing the measured or apparent heat resistance (Alderton *et al.*, 1974; Bradshaw *et al.*, 1977; Kim and Foegeding, 1992; Peck *et al.*, 1992; Scott and Bernard, 1985; Sebald *et al.*, 1972; Juneja *et al.*, 1995a). In a study by Sebald *et al.* (1972), when *C. botulinum* type E spores were heated in phosphate buffer for 10 min at 80°C, surviving spores were able to form colonies on a medium containing lysozyme, but not in its absence. Alderton *et al.* (1974) reported an estimated D-value of 1.30 min when type E spores were heated in phosphate buffer at 79.5°C and recovered on a medium without lysozyme. In the same study when the recovery medium contained lysozyme, the heating temperature had to be raised to obtain measurable spore destruction; the D-values were 13.50 and 3.80 min at 90.5°C and 93.3°C, respectively. Duncan *et al.* (1972) suggested that the heat alteration of the spore results in inactivation of the cortex lytic enzyme system, i.e. the system responsible for cortical degradation during germination. Lysozyme in the plating medium can replace the thermally inactivated spore germination enzymes (Scott and Bernard, 1985). The lysozyme is able to permeate the spore coat and degrade the cortex, allowing core hydration and, consequently, spore germination (Gould, 1989).

Peck *et al.* (1992) used the submerged tube procedure of Kooiman and Geers (1975) which gives instantaneous heating and subsequent rapid cooling of spores. In this study, when spores of non-proteolytic *Clostridium botulinum* were heated at 85°C and survivors were enumerated on a highly nutritive medium, a 5 decimal kill in less than 2 min was observed. However, enumeration of survivors on a medium supplemented with lysozyme showed that heating at 85°C for 5 min resulted in only an estimated 2.6 decimal kill of spores of strain 17B (type B); also, biphasic survivor curves, indicating heat-sensitive and heat-resistant population subfractions were observed. Only 0.2–1.4% of heated non-proteolytic *C. botulinum* spores of strain 17B (type B) and Beluga (type E) were permeable to lysozyme, but the proportion of heated spores permeable to lysozyme for Foster B96 (type E) strain was 20% (Peck *et al.*, 1993). Treatment with alkaline thioglycollate resulted in all heated spores being permeable to lysozyme and a biphasic heat-inactivation curve was converted to a logarithmic curve, the slopes of which were similar to the second part of the biphasic curves; the D-values are shown in Table 10.6. This treatment is known to rupture disulfide bonds in the spore coats, thereby making the spore coat permeable to lysozyme. These findings have implications for assessing heat treatments necessary to reduce risk of non-proteolytic *C. botulinum* survival and growth in REPFEDs. Lund and Peck (1994) presented electron micrographs showing sections through spores of *C. botulinum* 17B after heating at 85°C for 10 min and treatment with alkaline thioglycollate to demonstrate that the spore coats were disrupted.

Lysozyme cleaves the cell wall of vegetative cells and cortex of spores. Lysozymes are classified as muramidases (E.C. 3.2.1.17) that hydrolyze the β (1–4) linkages between *N*-acetylmuramic acid and *N*-acetylglucosamine which are parts of a sugar molecule (peptidoglycan) found in the outer membrane of bacterial cells (Proctor and Cunningham, 1988). The molecular weight ranges from 14 300 to 14 600 and the isoelectric point is pH 10.7. The enzyme is most active at pH ranging from 5 to 7 and the activity is generally measured by its ability to lyse *Micrococcus lysodeikticus*. The enzyme is relatively heat stable at pH values <7 (Lund and Peck, 1994).

Table 10.6 D-values of lysozyme-permeable spores of non-proteolytic *Clostridium botulinum* estimated based on enumeration of survivors on PYGS medium[a] supplemented with lysozyme

Strain	Temperature (°C)	Estimated D-values	
		From 2nd part of biphasic curve	After thioglycollate treatment[b]
17B	85	90	111
	90	18.1	19.4
	95	4.57	4.23
Beluga	85	48.3	42.8
	90	12.6	11.0
	95	3.17	2.34

[a]PGYS: peptone, yeast extract, glucose, starch.
[b]Heated spores were treated with alkaline thioglycollate before counting on PGYS.
Adapted from Peck *et al.* (1993).

Lysozyme is present in varying concentrations in a variety of foods of plant and animal origin such as eggs of birds and reptiles, mammalian tissue and milk, fish, plants (cauliflower, broccoli, cabbage, etc.), mollusks and crustaceans (Lund and Peck, 1994). The levels in foods of plant and animal origin are 1.8 to 27.6 µg/g and 20–160 µg/g, respectively. Since lysozyme is heat stable and is present in a variety of foods, the influence of this enzyme in the recovery of heat-damaged spores warrants further investigation to determine its effect on the efficacy of recommended heat processes. However, as yet, the protective effect has not been shown to occur in any (lysozyme-unsupplemented) food product.

Simulating the conditions in minimally processed refrigerated foods, Juneja and Eblen (1995) recovered heated spores on both Reinforced Clostridial Medium (RCM) with lysozyme and on RCM with lysozyme and the same salt levels as the heating menstruum. When the recovery medium contained no salt, D-values in turkey slurry containing 1% salt were 42.1, 17.1, 7.8 and 1.1 min at 75, 80, 85 and 90°C, respectively. Increasing levels (2 and 3%, w/v) of salt in the turkey slurry reduced the heat resistance as evidenced by reduced spore D-values (Table 10.7). The D-values were 27.4, 13.2, 5.0 and 0.8 min at 75, 80, 85 and 90°C, respectively, when both the turkey slurry and the recovery medium contained 1% salt (Table 10.7). Increasing levels (2–3%, w/v) of salt in turkey slurry resulted in parallel decrease in the D-values obtained from the recovery of spores on the media containing the same levels of salt as the heating menstruum (Table 10.7). The authors indicated that the decrease in D-values obtained from the recovery of heat-damaged spores on the media with added salt was due to the inability of heat-injured spores to recover in the presence of salt. The heat-injured spores were sensitive to salt in the recovery medium. These

Table 10.7 Heat resistance (75–90°C), expressed as D-values in min, for non-proteolytic *Clostridium botulinum* type B strain KAP B5 in turkey slurry

Heating menstruum	Recovery medium[a]	Temperature (°C)				z-value (°C)
		75	80	85	90	
Turkey + 1% salt	RCM + L	42.1	17.1	7.8	1.1	10.08
Turkey + 2% salt	RCM + L	25.7	15.1	5.5	0.6	8.82
Turkey + 3% salt	RCM + L	17.7	13.1	3.2	0.5	8.47
Turkey + 1% salt	RCM + L + 1% salt	27.4	13.2	5.0	0.8	9.86
Turkey + 2% salt	RCM + L + 2% salt	19.9	12.6	4.3	0.4	9.21
Turkey + 3% salt	RCM + L + 3% salt	16.9	8.2	2.6	0.3	9.51

[a]Reinforced Clostridial Medium (RCM) with lysozyme (L).

Adapted from Juneja and Eblen (1995).

data should assist food processors to design thermal processes that ensure safety against non-proteolytic *C. botulinum* type B spores in cook–chill foods while minimizing quality losses.

A wide range of REPFED are already produced and marketed successfully under commercial conditions where temperature abuse must have occurred at some point from production to consumption. These products have had an excellent safety record with regard to problems from *C. botulinum*; but the basis of the safety of these products needs to be explored.

The pasteurization of blue crab meat in hermetically sealed containers was introduced to the industry in the United States to increase refrigerated shelf life. Pasteurized crab meat has a refrigerated shelf life of at least 6 months. While spores of non-proteolytic *C. botulinum* have been isolated from the blue crab meat (Lynt *et al.*, 1977), there have been no reported illnesses attributed to this product. Based on the Tri-State Seafood Committee recommendations, a thermal process at 85°C for 1 min is applied to this product. The degree of hazard of this process is under debate. Lynt *et al.* (1977) reported that a 12D process would require 3 min at 85°C for type E spores. In earlier studies by Cockey and Tatro (1974), a thermal process at 85°C for 1 min reduced 10^8 viable spores per 100 g of crab meat to 6 or less, and the pasteurized meat remained non-toxic when stored at 4–6°C for up to 6 months. In these studies lysozyme was not incorporated in the recovery medium and thus the surviving spore numbers were underestimated.

Minimally processed oysters are processed at pasteurization temperatures of 55°C for 10 min and have a shelf life of 45 days (versus 10 to 14 days for fresh). When 1.0% salt and 0.13% sorbate is supplemented in oysters, the shelf life is increased to 60 days (Bucknavage *et al.*, 1990). The authors reported D-values at 55°C ranged from 1023 to 1259 min and at 80°C from 0.55 to 1.00 min. Based on their investigations, Bucknavage *et al.* (1990) expressed concerns that *C. botulinum* type E would survive in oysters that are minimally processed and suggested that the process temperatures would need to be much higher to eliminate the threat of botulism associated with type E spores. In another study, the heat resistance of spores of five type E *C. botulinum* strains was determined in oyster homogenates (Chai and Liang, 1992). The type E Minnesota strain was the most heat-resistant. D-values at 73.9, 75.0, 76.7, 79.4 and 82.2°C were 8.96, 5.28, 2.69, 1.03 and 0.43 min, respectively. The authors stated that current pasteurization methods are not sufficient to guarantee safety from type E *C. botulinum* spores. In these studies, the estimated heat resistance would have been higher if lysozyme was incorporated in the recovery medium. It is logical to characterize the conditions for growth and toxin production by adequate inoculated pack studies to elucidate the basis of safety of these products.

(c) Growth and toxin production

Laboratory media studies. Growth and toxin production by non-proteolytic *C. botulinum* has been thoroughly investigated in laboratory media (Graham and Lund, 1993; Graham *et al.*, 1996, 1997; Jensen *et al.*, 1987; Lund *et al.*, 1985, 1990), in meat and poultry (Abrahamsson *et al.*, 1966; Genigeorgis *et al.*, 1991; Meng and Genigeorgis, 1993, 1994), in fish and sea food (Lerke, 1973; Lindroth and Genigeorgis, 1986; Post *et al.*, 1985; Solomon *et al.*, 1977, 1982; Huss *et al.*, 1979a, b; Ikawa and Genigeorgis, 1987; Cann and Taylor, 1979; Cann *et al.*, 1965; Garcia *et al.*, 1987; Garren *et al.*, 1994) and in cooked vegetables (Carlin and Peck, 1995, 1996). The time to toxin production by non-proteolytic *C. botulinum* reported by a number of researchers is summarized in Table 10.8.

Quantitative information on growth rates from spores of 10 strains of type E was first reported by Ohye and Scott (1957). The authors reported mean doubling times of 42.6, 7.2 and 1.9 h at 5, 10 and 20°C, respectively. Graham and Lund (1993) studied growth rate, from vegetative cells rather than spore inocula, of a non-proteolytic type B strain (17B) and reported shorter doubling times (28.6, 5.7 and 1.3 h at 5, 10 and 20°C, respectively) than those reported for type E strains by Ohye and Scott (1957). It is possible that other strains of non-proteolytic type B will have an even shorter doubling time because the strain used by Graham and Lund (1993) may not be the fastest growing.

The ability of non-proteolytic *C. botulinum* type B spores to grow and form toxin in cooked meat medium at low temperatures was studied by Eklund *et al.* (1967). When the inoculum level was 5×10^6 spores, gas and toxin were produced within 17–18 days at 5.6°C and in 24–27 days at 4.4°C. At 3.3°C, gas was not detected during incubation for 109 days; however, toxin was detected after 85 days. Thus, increasing lengths of time were

Table 10.8 Time to toxin production by non-proteolytic *Clostridium botulinum* at low temperatures

C. botulinum type	Temperature (°C)	Substrate	Time (days)	Reference
B	3.3	Broth	>129	Eklund *et al.* (1967)
B	4.4	Broth	36	Eklund *et al.* (1967)
B	5.6	Broth	26	Eklund *et al.* (1967)
E	12	Broth	14	Eklund *et al.* (1967)
E	3.3	Beef stew	31	Schmidt *et al.* (1961)
E	10	Fish	8	Huss *et al.* (1979a)
E	15	Fish	1–2	Huss *et al.* (1979a)
E	10	Cooked potatoes	9	Notermans *et al.* (1981)
E	8	Broth	7–10	Lund *et al.* (1985)
E	12	Broth	3	Lund *et al.* (1985)
E	4.4	Crab meat	55	Cockey and Tatro (1974)
E	10	Crab meat	8	Cockey and Tatro (1974)
E	5	Cooked meat medium	56	Ohye and Scott (1957)
E	6	Chopped meat medium	21	Abrahamsson *et al.* (1966)
E	5	Raw herring	12	Hobbs (1982)
E	4.4	Crab meat	30	Lerke and Farber (1971)
F	3.3	Cooked meat medium	47	Eklund *et al.* (1967)
F	4.4	Cooked meat medium	22	Eklund *et al.* (1967)
F	5.6	Cooked meat medium	15	Eklund *et al.* (1967)
F	4	Broth	20	Solomon *et al.* (1982)

required to produce visible gas and toxin when temperature was decreased. Decreasing the inoculum level by one log (from 5×10^6 to 5×10^5) increased the length of incubation required for toxin and gas production. Thus, 21 days were needed for growth at 5.6°C, 33 days at 4.4°C and over 108 days at 3.3°C. Toxin was detected after 26 and 36 days of storage at 5.6 and 4.4°C, respectively.

The effect of temperature. Solomon *et al.* (1977) investigated the effect of incubation temperature on growth and toxin production by *C. botulinum* type E spores in crab meat and a TPGY broth at 4–26°C. The authors reported that unheated spores of type E grew and produced toxin in media at 8 and 12°C within 14 days and at 4°C in 52 days whereas heated spores grew and produced toxin at 12°C in 20 days and at both 8 and 4°C in 30 days. In crab meat, while unheated spores grew and produced toxin at 12°C in 14 days, the heated spores were unable to grow at temperatures 12°C and lower within 180 days (Table 10.9). In contrast, Cockey and Tatro (1974) reported toxin formation by type E spores in crab meat within 55 and 8 days at 4.4 and 10°C, respectively. In a similar study, Solomon *et al.* (1982) showed that for *C. botulinum* type B, the time required to grow and produce toxin increased with the decrease in incubation temperature; both heat and unheated spore inocula were unable to grow in crab meat at temperatures of 12°C and lower within the 180 days of the experiment; the unheated spores grew and produced toxin in TPGY broth at 12, 8 and 4°C in 6, 15 and 20 days, respectively. Heated spores grew and produced toxin at 12 and 8°C within 6 and 15 days; however, toxin was not detected at 4°C in 180 days (Table 10.9).

Published data based on studies conducted with low numbers of non-heat-shocked spores to simulate the natural conditions of growth in the REPFEDs clearly indicate high probability of toxin production in foods at low temperatures in a short incubation time. In a study by Ikawa and Geni-georgis (1987), when non-proteolytic type B spores were inoculated into heart infusion-cysteine broth and incubated for 60 days, the observed probability of growth initiation from a single spore was 0.0001% at 4°C and 1% at both 8 and 12°C. The probability of growth from a single spore or a vegetative cell in brain heart infusion broth after 28 days at 4, 8 and 12°C was 0.1% and 6.3%, 10 and 100%, and 100 and 100%, respectively (Jensen *et al.*, 1987). Based on these results, the calculated number of spores required to initiate growth was 1000, 10 and 1 at 4, 8 and 12°C, respectively, and the number of vegetative cells needed to initiate growth was 17 at 4°C and 1 at both 8 and 12°C. Lund *et al.* (1990) reported that the probability of growth from a vegetative cell was 1–100% after 20 days at 6°C, 10–100% after 11 days at 8°C and 100% after 2 days at 20°C. It should be noted that the use of vegetative cells in challenge studies gives an estimate of the maximum risk of growth of the organism. With regard to spores, a variety of factors including sporulation temperature and media, and storage conditions may effect the rates of germination.

The effect of pH. The minimum pH that permitted growth from spores of four type E strains at 30°C was between 5.10 and 5.21; however, minimum pH was 5.7–5.9 at 8°C (Segner *et al.*, 1966). Similar results were reported by Emodi and Lechowich (1969). According to them, the minimum pH values

Table 10.9 Influence of substrate on days to toxin production by spores of *Clostridium botulinum* type E and non-proteolytic type B

Substrate	Spores	Days			
		26°C	12°C	8°C	4°C
TPGY	E (Heated)	6	20	30	30
	E (Unheated)	3	14	14	52
	B (Heated)	3	6	15	>180
	B (Unheated)	3	6	15	20
Crab meat	E (Heated)	6	>180	>180	>180
	E (Unheated)	3	14	>180	>180
	B (Heated)	3	>180	>180	>180
	B (Unheated)	3	>180	>180	>180

Adapted from Soloman *et al.* (1977, 1982).

for the growth of six type E strains in broth was 5.4–5.6 and 6.2–6.4 at 15.6°C and 5°C, respectively. Lund *et al.* (1990) developed a mathematical model to predict the probability of growth of non-proteolytic type B spores as affected by varying combinations of pH (4.8–7.0), temperature (6–30°C) and sorbic acid concentrations up to 2270 mg/l in culture media. The authors reported that non-proteolytic type B had greater ability to grow at sub-optimal combinations of pH and temperature as compared to the report in the previous study by Emodi and Lechowich (1969).

Combined effects of heat treatment and storage temperatures. In a study on the effect of heat treatment and storage temperature on the survival and growth of non-proteolytic *C. botulinum* spores, spores of 5 type B, 5 type E and 2 type F strains were inoculated into tubes of an anaerobic meat medium containing lysozyme to obtain 6 $\log_{10}$ spores/tube. The tubes were heated at temperatures ranging from 65 to 95°C for up to 6 h (Peck *et al.*, 1995). The authors reported that heat treatment equivalent to maintaining 85°C for 19 min, combined with storage below 12°C for less than 28 days, would reduce the risk of spore growth by a factor of 10^6; it was pointed out that if foods are likely to be stored at temperatures above 10°C, growth and toxin production by proteolytic strains must also be prevented. The heat-treatment equivalent to maintaining 95°C for 15 min followed by incubation at 25°C for up to 60 days failed to prevent growth and toxin production by 10^6 spores of non-proteolytic *C. botulinum*, whereas incubation at 12°C or lower for 60 days resulted in inactivation/inhibition of 10^6 spores of non-proteolytic *C. botulinum*.

The effect of lysozyme. When anaerobic meat medium was supplemented with 10 µg/ml of lysozyme prior to heating at 90°C for 19.8 min, growth at 12°C was first observed after 51 days compared to 6 days if added after heating (Peck and Fernandez, 1995). When anaerobic meat medium was supplemented with 50 µg/ml of lysozyme, growth was first observed after 68 days at 8°C, 31 days at 12°C, 24 days at 16°C and 9 days at 25°C (Table 10.10). While a heat treatment equivalent to 90°C for 19.8 min was not sufficient to give a 6D inactivation for germination and outgrowth at 25°C, this heat treatment in conjunction with storage at temperatures less than 12°C prevented growth for 30 days. The authors indicated that if lysozyme is present at concentration up to 50 µg/ml in a refrigerated, processed food with an intended shelf life of 4 weeks, and the food is likely to be exposed to mild temperature abuse of up to 12°C, a heat treatment at 90°C for 19.8 min would be required to reduce the risk of growth of non-proteolytic *C. botulinum* by a factor of 10^6. However, if longer shelf life is expected, then higher heat treatment in conjunction with better control of temperature, or additional barriers would be required to ensure safety against neurotoxi-genesis by non-proteolytic *C. botulinum*. While this study was conducted with model food, it is logical to presume that other foods may have a greater or lesser protective effect on spores due to lysozymes. Other foods may contain a lesser or higher concentration of lysozyme or may contain other lytic enzymes with similar effects on spores (Lund and Peck, 1994). Peck and Fernandez (1995) closed their discussion by expressing concerns and cautioned that the addition of lysozyme, including a genetically modified lysozyme with increased heat resistance (with additional disulfide bonds), and other lytic enzymes as preservative factors (Cunningham *et al.*, 1991; Proctor and Cunningham, 1988; Nielsen, 1991; Gould, 1992), is likely to increase the risk of growth of non-proteolytic *C. botulinum*. However, at the present time, the effect of increased heat resistance has not been shown to occur in any (lysozyme-unsupplemented) food product (Gould, 1996) but more information is needed for procedures employing processes at 90°C/10 min equivalent lethality, and particularly for processes substantially below this, during which the presence and survival of lysozyme activity may be expected to be more likely, and at recovery temperatures relevant to food storage practices.

Studies in real foods. The potential hazard due to growth and toxin production by non-proteolytic *C. botulinum* in extended shelf life of refrigerated foods stored at temperatures above 3°C is confirmed by the following studies. Brown and Gaze (1990) and Brown *et al.* (1991) investigated the growth of non-proteolytic *C. botulinum* spores inoculated into carrot, cod and chicken homogenates processed using the *sous vide* method (vacuum packaged and then cooked at 70°C for 2 min) and stored at 5, 8 and 15°C. Type B toxin was detected in chicken within 6–8 weeks and cod within 3–6 weeks at 8°C. Type E toxin was formed between 3 and 4 weeks in chicken and 5 days to 3 weeks in cod. At 15°C, toxins were detected within 7 days

Table 10.10 Influence of lysozyme concentration on growth from spores of non-proteolytic *Clostridium botulinum* heated at 90°C for 19.8 min and then stored at low temperatures

Lysozyme concentration	Time to growth (days)			
	8°C	12°C	16°C	25°C
0	NG	NG	NG	NG
5	NG	65	63	33
10	NG	51	29	20
25	83	45	27	8
50	68	31	24	9
10*	13	6	3	2

*Lysozyme (10 μg/ml) added after heating.

Adapted from Peck and Fernandez (1995).

in chicken and cod. Notermans *et al.* (1990) studied the potential risk of botulism associated with REPFEDs. In their study, when fortified-egg-meat medium inoculated with non-proteolytic *C. botulinum* spores at low levels of 0.17 spore/ml was stored at 8°C, toxin production was not observed after approximately 16–20 days; however, with high levels (1.7–170 spores/ml) toxin production was observed after 12 days of incubation. In the same study, toxin was detected after 4 weeks in 2 of 11 commercially available *sous vide* products inoculated with non-proteolytic *C. botulinum* spores and then stored at 8°C. At this time it is worth reiterating that the storage temperature of chilled foods in retail and domestic refrigerators is often around 8°C and frequently above this (NFPA, 1988).

In general, vegetables are considered to be poor in nutrients for growth and toxin production by non-proteolytic *C. botulinum*. Carlin and Peck (1995) conducted studies to determine the growth and toxin production by 10^3 spores/g of non-proteolytic *C. botulinum* inoculated in cooked vegetables incubated anaerobically at 30°C. Growth and toxin production was observed within 4 days in the following vegetables: mushroom, spinach, potato, bean sprouts, cauliflower, asparagus, broccoli and kale. The vegetables which did not support the growth or toxin production were: garlic, butter squash, chicory, green beans, celery, fennel, brussels sprouts, parsnip, turnip, carrot, onion, green pepper, mange-tout and tomato. In another study by Carlin and Peck (1996), the authors assessed the growth and toxin production at refrigeration temperatures and reported that cooked cauliflower and mushrooms supported growth at 5, 8, 10 and 16°C whereas potatoes only at 16 and 8°C. The toxin was detected within 3 to 5 days at 16°C, 11 to 13 days at 10°C, 10 to 34 days at 8°C and 17 to 20 days at 5°C.

In contrast to this, botulinum toxin was not detected in 0.5 ml (10^4 cfu/ml, non-proteolytic *C. botulinum* type B and type E) inoculated samples of carrots, jacket potatoes, lasagne or smoked trout fillets after incubation at 4 and 10°C for 3 weeks; in inoculated whole trout, however, it was found that toxin production could occur within one week at 10°C but not at 4°C in a period of 3 weeks (Gibbs *et al.*, 1994). The authors concluded that with the exception of whole trout, toxin production is unlikely to occur within the current shelf life of these products if the temperature of 10°C is not exceeded. Hence although variable results have been achieved the potential for non-proteolytic botulinum in vegetables at chilled temperatures is a real one.

10.3.2 Clostridium perfringens

Clostridium perfringens is considered to be ubiquitous and is found in soil, dust, vegetation and a variety of animals that include cattle, poultry and humans. The optimum temperatures for *C. perfringens* growth range from 41°C to 45°C and a doubling time as short as 8–10 min has been reported.

The concern regarding this organism is increased following reports that some strains can grow at 6°C (Johnson, 1990). Optimum pH for growth is between 6.0 and 7.0 and the growth limiting pH ranges from 5.5–5.8 to 8.0–9.0. While most strains are inhibited by 5–6.5% salt, the organism has been observed to grow at up to 8% NaCl concentration (Johnson, 1990). Heat-resistant spores have D-values at 95°C ranging from 17.6 to 64 min (Ando *et al.*, 1985). Juneja and Majka (1995) reported D-values at 99°C that ranged from 23.33 min in the presence of 3% salt in beef at pH 7.0 to 13.99 in beef at pH 5.5 containing 3% salt. In another study, Juneja and Marmer (1996) reported that the D-values at 90°C ranged from 23.2 min when turkey was not supplemented with NaCl to 17.7 min in the presence of 3% NaCl. In a study by Juneja and Majka (1995), when outgrowth of *C. perfringens* spores in cook-in-bag beef products at 15°C was studied, growth occurred within 6 days in samples with pH 7.0, but was delayed until day 8 in the presence of 3% salt at pH 5.5. In another study, Juneja and Marmer (1996) assessed the growth of *C. perfringens* from spore inocula in *sous vide* turkey products and reported that growth at 15°C occurred at a relatively slow rate in the presence of 1–2% NaCl. These studies emphasize the importance of normal preservation factors of salt and low pH in foods. However, the organism is a continuing concern in view of the trend towards more natural less preserved foods. It should be noted that the risk due to *C. perfringens* is not as great as that of *C. botulinum* because food poisoning due to *C. perfringens* typically requires the ingestion of approximately 10^6 to 10^7 viable cells per gram of food; the ingested cells sporulate in the intestine and release a heat-labile enterotoxin known as *C. perfringens* enterotoxin (CPE), which is responsible for typical symptoms (Duncan, 1975; Stark and Duncan, 1971).

10.3.3 Bacillus cereus

Bacillus cereus is widely distributed in nature and causes two different types of food poisoning: the diarrheal type and the emetic type. The diarrheal type of food poisoning is caused by an enterotoxin(s) produced during vegetative growth of *B. cereus* in the small intestine (Granum, 1994). The emetic toxin is produced by cells growing in the food (Kramer and Gilbert, 1989). The organism has been isolated from a variety of foods including poultry products (Sooltan *et al.*, 1987), cook–chill meals (van Netten *et al.*, 1990) and many ready-to-serve meals (Harmon and Kautter, 1991). While the optimum temperature for growth of *B. cereus* ranges from 28 to 35°C (Gilbert, 1979), the psychrotrophic strains are able to grow at temperature as low as 4°C (van Netten *et al.*, 1990). The doubling time in a nutritive medium at optimum temperature is 18–27 min. The growth-limiting pH under ideal conditions ranges from pH 4.9 to 9.3 and the minimum water activity for growth is 0.95 (Johnson, 1990). It must be pointed out that the growth depends upon the interactive effects of all the environmental parameters. Based on model studies on cook/chill meals, van Netten *et al.* (1990) concluded that REPFEDs could present a risk of *B. cereus* intoxication unless stored at <4°C. Information on the influence of factors such as pH, a_w, etc., on growth is lacking in the scientific literature.

The heat resistance of *B. cereus* spores is a concern to the food industry and has been studied extensively. In general, the heat resistance is similar to that of other mesophilic spore-formers; however, some strains referred to as heat-resistant strains are 15–20-fold more heat resistant that the heat sensitive strains. *B. cereus* strains involved in food poisoning have D-values at 90°C ranging from 1.5 to 36 min (Johnson, 1990). Regarding the effect of antimicrobials on growth, 0.26% sorbic acid and 0.39% potassium sorbate at pH 5.5 and 6.6, respectively, are inhibitory (Raevuori, 1976). As with *C. perfringens*, the risk of food poisoning due to *B. cereus* is not high because of the high infective dose which ranges between 10^5 and 10^7 (total) for diarrheal type and 10^5–10^8 per gram of food for emetic syndrome.

10.4 Methods for control of spore-formers

The effectiveness of any control measures to guard against spore-formers is based on the probability that a spore will germinate, grow out and multiply and elaborate toxin in the preserved food product. Although the probability of growth increases as a function of the number of contaminating spores on raw materials, it is often difficult to control the number of spores on raw materials and almost impossible to guarantee absence. Therefore, the safety of such products must really depend not only on specifications of raw materials but also on combinations of preservation regimes that are designed to cope with normal levels of spores that will be present in REPFEDs.

Since the assurance of safety from psychrotrophic *C. botulinum* spores is a key factor in the success of the new generation of refrigerated food products, this section will concentrate mainly on this organism. Changes in the intrinsic properties of the food products, primarily pH, sodium chloride content, etc., are known to affect the ability of spores to survive thermal processing and on the ability of survivors to germinate, outgrow and produce toxin at temperatures >3.3°C. Thus, these intrinsic properties must be considered when designing adequate thermal processes or when trying to prevent toxin production during extended refrigerated storage. Moreover, it is often desirable to formulate foods and use combinations of these parameters that would allow for reduction of undesirable attributes such as high salt or acidity and design a reduced thermal process that would destroy the spores without negatively impacting the product quality. While a majority of REPFEDs are produced without any preservative factors or with few hurdles to prevent growth of non-proteolytic *C. botulinum*, it is very important to build/include factors to prevent toxin formation by *C. botulinum*. If intrinsic factors (low pH, reduced water activity, etc.) and extrinsic factors (temperature, time, etc.) are not adequate to prevent non-proteolytic spores from germination, heat processing must be sufficient to destroy spores if the food is to be safe. For example, while the heat process and storage temperature may be relied on in certain well-controlled food service operations it may be important to include additional hurdles to growth in relatively less well controlled retail operations. As a first approach, it is logical to know the minimum requirements for growth and the heat resistance of *C. botulinum* types A, B, E and F. Accordingly, researchers have sought to establish extremes of these parameters that would control *C. botulinum* growth in foods and they are given in Table 10.11.

The most critical step in the production of REPFEDs is the heating process for the inactivation of spores of non-proteolytic *C. botulinum* strains. Since these are less heat-resistant spores, it is practically feasible to inactivate these spores by heat. Heat resistance increases with decreasing water activity and decreases with decreasing the pH below 5 and increasing values above 9. Fat and protein also have a protective effect and increase spore heat resistance. While Juneja *et al.* (1995a) reported that contaminated turkey should be heated to an internal temperature of 80°C for at least 91.3 min to give a 6D process for type B spores, with the inclusion of 3% salt in turkey, 78.6 min at 80°C were sufficient to achieve a 6D process (Juneja and Eblen, 1995). Thus, it is suggested to incorporate low levels of salt while formulating foods and design a reduced thermal process that ensures safety against non-proteolytic *C. botulinum* type B in minimally processed foods while maintaining the desirable organoleptic attributes of foods.

Juneja *et al.* (1995b) assessed and quantified the effects and interactions of temperature, pH, salt and phosphate levels and found that the thermal inactivation of non-proteolytic *C. botulinum* spores is dependent on all four factors. Thermal resistance of spores can be lowered by combining these

Table 10.11 Properties of *Clostridium botulinum*: minimal requirements for growth and heat resistance

Property	*C. botulinum* Group I	*C. botulinum* Group II
Toxin types	A, B, F	B, E, F
Minimum temperature for growth	10°C	3.3°C
Minimum pH for growth	4.6	5.0
Minimum a_w for growth	0.94	0.97
Inhibitory brine (%)	10%	5%
Heat resistance of spores:		
D-value at 100°C	25 min	< 0.1 min
D-value at 121°C	<0.1–0.2 min	< 0.001 min

Adpated from Hauschild (1989).

intrinsic factors. The following multiple regression equation, developed in this study, predicts D-values for any combinations of temperature (70–90°C), salt (NaCl; 0.0–3.0%), sodium pyrophosphate (0.0–0.3%) and pH (5.0–6.5) that are within the range of those tested.

Log_e D-value =

$$-9.9161 + 0.6159(temp) - 2.8600 \, (pH) - 0.2190 \, (salt)$$
$$+ 2.7424 \, (phos) + 0.0240(temp)(pH) - 0.0041(temp)(salt)$$
$$- 0.0611(temp)(phos) + 0.0443(pH)(salt) + 0.2937(pH)(phos)$$
$$- 0.2705(salt)(phos) - 0.0053(temp)^2 + 0.1074(pH)^2$$
$$+ 0.0564(salt)^2 - 2.7678(phos)^2$$

Additionally, Juneja *et al.* (1995b) developed confidence intervals to allow microbiologists to predict the variation in the heat resistance of non-proteo-lytic *C. botulinum* spores. Using this predictive model, food processors should be able to design thermal processes for the production of a safe *sous vide* food with extended shelf life without substantially adversely affecting the quality of the product. Representative observed and predicted D-values at 70–90°C of non-proteolytic *C. botulinum* in ground turkey, at various pH levels (5.0–6.5) supplemented with salt (0.0–1.5% w/v) and sodium pyrophosphate (0.0–0.2% w/v) are given in Table 10.12.

Every effort should be made to extend the lag and generation time. By quantifying the effects and interactions of multiple food formulations, it is possible to provide secondary barriers to toxin production in cases of temperature abuse or failure of other primary preservative techniques. *C. botulinum* growth and subsequent toxin production can be prevented by combining several inhibitory parameters at sub-inhibitory levels during food formulation to produce a safe food product. This concept of 'microbial hurdles' describes pictorially the effects of various intrinsic food composition factors and extrinsic environmental conditions and has been described by Leistner (1985). Recently, a number of predictive models have been developed based on multifactorial design experiments, extensive data

Table 10.12 Observed and predicted D-values at 70–90°C of non-proteolytic *Clostridium botulinum* in ground turkey

Temperature (°C)	pH	% NaCl	% Phosphate	D-value observed (min)	D-value predicted (min)
70	6.50	0.0	0.00	57.7	66.0
70	6.50	1.5	0.15	40.1	46.5
75	6.25	1.0	0.10	39.1	42.3
75	6.25	1.0	0.20	32.9	38.6
90	5.00	0.0	0.00	5.0	6.3
90	5.00	1.5	0.15	3.1	4.8

Adapted from Juneja *et al.* (1995b).

collection and analysis. These models quantify the effects and interactions of intrinsic and extrinsic factors and describe the growth responses of spore-formers. Food processors can optimize product formulation by the use of these predictive models.

Meng and Genigeorgis (1993) assessed the interactive effects of NaCl (0–2%), 0–3% sodium lactate (NaL), temperature (4–30°C) and spore inoculum (10^2–10^4/g) on the length of lag phase of non-proteolytic *C. botulinum* type B and E spores. The spores were inoculated in cooked turkey and chicken meats without nitrite and stored under vacuum for up to 120 days. In this study, toxin was not detected at 4°C in the presence of NaL and an inoculum of 10 spores/3g sample. The authors developed the following predictive regression model for the lag phase duration as affected by NaCl, NaL, inoculum (*I*) and temperature (*T*) of 8–30°C and their interactions.

$$\text{Log } (1/\text{LP}) = -2.29 - 0.123(\text{NaCl}) + 0.22(\text{NaL}) + 0.439(T) \\ + 0.02(T)(I) \text{ with } R^2 = 0.945$$

where *T* is the square root of temperature.

The study demonstrated that lag phase can be extended to >38 days at <8°C in the presence of 2% NaL and 1% NaCl and an inoculum of 100 spores/sample. Increasing NaCl concentration to 2% extended the lag phase to >55 days. At a mild temperature abuse of 12°C, incorporation of 3% NaL and 2% NaCl was required to prevent toxin production for at least 36 days in turkey meat containing 100 spores (Table 10.13). In a previous study, Genigeorgis *et al.* (1991) found that turkey meat without added NaL but with 2.2% brine became toxic after 130, 10 and 9 days at 4, 8 and 12°C, respectively. From the above model, Meng and Genigeorgis (1993) were able to determine that temperature was the most influential factor on toxigenesis. Storage temperature explained 65% of the variation in results, followed by lactate (21.2%), the interaction of inoculum and temperature (4.9%) and NaCl (3.4%). The R^2 values (percentage of result variation) were used as a method to quantify barriers to toxicity (hurdle concept). In another study, Meng and Genigeorgis (1994) inoculated processed 'sous vide' products (beef and salmon homogenates) containing 0, 2.4 and 4.8% (w/w) NaL and chicken containing 0, 1.8 and 3.6% (w/w) lactate, with 4 $\log_{10}$ non-proteolytic *C. botulinum* spores/3g sample, vacuum packaged and stored at 4, 8, 12 and 30°C for up to 90 days. Lactate at ≥ 2.4% in beef and ≥ 1.8% in chicken delayed toxigenesis for ≥ 40 d at ≤ 12°C. In this study, it was interesting to note that salmon was the most conducive to toxigenesis and required 4.8% lactate for delaying toxigenesis for a similar time period. Increased lactate levels and decreased temperatures resulted in a parallel delay in toxigenesis in the 'sous vide' products. The authors concluded that the use of ≥ 2.4% lactate and storage at ≤ 12°C can ensure inhibition of toxigenesis for time periods well beyond the expected shelf life of 3–6 weeks.

Table 10.13 Influence of salt and sodium lactate levels on non-proteolytic *Clostridium botulinum* time to toxin production in cooked turkey meat inoculated with 100 spores and incubated at 8 and 12°C for up to 80 days

Temperature (°C)	Salt (%)	Sodium lactate (%)			
		0	1.2	2	3
8	0	8	>28	>34	>40
	1	14	>28	>38	>55
	2	>28	>32	>55	>80
12	0	7	10	10	24
	1	7	>18	>22	>36
	2	13	22	26	>36

Adapted from Meng and Genigeorgis (1993).

Graham *et al.* (1996) assessed the effects of pH (5.0–7.3), NaCl concentration (0.1–5.0%) and temperature (4–30°C) on growth of non-proteolytic *C. botulinum* in laboratory media; growth curves were fitted by the Gompertz (Gibson *et al.*, 1987) and Baranyi models (Baranyi and Roberts, 1994), and parameters derived from the curve-fit were modeled. Both models predicted that the optimal conditions for growth were temperatures between 25 and 28°C, pH between 6.6 and 6.7 and a NaCl concentration less than 1%. Growth was not observed at pH less than 5.1 or at 5% NaCl. Experimental and predicted values for doubling time and lag time are shown in Table 10.14.

For *B. cereus* and *C. perfringens* control, the most effective control measure is to ensure that the rate and extent of cooling to < 4°C is rapid to prevent potential spore germination and growth of vegetative cells. Juneja *et al.* (1994) conducted studies to determine a safe cooling rate for cooked beef and observed that *C. perfringens* spores germinated and grew from an inoculum of approximately 1.5 $\log_{10}$ to about 6.0 $\log_{10}$ cfu/g when the cooling time to achieve 7.2°C was extended to 18 h. The authors reported that the pasteurized cooked beef must be cooled to 7.2°C in 15 h or less to prevent growth from spores of *C. perfringens*. If cooling rate is not rapid, sufficient reheating before consumption can kill large numbers of vegetative cells.

The effects and interactions of temperature (12, 19, 28, 37, 42°C), initial pH (5.5, 6, 6.25, 6.5, 7), sodium chloride content (0, 0.1, 1.5, 2, 3%), and sodium pyrophosphate concentration (0, 0.1, 0.15, 0.2, 0.3) on the growth of a three strain mixture of *C. perfringens* vegetative cells indicated that the growth kinetics of *C. perfringens* was dependent on the interaction of the four variables, particularly in regard to exponential growth rates and lag phase duration (Juneja *et al.*, 1996). The authors concluded that sodium pyrophosphate can have significant bacteriostatic activity against *C. perfringens* and

Table 10.14 Representative doubling time and lag time of non-proteolytic *Clostridium botulinum*: Effect of temperature, pH and sodium chloride

Temp. (°C)	pH	NaCl (%)	Doubling time (h)		Lag time (h)	
			Obs.	Pred.	Obs.	Pred.
5.0	6.04	1	22	20	246	242
5.0	5.98	0.1	28	20	314	259
5.0	6.12	2	32	24	290	267
7.0	5.89	2	14	16	180	213
7.9	5.33	1	11	23	596	561
8.2	6.81	0.1	4.7	5.1	59	62
9.9	6.52	0.1	4.4	3.1	46	36

Adapted from Graham *et al.* (1996).

may provide processed meats with a degree of protection against this microorganism, particularly if employed in conjunction with a combination of acidic pH, high salt concentrations, and adequate refrigeration.

Other interventions to the ability of surviving spores to grow out in pasteurized, refrigerated foods may include:

(a) Addition of competing microflora: These microorganism can have antagonistic action on germination, growth and toxin production of sporeformers. Mossel and Struijk (1991) suggested the addition of *Sporolactobacillus inulus* to REPFEDs. The spores of this organism survive pasteurization temperatures and the growth from spores does not occur at temperatures <5°C (Botha and Holzapfel, 1988). At mild temperature abuse conditions, spores will germinate, grow out and produce acid which can warn the consumer that the product is no longer edible. Also, lactic acid bacteria may be incorporated in foods. These bacteria may survive processing and grow when there is temperature abuse, and produce acid and bacteriocins, make the product inedible and warn the consumer of possible hazard. However, further research needs to be done in this area.

(b) The use of bacteriolytic enzymes (lysozyme): As discussed earlier in the chapter, lysozyme is present in a variety of foods of both plant and animal origin and is relatively heat stable, particularly under acidic conditions. The implications on enhanced germination and outgrowth of spores in REPFEDs during their storage must be considered while designing the heat treatment and assessing the safety of such foods (Lund and Notermans, 1993).

(c) Use of time–temperature indicators: Rigorous control of temperature during transportation, distribution, retail storage or handling before consumption is extremely important. Since temperature abuse is a common occurrence at both the retail and consumer levels, producers should not rely on low-temperature storage and should use time–temperature indicators (TTI) to track the time and temperature history of the products from production to consumption. It is recommended that TTI devices be added to individual food packages or cartons to indicate temperature abuse and an increased risk of food poisoning. Reviews on TTIs have been published (Wells and Singh, 1988; Taoukis and Labuza, 1989; Selman, 1990). Time–temperature indicators should be designed in such a way that their expiration, as affected by storage temperature and time, will follow bio-chemical or physical kinetics slightly faster than the rate of growth of pathogens or toxin production.

(d) An integral part of preventing hazards associated with spore-formers in REPFEDs is education. Workers in food-processing, food-distribution, and food-service establishments should be exposed to continuing training for safe food production including the consequences of temperature abuse of these food products. Additionally, consumers must also be made aware of the potential hazards associated with these products and they must receive the knowledge regarding handling and storage of these products.

10.5 Regulations

In most countries, food industry and retail food establishments are required to comply with the published regulations for hygienic production, distri-bution and sale of REPFEDs. Also, a Code of Practice, which is advisory rather than prescriptive, has been developed in some countries. In com-parison with legislation, the main advantage of the codes is that they can be more easily updated to take into account the changing industry practices in response to consumer demands.

Health Canada (Health and Welfare Canada, 1992) published guidelines on 'refrigerated prepackaged foods with extended shelf life' in a document entitled 'Guidelines for the production, distribution, retailing and use of refrigerated prepackaged foods with extended shelf life'. The REPFEDs in-clude a range of products which share the following characteristics: (1) rely largely on refrigeration for microbiological safety after processing; (2) are prepackaged using a packaging technology which does not adequately inhibit microbial growth; (3) are not commercially sterile; and (4) have a shelf life equal to or greater than 10 days. It is recommended for the proces-sors to adopt the principles and practices of the Hazard Analysis and Criti-cal Control Point (HACCP) system, and products should be formulated and processed to reduce the microbial loads and limit their potential for growth, in case of occurrence of temperature abuse. Significant emphasis is given on record keeping particularly at steps of process parameters, changes in formulation, and microbiological challenge studies carried out to determine the shelf life of a particular product with regard to spoilage versus relative growth of foodborne pathogens. It is recommended in these guidelines that the labels on the packages should clearly indicate: (1) refrigerate at all times; (2) 'Use By' date in addition to a 'Best Before' date. If relevant, preparation and cooking instruction should be given. It is also recommended that the processors should use time–temperature indicators to indicate that temperature abuse has occurred leading to increased risk of food poisoning. Products must be continuously maintained at appropriate refrigeration temperatures at all points along the food chain, including retail display. Also stressed is the fact that retailers should have operating procedures in place regarding temperature-abused or out-dated products.

For the retail sector, it is recommended that retailers follow the principles outlined in the document entitled *Canadian Code of Recommended Handling Practices for Chilled Food,* issued by the Food Institute of Canada (FIOC, 1990). While the main objective of the chilled food industry is to maintain temperatures between –1 and 4°C and to ensure proper care and handling of chilled foods from production to consumption, a few significant points covered in the code include: (1) preparation of raw materials should be carried out in an area maintained at <10°C; (2) all products should be rapidly cooled to between –1 and 4°C; (3) all refrigerated storage facilities, transportation vehicles and warehouses should be monitored with a suitable time–temperature monitoring device and maintained at temperatures between –1 and 4°C; (4) additional hurdles or barriers must be built into the processed refrigerated products for an additional degree of safety; (5) receivers of refrigerated food should check that products received are maintained at –1 to 4°C. (6) A 'Keep Refrigerated' between –1 and 4°C should be clearly displayed on the shipping carton, and on the inner carton or packaging material a 'Keep Refrigerated' and 'Use By' date should be clearly visible.

The *Canadian Code of Recommended Manufacturing Practices for Pasteurized/Modified Atmosphere Packaged/Refrigerated Food* was compiled by the Agri-Food Safety Division (Agriculture Canada, 1990). The objective of this document is to ensure microbiological safety of the foods that are pasteurized either before or after being packaged under vacuum or in modified atmospheres in hermetically sealed containers and require refrigeration at –1 to 4°C throughout their shelf life. Overall, the approach of this document is based on the principles of HACCP and as such is divided into three main sections, with the first section dealing with product safety and identifying factors influencing microbiological contamination, i.e. process development and a thorough risk analysis. The second section deals with the recommended manufacturing practices and covers the following factors affecting microbiological contamination: raw ingredients, packaging, hygiene, time/temperature relationships and extra hurdles. The third section deals with monitoring the parameters and designing acceptance limits on the critical control points. The product assembly or preparation areas should be maintained at <10°C. There is no recommended pasteurization process but it is stated that the process should destroy a defined level of identified target microorganisms and reference is made to the French recommendations, and a similar relationship between 'pasteurization value' and shelf life is adopted. Rate and extent of cooling should be rapid to achieve <4°C within 2 h of end of pasteurization. Storage and distribution of the finished products must be done at −1 to 4°C. While recommendations on the shelf life are not stated, reference is made to French legislation as an example of shelf life. The use of extra hurdles/barriers to the growth of microorganisms, such as acidification, the addition of preservatives or the reduction of water activity is 'strongly encouraged'.

In the United States, the document produced by the National Advisory Committee on Microbiological Criteria for Foods (NACMCF, 1990) includes information on areas: (1) epidemiological and microbiological considerations, (2) processing and packaging, (3) recommendations. The committee recommended that a HACCP system based on sound scientific principles following a thorough risk assessment be used in the production and distribution of refrigerated foods. Thermal treatments must be designed to achieve a 4D process for *Listeria monocytogenes*. In addition, the committee recommended that it is the processor's responsibility to conduct challenge studies to document that botulinal toxin is not elaborated from the time of production to consumption. Guidelines for *C. botulinum* challenge studies to evaluate the potential risk of REPFEDs state that the food product after inoculation be incubated at 12.8°C for one and one-half times the product's shelf life or for up to the time when the product is organoleptically spoiled and is unfit for human consumption. The mouse bioassay procedure as described in the *US Food and Drug Administration Bacteriological Analytical Manual* is the recommended method for botulinal toxin testing. If botulinal toxin formation occurs within the time frame specified in the guidelines, then the processors must either reformulate the product to prevent toxigenesis or design a strategy to ensure that the consumer can identify that the product is a botulinal hazard before toxin may be produced. Secondary barriers or hurdles are also suggested as an effective alternative to adequate thermal processing. Processors must have a verified HACCP plan prior to production and marketing.

American recommendations on the safety of vacuum or modified atmosphere packaging for refrigerated raw fishery products have been produced by the National Advisory Committee on Microbiological Criteria for Foods (NACMCF, 1992). The committee concluded that temperature control at or below 3.3°C from packaging through preparation is the primary preventive measure to guard against *C. botulinum* hazard. The committee recommended that the production and marketing of raw fishery products packaged under vacuum or modified atmosphere be permitted only if certain conditions are fulfilled. These include use of high-quality raw fish, storage of packaged product at or below 38°F, spoilage precedes toxigenesis, product labels prominently indicate storage temperature, shelf life and cooking requirements, and the products are packaged under an established HACCP plan.

Rhodehamel (1992) at the Food and Drug Administration expressed potential health hazards concerns associated with *sous vide* processing. He recommended that: (1) *sous vide* products must be produced and distributed using a HACCP approach; HACCP offers an effective, rational and systematic approach that involves shifting the emphasis from finished-product testing to raw material and process control, thereby ensuring the safety and quality of foods; (2) processors must follow Good Manufacturing Practices sanitation guidelines; (3) while refrigeration is the primary barrier, additional hurdles or barriers must be built into a particular product; (4) inoculated pack or challenge studies must be conducted to validate the efficacy of the multiple barriers; and (5) since temperature abuse is a common occurrence at both the retail and consumer levels, *sous vide* processors should use time–temperature indicators (TTI) to track the time and temperature history of the products from production to consumption. Also recommended is that TTI devices be included on each food package to indicate if temperature abuse had occurred leading to increased risk of food poisoning.

In the United Kingdom, a document entitled *Chilled and Frozen Guidelines on Cook–Chill and Cook–Freeze Systems* (DoH, 1989) provides information on hygienic practices or guidelines for the manufacture of pre-cooked chilled foods and apply only to catering operations. The prepared foods should be stored at <10°C and thermally processed to an internal temperature of 70°C for 2 min to achieve a shelf life of 5 days including day of cooking and consumption. This recommended pasteurization treatment is aimed at the destruction of *L. monocytogenes*. Recommendations are also given for cooling of foods after thermal processing: chilling should begin within 30 min of end of cooking; the rate and extent of cooling should be rapid enough to achieve core temperature of 0–3°C within 90 min; these temperatures should be maintained during storage and distribution. In case of scenarios where the specified temperatures are exceeded, corrective actions are provided. If the food temperature is between 5 and 10°C, the food must be consumed within 12 h, and if this time period has lapsed, then the food should be discarded. Also, the food should be discarded immediately if the food temperature exceeds 10°C. Reheating of chilled foods should begin within 15 min of removal from the chiller and an internal temperature of 70°C for 2 min should be achieved. Reheated foods should be served within 15 min and the temperature should not fall below 63°C; reheated foods, if not consumed, must be discarded.

To prevent the growth and toxin production in chilled foods with a shelf life of 10 days or longer, the UK Advisory Committee on the Microbiological Safety of Food (ACMSF, 1992) recommended that in addition to storage at refrigeration temperatures (<3°C), the safety of minimally processed foods could be ensured by the following factors, used singly or in combination: (1) a heat treatment of 90°C for 10 min or equivalent lethality; (2) a pH of <5 throughout the food; (3) a minimum salt level of 3.5% in the aqueous phase of all components of complex foods; (4) an a_w of 0.97 or less throughout all components of complex foods; 5) a combination of heat and preservative factors which can be shown consistently to prevent growth and toxin production by non-proteolytic *C. botulinum* at temperatures up to 10°C.

A document entitled *A Code of Practice for* Sous-vide *Catering Systems* has been produced by *Sous-Vide* Advisory Committee (SVAC, 1991) and provides recommendations for process parameters at various stages of processing from raw materials to finished product. The code of practice is designed to be used by 'In-house' catering and a maximum shelf life of 8 days has been recommended at storage temperatures of <3°C. It is recommended that the raw ingredients be stored between 0 and +3°C except fresh fruits and vegetables which should be stored at <8°C. The pre-cooked foods should be held at <10°C and thermally processed within a maximum of 2 hours after preparation. The recommended pasteurization treatment should be aimed to achieve a 6 log reduction of non-proteolytic *C. botulinum*. The recommended time and temperatures to achieve an equivalent lethal effect are given in Table 10.15. It is interesting to note that these heat treatments would only achieve a 4 log reduction of non-proteolytic *C. botulinum* type B based on the study of Gaze and Brown (1990) in which strain 25765 was heated in cod. Recommendations for post-pasteurization cooling of foods are the same as given in the DoH guidelines. Reheating of chilled foods should begin within 30 min of leaving chiller and heating to a minimum core temperature of 75°C should be achieved. Reheated foods should be served within 15 min and the temperature should not fall below 65°C; reheated foods, if not consumed, must be discarded.

French regulations for pre-cooked ready-to-eat, refrigerated foods provide guidelines regarding the preparation, distribution and sale of ready-to-eat meals (Anon., 1974) and the necessary changes in manufacturing protocol for extending the shelf life of pre-cooked, ready-to-eat chilled meals (Anon., 1988). The time/temperature treatments and the proposed shelf life are given in Table 10.16; these processes are based on the thermal destruction of *Enterococcus* strain which has a D-value at 70°C of 2.95 min ($z = 10°C$). According to the guidelines, cooked products must be cooled rapidly to achieve a core temperature of <10°C within 120 min. The cooled products must be stored at <3°C; for reheating prior to consumption, an internal temperature of 65°C must be achieved and maintained for 60 min. The food must not fall below this temperature until served.

In the document entitled *Guidelines of Good Hygienic Practices for Prepared Refrigerated Meals* (SYNAFAP, 1989) produced in France, there are many recommendations given for the various stages of processing. It is

Table 10.15 The recommended time and temperature to achieve a six log reduction of non-proteolytic *Clostridium botulinum* for *sous vide* cook–chill foods with a maximum shelf life of 8 days at 0–3°C

Temperature (°C)	Time (min)
80	26
85	11
90	4.5
95	2

Adapted from SVAC (1991).

recommended that HACCP plans are implemented to control microbiological hazards. While a temperature of 8 to 12°C should be maintained in preparation areas, perishable raw materials should be stored chilled at 0 to 3°C or frozen at –18°C. The recommended pasteurization process and post-pasteurization cooling and reheating requirements are in accordance with the French regulations. It is recommended that secondary packaging be done at a temperature of <12°C and storage, distribution and retail display is at 0–3°C.

The Botulinum Working Party was set up by the European Chilled Food Federation (Gould, 1996) to consider the safety of chilled food products which have been mildly heated in hermetically sealed packages, or heated and packaged without recontamination. There was general agreement that for products intended to have a short shelf life, a heat process less than that known to deliver a 6 decimal reduction of psychrotrophic *C. botulinum* spores is acceptable, provided the temperatures of storage are sufficiently low and the shelf lives are sufficiently short to prevent growth from spores that may be present. For products intended to have a long shelf life, heat processing sufficient to achieve at least a 6 decimal reduction of psychrotrophic *C. botulinum* is acceptable provided the temperature is maintained below that at which mesophilic strains of *C. botulinum* can grow (10°C) during storage, distribution, retail sale and storage. If milder heat treatments are given, evidence must be provided that other food preservation factors such as low pH, water activity, strict temperature control

Table 10.16 Recommended core temperature, heating time and the expected shelf life of *sous vide* pre-cooked foods

Temperature (°C)	Time (min)	Shelf life (days)
70	40	6
70	100	21
70	1000	42

Adapted from Anon. (1974, 1988), French regulations.

below 3.3°C, etc., are present and will inhibit growth from surviving spores in model studies and inoculated pack/challenge tests.

10.6 Conclusions and outlook to the future

Survival of spore-formers and the occurrence of temperature abuse throughout distribution, in retail markets and home refrigerators is a challenge for innovative interventions. The safety of REPFEDs cannot be considered to rely on only one single factor. While many guidelines have been proposed, it appears that all of them lack a sound scientific basis and there are insufficient data to suggest a change from the current guidelines. Recent research has assessed and quantified the combination of hurdles to decrease the heat-processing requirements and control subsequent germination of surviving spores during storage. Further research on complex multifactorial experiments and analysis to quantify the effects and interactions of additional intrinsic and extrinsic factors and development of 'enhanced' predictive models, followed by validation by challenge tests, is warranted to ensure the safety. In view of the continued interest that exists in lowering the heat treatment it would be logical to define a specific lethality at low temperatures. Also, it would be useful to determine the possible effects of injury to spores that may result from mild heat treatments and factors in foods that influence the recovery of spores heated at these low temperatures. Presently, lysozyme appears to play a crucial role in the repair process of heat-injured spores; the influence of this enzyme must be considered while designing thermal processes for safe production of REPFEDs. However, the effect of indigenous lysozyme in foods on the measured heat resistance of spores needs to be investigated. Development of quantitative risk assessment models based on product composition/formulation, processing and storage, in conjunction with implementation of HACCP plans, should provide an adequate degree of protection against non-proteolytic *C. botulinum*.

Finally, all establishments within the food chain also should have HACCP systems in place to ensure that *C. botulinum* controls are properly executed and maintained for the entire life of high-risk products produced. Actually, HACCP should be in place to minimize the risk of all pathogens, not only *C. botulinum*.

References

Abrahamsson, K., Gullmar, B. and Molin, N. (1966) The effect of temperature on toxin formation and toxin stability of *Clostridium botulinum* type E in different environments. *Canadian Journal of Microbiology*, **12**, 385–393.

Advisory Committee on the Microbiological Safety of Food (ACMSF) (1992) *Report on Vacuum Packaging and Associated Processes*, HMSO, London, UK.

Agriculture Canada (1990) *Canadian Code of Recommended Manufacturing Practices for Pasteurized/Modified Atmosphere Packaged/Refrigerated Food*, Agri-Food Safety Division, Agriculture Canada, Ottawa, Canada.

Alderton, G., Chen, J.K. and Ito, K.A. (1974) Effect of lysozyme on the recovery of heated *Clostridium botulinum* spores. *Applied Microbiology*, **27**, 613–615.

Ando, Y., Tsuzuki, T., Sunagawa, H. and Oka, S. (1985) Heat resistance, spore germination, and enterotoxigenicity of *Clostridium perfringens*. *Microbiology and Immunology*, **29**, 317–326.

Anon. (1989) Temperature abuses of food. *Audits International Monthly*, April 1989. Audits International, Highland Park, IL.

Anon. (1974) Réglementation des conditions d'hygiène relatives à la préparation, la conservation, la distribution et la vente des plats cuisinés à l'avance. Arrêté du 26 Juin 1974. République Française Ministère de l'Agriculture.

Anon. (1988) Prolongation de la durée de vie des plats cuisinés à l'avance, modification du protocole permettant d'obtenir les autorisations. Note de Service DGAL/SVHAIN 881 No. 8106, du 31 Mai 1988. République Française Ministère de l'Agriculture.

Baranyi, J. and Roberts, T.A. (1994) A dynamic approach to predicting bacterial growth in food. *International Journal of Food Microbiology*, **23**, 277–294.

Bohrer, C.W., Denny, C.B. and Yao, M.G. (1973) Thermal destruction of type E *Clostridium botulinum*. Final Report on RF 4603, National Canners Association Research Foundation, Washington, DC.

Botha, S.J. and Holzapfel, W.H. (1988) Effect of reduced water activity on vegetative growth of cells and on germination of endospores of *Sporolactobacillus*. *International Journal of Food Microbiology*, **6**, 19–24.

Bradshaw, J.G., Peeler, J.T. and Twedt, R.M. (1977) Thermal inactivation of ileal loop-reactive *Clostridium perfringens* type A strains in phosphate buffer and beef gravy. *Applied and Environmental Microbiology*, **34**, 280–284.

Brown, G.D. and Gaze, J.E. (1990) Determination of the growth potential of *Clostridium botulinum* types E and nonproteolytic B in 'sous-vide' products at low temperatures. *Technical Memorandum No. 593*, Campden Food and Drink Research Association, Chipping Campden, Glos., UK.

Brown, G.D., Gaze, J.E. and Gaskell, D.E. (1991) Growth of *Clostridium botulinum* nonproteolytic type B and type E in 'sous-vide' products stored at 2–15°C. *Technical Memorandum No. 635*, Campden Food and Drink Research Association, Chipping Campden, Glos., UK.

Brown, M.H. and Gould, G.W. (1992) Processing, in *Chilled Foods: A Comprehensive Guide* (eds C. Dennis and M.F. Stringer), Ellis Horwood, London, pp. 111–146.

Bryan, F.L., Seabolt, L.A., Peterson, R.W. and Roberts, L.M. (1978) Time-temperature observations of food and equipment in airline catering operations. *Journal of Food Protection*, **41**, 80–92.

Bucknavage, M.W., Pierson, M.D., Hackney, C.R. and Bishop, J.R. (1990) Thermal inactivation of *Clostridium botulinum* type E spores in oyster homogenates at minimal processing temperatures. *Journal of Food Science*, **55**, 372–373.

Cann, D.C. and Taylor, L.Y. (1979) The control of the botulism hazard in hot-smoked trout and mackerel. *Journal of Food Technology*, **14**, 123–129.

Cann, D.C., Wilson, B.B., Hobbs, G. and Shewan, J.M. (1965) The growth and toxin production of *Clostridium botulinum* type E in certain vacuum packed fish. *Journal of Applied Bacteriology*, **28**, 431–436.

Carlin, F. and Peck, M.W. (1995) Growth and toxin production by non-proteolytic and proteolytic *Clostridium botulinum* in cooked vegetables. *Letters in Applied Microbiology*, **20**, 152–156.

Carlin, F. and Peck, M.W. (1996) Growth and toxin production by nonproteolytic *Clostridium botulinum* in cooked pureed vegetables at refrigeration temperatures. *Applied and Environmental Microbiology*, **62**, 3069–3072.

Chai, T.J. and Liang, K.T. (1992) Thermal resistance of spores from five E *Clostridium botulinum* strains in eastern oyster homogenates. *Journal of Food Protection*, **55**, 118–22.

Cockey, R.R. and Tatro, M.C. (1974) Survival studies with spores of *Clostridium botulinum* type E in pasteurized meat of the blue crab, *Calinectes sapidus*. *Applied Microbiology*, **27**, 629–633.

Conner, D.E., Scott, V.N., Bernard, T. and Kautter, D.A. (1989). Potential *Clostridium botulinum* hazards associated with extended shelf life refrigerated foods: a review. *Journal of Food Safety*, **10**, 131.

Crisley, F.D., Peeler, J.T. Angelotti, R. and Hall, H.E. (1968) Thermal resistance of spores of five strains of *Clostridium botulinum* type E in ground whitefish chubs. *Journal of Food Science*, **33**, 411–416

Cunningham, F.E., Proctor, V.A. and Goetsch, S.J. (1991) Egg-white lysozyme as a food preservative: an overview. *World's Poultry Science Journal*, **47**, 141–163.

Daniels, R.W. (1991) Applying HACCP to new-generation refrigerated foods at retail and beyond. *Food Technology*, **45** (6), 122–124.

DoH (1989) *Chilled and frozen Guidelines on Cook-Chill and Cook-Freeze Systems*. HMSO, ISBN 0113211619.

Duncan, C.L. (1975) Role of clostridial toxins in pathogenesis, in *Microbiology* (ed. D. Schlessinger), American Society for Microbiology, Washington, DC, pp. 283–291.

Duncan, C.L., Labbe, R.G. and Reich, R.R. (1972) Germination of heat- and alkali-altered spores of *Clostridium perfringens* type A by lysozyme and an initiation protein. *Journal of Bacteriology*, **109**, 550–559.

Eklund, M.W., Wieler, D.I. and Poysky, F.T. (1967) Outgrowth and toxin production of nonproteolytic type B *Clostridium botulinum* at 3.3 to 5.6°C. *Journal of Bacteriology*, **93**, 1461–1462.

Eklund, M.W., Poysky, F.T. and Wieler, D.I. (1967) Characteristics of *Clostridium botulinum* type F isolated from the pacific coast of the United States. *Applied Microbiology*, **15**, 1316–1323.

Emodi, A.S. and Lechowich, R.V. (1969) Low temperature growth of type E *Clostridium botulinum* spores. 1. Effects of sodium chloride, sodium nitrite and pH. *Journal of Food Science*, **34**, 78–81.

FIOC (Food Institute of Canada). (1990) *The Canadian Code of Recommended Handling Practices for Chilled Food*, The Food Institute of Canada, Ottawa, Ontario, Canada.

Foegeding, P.M. and Busta F.F. (1981) Bacterial spore injury – an update. *Journal of Food Protection*, **44**, 776–786.

Garcia, G.W., Genigeorgis, C.A. and Lindroth, S. (1987) Risk of growth and toxin production by *Clostridium botulinum* nonproteolytic type B, E, and F in salmon fillets stored under modified atmospheres. *Journal of Food Protection*, **50**, 390–397.

Garren, D.M., Harrison, M.A. and Huang, Y.W. (1994) *Clostridium botulinum* type E outgrowth and toxin production in vacuum-skin packaged shrimp. *Food Microbiology*, **11**, 467–472.

Gaze, J.E. and Brown, G.D. (1990) Determination of the heat resistance of a strain of nonproteolytic *Clostridium botulinum* type B and a strain of type E, heated in cod and carrot homogenate over a temperature range 70–90C. *Technical Memorandum No. 592*, Campden Food and Drink Research Association, Chipping Campden, Glos., UK.

Genigeorgis, C.A., Meng, J. and Baker, D.A. (1991) Behavior of nonproteolytic *Clostridium botulinum* type B and E spores in cooked turkey and modeling lag phase and probability of toxigenesis. *Journal of Food Science*, **56**, 373–379.

Gibbs, P.A., Davies, A.R. and Fletcher, R.S. (1994) Incidence and growth of psychrotrophic *Clostridium botulinum* in foods. *Food Control*, **5**, 5–7.

Gibson, A.M., Bratchell, N. and Roberts, T.A. (1987) The effect of sodium chloride and temperature on the rate and extent of growth of *Clostridium botulinum* type A in pasteurized pork slurry. *Journal of Applied Bacteriology*, **62**, 479–490.

Gilbert, R.J. (1979) In *Foodborne Infections and Intoxications*, 2nd edn (eds H. Reimann and F.L. Bryan), Academic Press, New York, pp. 495–518.

Gould, G.W. (1989) Heat-induced injury and inactivation, in *Mechanisms of Action of Food Preservation Procedures* (ed. G.W. Gould), Elsevier, London, pp. 11–42.

Gould, G.W. (1992) Ecosystem approach to food preservation. *Journal of Applied Bacteriology Symposium Supplement*, **73**, 58S-68S.

Gould, G.W. (1996) Conclusions of the ECFF Botulinum Working Party, in *Proceedings: Second European Symposium on Sous- vide*, 10–12 April, 1996, pp. 174–180.

Graham, A. and. Lund, B.M. (1993) The effect of temperature on growth of non-proteolytic type B *Clostridium botulinum*. *Letters in Applied Microbiology*, **16**, 158–160.

Granum, P.E. (1994) *Bacillus cereus* and its toxins. *Journal of Applied Bacteriology Symposium Supplement*, **76**, 61S-66S.

Graham, A.F., Mason, D.R. and Peck, M.W. (1996) Predictive model of the effect of temperature, pH and sodium chloride on growth from spores of non-proteolytic *Clostridium botulinum*. *International Journal of Food Microbiology*, **31**, 69-85.

Graham, A.F., Mason, D.R., Maxwell, F.J. and Peck, M.W. (1997) Effect of pH and NaCl on growth from spores of non-proteolytic *Clostridium botulinum* at chill temperatures. *Letters in Applied Microbiology*, **24**, 95-100.

Harmon, S.M. and Kautter, D.A. (1991) Incidence and growth potential of *Bacillus cereus* in ready-to-serve foods. *Journal of Food Protection*, **54**, 372-374.

Health and Welfare Canada (1992) *Guidelines for the Production, Distribution, Retailing and Use of Refrigerated Prepackaged Foods with Extended Shelf Life*, Guideline #7, Food Directorate, Health Protection Branch, Health Canada.

Hobbs, G. (1982) The ecology and taxonomy of psychrotrophic strains of *Clostridium botulinum*, in *Psychrotrophic Microorganisms in Spoilage and Pathogenicity* (eds T.A. Roberts, G. Hobbs, J.H. B.Christian and N. Skovgaard), Academic Press, NY, pp. 449-462.

Hauschild, A.H.W. (1989) *Clostridium botulinum*, in *Foodborne Bacterial Pathogens* (ed. M.P. Doyle), Marcel Dekker, New York, p. 111.

Huss, H.H., Schaeffer, I., Pedersen, A. and Jespen, A. (1979a) Toxin production by *Clostridium botulinum* type E in smoked fish in relation to the measured oxidation reduction potential (Eh), packaging method and the associated microflora, in *Advances in Fish Science and Technology* (ed. J.J. Connel), Fishing News Books Ltd, Farnham, UK, pp. 476-479.

Huss, H.H., Schaeffer, I., Petersen, E.R. and Cann, D.C. (1979b) Toxin production by *Clostridium botulinum* type E in fresh herring in relation to the measured oxidation reduction potential (Eh). *Nord. Vetinary-medicine*, **31**, 81-86.

Hutton, M.T., Dhehak, P.A. and Hanlin, J.H. (1991) Inhibition of botulinum toxin production by *Pedicoccus acidilacti* in temperature abused refrigerated foods. *Journal of Food Safety*, **11**, 255-267.

Ikawa, J.Y. and Genigeorgis, C. (1987) Probability of growth and toxin production by non-proteolytic *Clostridium botulinum* in rockfish fillets stored under modified atmospheres. *International Journal of Food Microbiology*, **4**, 167-181.

Insalata, N.F., Witzeman, S.F., Fredericks,G.J. and. Sunga, F.C.A. (1969) Incidence study of spores of *Clostridium botulinum* in convenience foods. *Applied Microbiology*, **17**, 542-544.

Ito, K.A., Seslar, D.J, Mercer, W.A. and. Meyer, K.F. (1967) The thermal and chlorine resistance of *Clostridium botulinum* types A, B, and E spores, in *Botulism 1966* (eds M. Ingram and T.A. Roberts), Chapman & Hall, London, pp. 108-122.

Jensen, M.J., Genigeorgis, C. and Lindroth, S. (1987) Probability of growth of *Clostridium botulinum* as affected by strain, cell and serologic type, inoculum size and temperature and time of incubation in a model broth system. *Journal of Food Safety*, **8**, 109-126.

Johnson, E.A. (1990) *Clostridium perfringens* food poisoning, in *Foodborne Diseases* (ed. D.O. Cliver), Academic Press, California, pp. 229-240.

Juneja, V.K. and Eblen, B.S. (1995) Influence of sodium chloride on thermal inactivation and recovery of non-proteolytic *Clostridium botulinum* type B strain KAP B5 spores. *Journal of Food Protection*, **58**, 813-816.

Juneja, V.K. and Majka, W.M. (1995) Outgrowth of *Clostridium perfringens* in cook-in-bag beef products. *Journal of Food Safety*, **15**, 21-34.

Juneja, V.K. and Marmer, B.S. (1996) Growth of *Clostridium perfringens* from spore inocula in *sous-vide* turkey products. *International Journal of Food Microbiology*, **32**, 115-123.

Juneja, V.K. and Snyder, O.P. (1994) Influence of cooling rate on outgrowth of *Clostridium perfringens* spores in cooked ground beef. *Journal of Food Protection*, **57**, 1063-1067.

Juneja, V.K., Marmer B.S., Philips, J.G. and Miller, A.J. (1995b) Influence of the intrinsic properties of food on thermal inactivation of spores of nonproteolytic *Clostridium botulinum*: Development of a predictive model. *Journal of Food Safety*, **15**, 349-364.

Juneja, V.K., Marmer, B.S., Philips, J.G. and Palumbo, S.A. (1996) Interactive effects of temperature, initial pH, sodium chloride, and sodium pyrophosphate on the growth kinetics of *Clostridium perfringens*. *Journal of Food Protection*, **59**, 963-968.

Juneja, V.K., Eblen, B.S., Marmer, B.S., Williams, A.C., Palumbo, S.A. and Miller, A.J. (1995a)

Heat Resistance of Non-proteolytic Type B and Type E *Clostridium botulinum* spores in phosphate buffer and turkey slurry. *Journal of Food Protection*, **58**, 758–763.

Kim, J. and Foegeding, P.M. (1992) Principles of control in *Clostridium botulinum*: *Ecology and Control in Foods* (eds A.H.W. Hauschild and K.L. Dodds), Marcel Dekker, New York, pp. 279–303.

Kooiman, W.J. and Geers, J.M. (1975) Simple and accurate technique for the determination of heat resistance of bacterial spores. *Journal of Applied Bacteriology*, **38**, 185–189.

Kramer, J.M. and Gilbert, R.J. (1989) *Bacillus cereus* and other *Bacillus* species, in *Foodborne Bacterial Pathogens* (ed, M.P. Doyle), Marcel Dekker, New York, pp. 21–70.

Lee, W.H. and Riemann, H. (1970) Correlation of toxic and non-toxic strains of *Clostridium botulinum* by DNA composition and homology. *Journal of General Microbiology*, **60**, 117–123.

Leistner, L. (1985) Hurdle technology applied to meat products of the shelf-stable product and intermediate moisture food types, in *Properties of Water in Foods* (eds D.Simatos and J.L. Multon), Martinus Nijhoff, Dordrecht, pp. 309–329.

Lerke, P. (1973) Evaluation of potential risk of botulism from seafood cocktails. *Applied Microbiol*, **25**, 807–810.

Lerke, P. and Farber, L. (1971) Heat pasteurization of crab and shrimp from the pacific coast of the United States: Public health aspects. *Journal of Food Science* **36**, 277–279.

Licciardello, J.J. (1983) Botulism and heat processed seafoods. *Marine Fisheries Review*, **45**(2), 1–7.

Lindroth, S.E. and Genigeorgis, C.A. (1986) Probability of growth and toxin production by non-proteolytic *Clostridium botulinum* in rockfish stored under modified atmospheres. *International Journal of Food Microbiology*, **3**, 167–181.

Livingston, G.E. (1985) Extended shelf life chilled prepared foods. *Journal of Foodservice Systems*, **3**, 221–230.

Lund, B.M. and Notermans, S.H. W. (1993) Potential hazards associated with REPFEDs, in *Clostridium botulinum*: *Ecology and Control in Foods* (eds A.H.W. Hauschild and K.L. Dodds), Marcel Dekker, New York, pp. 279–303.

Lund, B.M. and Peck, M.W. (1994) Heat resistance and recovery of spores of non-proteolytic *Clostridium botulinum* in relation to refrigerated, processed foods with an extended shelf life. *Journal of Applied Bacteriology, Symposium Suppl*ement, **76**, 115S–128S.

Lund, B.M., Wyatt, G.M. and Graham, A.F. (1985) The combined effect of low temperature and low pH on survival of, and growth and toxin formation from spores of *Clostridium botulinum*. *Food Microbiol*ogy, **2**, 135–145.

Lund, B.M., Graham, A.F., George, S.M. and Brown, D. (1990) The combined effect of incubation temperature, pH and sorbic acid on the probability of growth of non-proteolytic type B *Clostridium botulinum*. *Journal of Applied Bacteriology*, **69**, 481–492.

Lynt, R.K., Solomon, H.M. and Kautter, D.A. (1971) Immunofluorescence among strains of *Clostridium botulinum* and other *Clostridium* spp. by direct and indirect methods. *Journal of Food Science*, **36**, 594–599.

Lynt, R.K., Solomon, H.M., Lilly, T. and Kautter, D.A. (1977) Thermal death time of *Clostridium botulinum* type E in meat of the blue crab. *Journal of Food Science*, **42** (4), 931–934.

Lynt, R.K., Kautter, D.A. and Solomon, H.M. (1982) Differences and similarities among proteolytic and nonproteolytic strains of *Clostridium botulinum* types A, B, E, and F: a review. *Journal of Food Protection*, **45**, 466–474.

Meng, J. and Genigeorgis, C.A. (1994) Delayed toxigenesis of *Clostridium botulinum* by sodium lactate in 'sous-vide' products. *Letters in Applied Microbiol*ogy, **19**, 20–23.

Meng, J. and Genigeorgis, C.A. (1993) Modeling lag phase of nonproteolytic *Clostridium botulinum* toxigenesis in cooked turkey and chicken breasts as affected by temperature, sodium lactate, sodium chloride and spore inoculum. *International Journal of Food Microbiology*, **19**, 109–122.

Mossel, D.A.A. and Struijk, C.B. (1991) Public health implication of refrigerated pasteurized ('sous-vide') foods. *International Journal of Food Microbiology*, **13**, 187–206.

Mossel, D.A.A. and Thomas, G. (1988) Microbiological safety of refrigerated meals: Recommendations for risk analysis, design and monitoring of processing. *Microbiologie-Aliments-Nutrition*, **6**, 289–309.

NACMCF (National Advisory Committee on Microbiological Criteria for Foods) (1990)

Recommendations for Refrigerated Foods containing Cooked, Uncured Meat or Poultry Products that are Packed for Extended, Refrigerated Shelf Life and that are Ready-to-Eat or Prepared with Little or No Additional Heat Treatment, adopted 31 January 1990, Washington, DC.

NACMCF (National Advisory Committee on Microbiological Criteria for Foods) (1992) *Vacuum or Modified Atmosphere Packaging for Refrigerated Raw Fishery Products*, adopted 20 March 1992, Washington, DC.

Narayan, K.G. (1967) Incidence of the food poisoning clostridia in meat animals. *Zentralbl. Bakteriol. I. Orig*, **204**, 265.

NFPA (National Food Processors Association) Refrigerated Foods and Microbiological Criteria Committee (1988). Factors to be considered in establishing good manufacturing practices for the production of refrigerated foods. *Dairy Food Sanitation*, **8**, 5–7.

Nielsen, H.K. (1991) Novel bacteriolytic enzymes and cyclodextrin glycosyl transferase for the food industry. *Food Technology*, **45**, 102–104.

Notermans, S., Dufrenne, J. and. Keijbets, M.J.H. (1981) Vacuum-packed cooked potatoes: Toxin production by *Clostridium botulinum* and shelf life. *Journal of Food Protection*, **44**, 572–575, 579.

Notermans, S., Dufrenne, J. and Lund , B.M. (1990) Botulism risk of refrigerated, processed foods of extended durability. *Journal of Food Protection*, **53**,1020–1024.

Ohye, D.F. and Scott, W.J. (1957) Studies on the physiology of *Clostridium botulinum* type E. *Australian Journal of Bioogical Sciences*, **10**, 85–94.

Peck, M.W., Fairbairn, D.A. and. Lund, B.M. (1992) The effect of recovery medium on the estimated heat resistance of spores of non-proteolytic *Clostridium botulinum*. *Letters in Applied Microbiology*, **15**, 146–151.

Peck, M.W., Fairbairn, D.A. and Lund, B.M. (1993) Heat resistance of spores of non-proteolytic *Clostridium botulinum* estimated on medium containing lysozyme. *Letters in Applied Microbiology*, **16**, 126–131.

Peck, M.W., Lund, B.M., Fairbairn, D.A., Kaspersson, A.S. and Undeland, P.C. (1995) Effect of heat treatment on survival of, and growth from, spores of non-proteolytic *Clostridium botulinum* at refrigeration temperatures. *Applied and Environmental Microbiology*, **61**, 1780–1785.

Peck, M.W. and Fernandez, P.S. (1995) Effect of lysozyme concentration, heating at 90°C, and then incubation at chilled temperatures on growth from spores of non-proteolytic *Clostridium botulinum*. *Letters in Applied Microbiology*, **21**, 50–54.

Post, L.S., Lee, D.A., Solgberg, M., Furgang, D., Specchio, J. and Graham.C. (1985) Development of botulinal toxin and sensory deterioration during storage of vacuum and modified atmosphere packaged fish fillets. *Journal of Food Science*, **50**, 990–996.

Proctor, V.A. and Cunningham, F.E. (1988) The chemistry of lysozyme and its use as a food preservative and a pharmaceutical. *CRC Critical Review in Food Science and Nutrition*, **26**, 359–395.

Raevuori, M. (1976) Effect of sorbic acid and potassium sorbate on growth of *Bacillus cereus* and *Bacillus subtilis* in rice filling of karelian pastry. *European Journal of Applied. Microbiology*, **2**, 205–213.

Rhodehamel, E.J. (1992) FDA's concerns with *sous-vide* processing. *Food Technology*, **46**, 73–76.

Rhodehamel, E.J., Solomon, H.M., Lilly, T. Jr, Kautter, D.A. and Peeler, J.T. (1991) Incidence and heat resistance of *Clostridium botulinum* type E spores in menhaden surimi. *Journal of Food Science*, **56**, 1562–1563, 1592.

Roberts, T.A. and Ingram, M. (1965) The resistance of *Clostridium botulinum* type E to heat and radiation. *Journal of Applied Bacteriology*, **28**, 125–141.

Rosset, R. and Poumeyrol, G. (1986) Modern processes for the preparation of ready-to-eat meals by cooking before or after sous-vide packaging. *Science Aliments 6 H.S*, **VI**, 161–167.

Russell, A.D. (1982) *The Destruction of Bacterial Spores*, Academic Press, London.

Schmidt, C.F. (1964) Spores of *Clostridium botulinum*: Formation, resistance, and germination. In *Botulism* (eds K.H. Lewis and K. Cassel, Jr), US Public Health Service, Cincinnati, OH, p. 69.

Schmidt, C.F., Lechowvich, R.V. and Folinazzo, J.F. (1961) Growth and toxin production by type E *Clostridium botulinum* below 40°F. *Journal of Food Science*, **26**, 626–630.

Scott, V.N. and Bernard, D.T. (1982) Heat resistance of spores of non-proteolytic type B *Clostridium botulinum. Journal of Food Protection*, **45**, 909–912.

Scott, V.N. and Bernard, D.T. (1985) The effect of lysozyme on the apparent heat resistance of non-proteolytic type B *Clostridium botulinum. Journal of Food Safety*, **7**, 145–154.

Sebald, M., Ionesco, H. and. Prevot, A.R. (1972) Germination IzP-dependante des spores de *Clostridium botulinum* type E. *E.C.R. Acad. Sci. Paris* (Serie D) **275**, 2175–2182.

Segner, W.P., Schmidt, C.F. and Boltz, J.K. (1966) Effect of sodium chloride and pH on the outgrowth of spores of type E *Clostridium botulinum* at optimal and suboptimal temperatures. *Applied Microbiology*, **14**, 49–54.

Selman, J.D. (1990) Time-temperature indicators: how they work. *Food Manufacture*, **65**, 30–34

Simunovic, J., Oblinger, J.L. and Adams, J.P. (1985) Potential for growth of non-proteolytic types of *Clostridium botulinum* in pasteurized, restructured meat products: A review. *Journal of Food Protection*, **48**, 265–276.

Smelt, J.P.P.M. and Haas, H. (1978) Behavior of proteolytic *Clostridium botulinum* types A and B near the lower temperature limits of growth. *European Journal of Applied. Microbiology Biotechnology*, **5**, 143–154.

Smith, L.D.S. and Sugiyama, H. (1988) *Botulism: The Organism, its Toxins, the Disease*, Springfield, IL: Charles C. Thomas.

Solomon, H.M., Kautter, D.A. and Lynt, R.K. (1982) Effect of low temperatures on growth of nonproteolytic *Clostridium botulinum* type B and F and proteolytic type G in crabmeat and broths. *Journal of Food Protection*, **45**, 516–518.

Solomon, H.M., Lynt, R.K., Lilly, T. Jr and Kautter, D.A. (1977) Effect of low temperatures on growth of *Clostridium botulinum* spores in meat of the blue crab. *Journal of Food Protection*, **40**, 5–7.

Sooltan, J.R.A., Mead, G.C. and Norris, A.P. (1987) Incidence and growth potential of *Bacillus cereus* in poultry meat products. *Food Microbiology*, **4**, 347–351.

Stark, R.L. and Duncan C.L. (1971) Biological characteristics of *Clostridium perfringens* type A: in vitro system for sporulation and enterotoxin synthesis. *Journal of Bacteriology*, **144**, 306–311.

SVAC (*Sous-Vide* Advisory Committee). (1991) *Code of Practice for Sous-Vide Catering Systems*, SVAC, Tetbury, Gloucestershire, UK.

SYNAFAP. (1989) *Guidelines of God Hygienic Practices for Prepared Refrigerated Meals*, 44 rue d'Alesia – 75682 PARIS CEDEX 14.

Taclindo, C.T., Midura, G., Nygaard, S. and Bodily, H.L. (1967) Examination of prepared foods in plastic packages for *Clostridium botulinum. Applied Microbiology*, **15**, 426–430.

Taoukis, P.S. and Labuza, T.P. (1989) Applicability of time-temperature indicators as shelf life monitors of food products. *Journal of Food Science*, **54**, 783–788.

Tui-Jyi Chai and Kuang T. Liang (1992) Thermal resistance of spores from five type E *Clostridium botulinum* strains in Eastern oyster homogenates. *Journal of Food Protection*, **55** (1), 18–22.

van Netten, P., van de Moosdijk, A., van Hoensel, P., Mossel, D.A.A. and Perales, I. (1990) Psychrotrophic strains of *Bacillus cereus* producing enterotoxin. *Journal of Applied Bacteriology*, **69**, 73–79.

van Grade, S.J. and Woodburn, M.J. (1987) Food discard practices of householder. *Journal of the American Dietary Association*, **87**, 322–329.

Wells, J.H. and Singh, R .P. (1988) Application of time-temperature indicators in monitoring changes in quality attributes of perishable and semi-perishable foods. *Journal of Food Science*, **53**, 148–152.

Whiting, R.C. (1993) Modeling bacterial survival in unfavorable environments. *Journal of Industrial Microbiology*, **12**, 240–246.

Wyatt, L.D. and Guy, V. (1980) Relationships of microbial quality of retail meat samples and sanitary conditions. *Journal of Food Protection*, **43**, 385–389.

11 Application of combined factors technology in minimally processed foods

STELLA MARIS ALZAMORA

11.1 Introduction

Safety of mild preservation technologies, introduced as a response to increasing consumers' demands for fresher, more natural and less heavily processed foods as well as for convenience foods, are a special problem for the food industry. These new trends have generally implied an important reduction in the intrinsic and extrinsic preservation factors compromising the safety and wholesomeness of the food (Gould, 1995). Among these minimal processing technologies, chilled foods, in particular cook–chill and *sous vide* foods, are considered as the nearest alternative to home-prepared meals and are attracting considerable interest for applications in institutional foodservice, catering and in-home convenience foods in North America and Europe (Bond, 1992).

The cook–chill preservation method comprises a step of cooking-pasteurization followed by rapid chilling and chilled storage of the food below 3°C before controlled reheating and service, while *sous vide* technology includes three categories of processes, the first one being usually referred to when speaking of '*sous vide*' (Bailey, 1995):

- 'la cuisson *sous vide*': cooking a food product inside a hermetically sealed vacuum package, rapid chilling and chilled storage of end product;
- 'cook–chill': vacuum packaging a product that has already been fully cooked and pre-cooled, pasteurization, rapid chilling and chilled storage of final product; and
- 'hot fill': vacuum packaging a cooked product while it is hot, rapid chilling and chilled storage of end product.

Depending on the severity of the pasteurization stage and the storage temperature, typical shelf lives of these products range from less than one week to 42 days or more (in the case of REPFED products).

These methodologies may pose a considerable threat to public safety if they are not properly used. Causes for major concern have centered around:

- insufficient thermal processing to inactive vegetative cells of pathogenic bacteria;

- post-processing contamination by organisms that can be introduced into or onto the food from a number of sources (air, surfaces, humans, other foods, etc.);
- abused temperature conditions, especially the weak links in the chill chain during distribution, retail sale and in the home; as these products usually have a pH $\geq$ 4.6, $a_w \geq$ 0.93 and contain no preservatives, abuse temperatures can lead to germination and outgrowth of spore-forming pathogenic (i.e. *Clostridium botulinum*) as well as to repairing thermally injured cells.

Due to their 'potentially hazardous' character, specific documents dealing with guidelines, recommendations and codes of good hygiene practice on the manufacture, handling, transport and storage temperatures and retail sale of these type of products have been published by industry associations and regulatory agencies of many countries (Turner, 1992; Farber, 1995; Schellekens, 1996). Moreover, because of their reduced microbiological safety margins, work is now being focused not only on the application of strict hygiene and Good Manufacturing Practices (GMP) and implementation of Hazard Analysis and Critical Control Points (HACCP) principles, but also on the design of the preservation system with additional hurdles so that imperfect processing and/or packaging, distribution and storage can still guarantee safe products. In this way, the UK Advisory Committee on the Microbiological Safety of Food (1992) has recommended, in addition to chill storage, the following preservation factors to be used singly or in combination: (1) heat treatment at 90°C for 10 min or equivalent process; (2) a pH $\leq$ 5.0 throughout all components of complex foods; (3) a minimum salt concentration of 3.5% in the aqueous phase of all components; (4) an $a_w \leq$ 0.97 throughout all components of complex foods; and (5) a combination of heat and preservative factors that have been shown to prevent growth of and production of toxin by psychrotrophic *C. botulinum*. Incorporation of multiple barriers in addition to chill temperatures have been also recommended by the US Food and Drug Administration (Rhodehamel, 1992) to inhibit *C. botulinum* growth and the USDA (NACMCF, 1990) suggested secondary barriers formulated into the product as an alternative to full processing for cooked meat products. As Hanlin and Evancho (1992) recognized, the use of GMP and HACCP are efficacious tools in controlling and minimizing the risks during manufacturing but distributors', retailers' and customers' food-handling practices are beyond the control of the food processors. It is mainly in these last stages of the food chain that hurdle technology must play its role.

Many possible additional hurdles proposed by various authors for being used in combination with *sous vide* process have been critically reviewed by Grant and Patterson (1995) and by Schellekens (1996). Biopreservation or the use of sodium lactate and irradiation have been suggested as the most promising potential hurdles by these authors, respectively. However, it was also recognized that the commercial potential of these hurdles has to be fully evaluated as there are very few practical applications to date.

This chapter intends to analyze the prospects of enhancing the microbiological safety of *sous vide* and cook–chill products by using the combined factors approach. Some facts that might compromise the success of the combined technologies (e.g. the stress factor – food matrix interaction and the induced tolerance of microorganisms when exposed to sublethal stresses) are also outlined. The aim of this chapter is not to present an exhaustive compilation of information about potential factors used in combination to reduce microbial hazards in these foods, but to discuss some ideas for a rational design of the combined preservation systems.

11.2 Combined factors technique

Homeostasis or internal media stability (composition and volume of fluids) is vital for survival and growth of microorganisms. Preservation procedures are effective when they overcome, temporarily or permanently, the various homeostatic reactions that microorganisms have evolved in order to resist stresses (Gould *et al.*, 1995). Table 11.1 shows, as mode of example, the mechanisms of action of some environmental stresses in foods or during processing and the microbial homeostatic reactions produced by their application.

Homeostatic mechanisms that vegetative cells have evolved in order to survive extreme environmental stresses are energy-dependent and allow microorganisms to keep functioning. In contrast, homeostasis in spores is passive, acting to keep the central protoplast in a constant low-water-level environment, this being the prime reason for the extreme metabolic inertness or dormancy and resistance of these cells. It seems likely that the structural arrangement of the cortex of the spore is responsible for the dehydration of the protoplast (Gould *et al.*, 1983; Gould, 1977).

In foods preserved by combined methods, the active homeostasis of vegetative microorganisms and the passive refractory homeostasis of spores are disturbed by a combination of gentle antimicrobial factors at a number of sites or in a cooperative manner (Gould and Jones, 1989; Leitsner, 1995a, b). For example, for vegetative cells (where homeostasis is energy dependent), the goal is to reduce the availability of energy (removing O_2, limiting nutrients, and reducing the temperature) and/or to increase the demand for energy (reducing a_w, reducing pH, and adding membrane-active compounds). For spores (where homeostasis is non-energetic and depends on the structures of the organism), the goal is to damage key structures (by chemical, enzymic, or physical attacks on coats, cortex, etc.) or to release spores from dormancy (initiating germination with natural germinants or with false triggers, or applying high pressures).

Table 11.1 Modes of action of some stress factors and homeostatic response in microorganisms

Stress factor	Principal effect	Homeostatic mechanisms
Vegetative cells		
Low water activity	Loss of water and consequent reduction or prevention of metabolism.	Accumulation by synthesis or active transport of compatible cytoplasmic solutes to balance external osmolality avoiding excessive loss of water from cell and of membrane turgor; changes in membrane lipids.
Low pH		
Strong acids	Denaturation of enzymes on cell surface and lowering of cytoplasmic pH due to proton permeation through membrane in response to the increased pH gradient.	Control of the activity of ion transport systems that facilitate proton entry to maintain the intracellular pH within a narrowed range (mechanisms not fully understood).
Lipophilic weak acids	Lowering of cytoplasmic pH due to undissociated acid passage through the membrane; specific effects of acid anion on metabolism; extensive protein denaturation and DNA damage in some organisms.	Increase of ATPase activity, synthesis of new proteins and synthesis of positively charged aminophospholipids that impart to the membrane surface a net positive charge that acts as a barrier to protons, etc.

Table 11.1 Continued

Stress factor	Principal effect	Homeostatic mechanisms
Low temperature	Reduction or inhibition of growth rate; alteration of metabolic activity and substrate uptake (i.e. permeases unable to combine with substrates due to changes in the molecular architecture of lipid bilayer of the cytoplasmic membrane and arrest of active transport of solute across membrane due to insufficient energy in mesophilic organisms).	Depending on the organisms involved, alterations in fatty acid and phospholipid composition of membrane so as to obtain lipids with a lowered-gel-to-liquid-crystalline state transition temperature maintaining the membrane fluidity; increased enzyme synthesis in psychrophiles to compensate for reduced enzyme activity.
Thermal treatment	Single- and double-strand breaks in DNA, inactivation of specific enzymes, RNA degradation, leakiness of membranes and heat-induced damage of the various chemiosmotic and transport functions of the cell.	Repair of DNA; synthesis of 'heat-shock' proteins leading to a rise in heat resistance when long, slow heating is involved.
Ionizing radiation	Damage to DNA	Development of enzymic mechanisms for repairing ionizing radiation-induced lesions in DNA.
Nisin	Pore formation in cytoplasmic membrane; leakage of ATP from target cells.	Synthesis of nisinase. Changes in the bacterial membrane structure to decrease membrane fluidity.
Spores Thermal treatment	Depurination and depyrimidation of spore DNA; enzyme attack on apurinic DNA after germination; destruction of the activity of the enzyme(s) on the germination pathway.	Resistance/dormancy mechanisms comprise low core water content by some osmotic and/or pressure-generating function of the cortex and 'mineralization' and immobilization of ions in the core of the spore or adaptation ('heat shock response') in vegetative cells.

Adapted from Gould (1989), Gould et al. (1995), Abee et al. (1995) and Tapia de Daza et al. (1996).

Antimicrobial preservation of foods by combined methods should be considered not only as interference of the homeostasis by additive or synergistic hurdles on the same microorganism, but also as the selective application of preservation factors that may be effective against a specific organism or group of organisms, while not against others. All preservation techniques are primarily based on prevention of the growth of food spoilage and poisoning organisms, but the preservation of other attributes of foods is of additional concern. The combined technology concept refers not only to microbiological stability, but also to total quality of foods, comprising the application of preservation factors to inhibit or delay physicochemical and biochemical reactions deleterious to color, texture, flavor, and nutritive value of foods (Tapia de Daza et al., 1996).

11.3 Examples of potential safety hurdles for *sous vide* and cook–chill foods

Over the last 10 years, the popularity of the hurdle concept has dramatically increased and numerous papers dealing specifically with its application for preventing growth of hazardous microorganisms in minimally processed foods have been published. Some of the reported examples of additional hurdles intended to *sous vide* and cook–chill foods to be microbiologically safe as well as of other barriers of potential use are summarized in Table 11.2. This list is far from being exhaustive, as *sous vide* and cook–chill foods represent an extremely wide range of food types and shelf lives, but gives a clear picture of the variety of possibilities that are being considered for the design of the combined techniques. It is stressed than more than 50 potential traditional and emerging hurdles of significance for the safety and/or quality of foods of animal or plant origin have been identified by Leitsner (1994).

An analysis of the large amount of information on this subject highlights the following aspects:

- In many studies, the intensity of the hurdles is detrimental to the food quality. Effective control of microbial growth is a necessary effect of a hurdle factor; however, to be useful, that barrier also must not cause deleterious effects on flavor, color or odor.
- Most work has been focused on the response of key microorganisms to stresses in laboratory media and in model systems. Few studies have used actual foods, taking into account the interactions between the intrinsic factors, the food matrix and the stresses applied. Moreover, when real foods were assayed, native flora had generally been previously inactivated. The ecology of pathogens and non-pathogens in foods is very complex and their interactions are difficult to predict.

Table 11.2 Examples of barriers suggested for using in combination in the formulation of *sous vide* and cook-chill foods

Stress factors	Target microorganism(s) and food or test medium	Critical control point	References
SV[a] + sodium lactate	C. botulinum (non-proteolytic) beef, chicken breast, salmon, comminuted turkey meat	Storage (mild temperature-abuse conditions)	Meng and Genigeorgis (1994); Maas et al. (1989)
SV + a_w reduction + lowering of pH	C. botulinum (proteolytic types A and B) spaghetti and meat sauce	Storage (mild temperature-abuse conditions)	Simpson et al. (1995)
CC[b] + pediocin AcH	L. monocytogenes Scott A raw and cooked chicken	Storage (mild temperature-abuse conditions)	Goff et al. (1996)
CC + Carnobacterium piscicola LKS	L. monocytogenes Scott A cooked meat, pasteurized crabmeat	Storage (mild temperature-abuse conditions)	Buchanan and Klawitter (1992)
CC + Lactobacillus bavaricus MN	L. monocytogenes beef and beef in gravy	Storage (mild temperature-abuse conditions)	Winkowski et al. (1993)
SV + NaCl + sodium pyrophosphate + reduction of pH (5.5)	C. perfringens ground beef	Storage (mild temperature-abuse conditions)	Juneja and Majka (1995)
SV + dipping in 1% acetic acid	C. sporogenes pork chops	Storage (temperature-abuse conditions)	Prabhu et al. (1988)
SV + sodium lactate + Brifisol 414™	L. monocytogenes, C. sporogenes beef roasts	Storage (temperature-abuse conditions)	Unda et al. (1991)
SV + Brifisol 414™ + glyceryl monolaurin	L. monocytogenes, C. sporogenes beef roasts	Storage (temperature-abuse conditions)	Unda et al. (1991)
SV or CC + slight reduction of pH + sodium lactate	Various food spoilage organisms and pathogens, laboratory medium	Storage (temperature-abuse conditions)	Houtsma et al. (1996)
SV or CC + slight reduction of a_w (0.97–0.98) and pH (5.3–5.5) (HCl)	C. botulinum type E and type B laboratory medium	Storage (temperature-abuse conditions)	Baird-Parker and Freame (1967); Emodi and Lechowich (1969)

SV or CC + NaCl (1.5%) + sage	*B. cereus* laboratory medium, rice, meat	Storage (temperature-abuse conditions)	Shelef *et al.* (1984)
SV or CV + slight reduction of pH (citric acid)	*C. botulinum* (type A, proteolytic and non-proteolytic type B, type E) laboratory medium	Storage (temperature-abuse conditions)	Graham and Lund (1987)
SV or CC + monolaurin + reduction of pH + NaCl	*L. monocytogenes* Scott A culture medium	Storage (temperature-abuse conditions)	Bal'a and Marshall (1996)
SV or CC + NaCl + potassium sorbate + sodium acid pyrophosphate + reduction of pH	*C. botulinum* (types A and B) laboratory medium	Storage (temperature-abuse conditions)	Wagner and Busta (1984)
CC + slight reduction of a_w and pH (5.6–6.2)	*E. coli*, salmonellae laboratory medium	Storage (temperature-abuse conditions and recontamination)	Gibson and Roberts (1986)
SV + low-dose irradiation	*L. monocytogenes* chicken breast meat	Thermal treatment step and storage (mild temperature-abuse conditions)	Shamsuzzaman *et al.* (1992)
Low-dose irradiation + heat treatment	*C. botulinum* spores bacon	Thermal treatment step	Rowley *et al.* (1983)
Slight reduction of pH + NaCl + sodium pyrophosphate + heat treatment	*C. botulinum* (non proteolytic types B and E) turkey slurry turkey	Thermal treatment step	Juneja and Majka (1995)
Polyphosphates + heat treatment	Salmonellae laboratory medium	Thermal treatment step	Seward *et al.* (1986)

[a]SV: *sous vide* process.
[b]CC: cook–chill process.

- Even in model systems, studies on the effectiveness of the interaction of various factors for inhibiting microbial growth used at very low concentrations or in very low doses (i.e. $a_w \geq 0.97$; pH ≥ 5.0–5.3; concentrations of antimicrobials below the sensorial threshold) are lacking. Only such minute intensities of additional hurdles would allow the optimum quality for marketing of these chilled foods. In the same way, there are very few systematic studies concerning the influence of gentle stresses used in combination on the lag phase of growth of pathogenic microorganisms, either at chilled or refrigerated temperatures or in temperature abuse conditions. This investigation is very important since these techniques usually work, in the case of spore-forming pathogens, by increasing the lag phase.
- Most proposed hurdles had been selected for inhibiting pathogenic and spoilage organisms during storage of the food subjected to temperature abuse. But the effect of those hurdles on the effectiveness of the other stages of the preservation process had not been evaluated. For instance, microbial sensitivity to heat is increased as the pH becomes more acidic, as some antimicrobials are included or as when a low irradiation dose is applied; while thermal resistance may be increased or decreased as a_w is lowered, depending on the particular microorganism and on the solutes present.

Based on these considerations, much remains to be done regarding the contribution of the hurdle approach to the safety of *sous vide* and cook–chill foods. In addition, as Gould *et al.* (1995) pointed out, the great progress done on the physiology of the most important target microorganisms in relation to factors affecting growth and survival has not yet been sufficiently exploited in the control of food-poisoning microflora in a practical way for commercial interest.

11.4 Issues for concern in the application of hurdle technology

The success of hurdle technology in increasing the safety of minimally processed foods can be compromised in many ways. Two phenomena that can significantly affect the efficiency of the combined approach are addressed below.

11.4.1 Adaptation of microorganisms to sublethal stresses

Spore-forming and non-spore-forming microorganisms sometimes react or adapt to mild stress factors by developing some mechanisms to repair the damages and to become even more resistant, surviving more severe homologous or heterologous stresses (Gould *et al.*, 1995; Knochel and Gould,

1995). Much of the research on the so-called Global Stress Response has only focused on the heat stress response and on acid adaptation, the physiological basis involved being not yet fully understood. For example, bacteria which have been exposed to mildly acidic conditions acquire the ability to survive not only normally lethal pH values, but also other stresses that microorganisms may meet in foods (Hill *et al.*, 1995; Lou and Yousef, 1996).

Foster (1995) has explored inducible acid survival of *Salmonella typhimurium*. Lag phase cells adapted to pH 5.8 (pre-acid shock) prior to an acid challenge of pH 3.3 (post-acid shock) survived this last condition better than unadapted cells that were grown at pH 7.7. It seems that an elevated pH homeostasis induced at pH 5.8 allows synthesis of a set of 50 proteins (called 'acid shock proteins') required for maximum protection against pH. These proteins are not synthesized if the pre-acid shock is not performed. Acid adaptation of *S. typhimurium* cells also increased their survival in cheese and their resistance to heat, osmotic stress, organic acids, lactoperoxidase system and hydrophobic and surface-active compounds (Leyer and Johnson, 1992; Lou and Yousef, 1996). Hill *et al.* (1995) found that the ability of *Listeria monocytogenes* L028 to survive at pH 3.5 was dramatically enhanced by prior induction for 90 min at pH 5.0, although no growth was observed. They also reported that acid-tolerant strains of *L. monocytogenes* were better adapted to survive in fermented dairy products. These authors concluded that the acid tolerance response and the cross-induction of other stress responses are extremely significant in minimally processed foods, in which low levels of different stresses are applied and pointed out the high risks involved if these stresses are employed in sequence.

In recent work, Archer (1996) focused on the ability of environmental stresses to increase bacterial virulence promoting adaptive mutations that may select strains of invasive organisms that are even more virulent. This enhanced virulence is extremely serious in immuno-compromised consumers.

The thermotolerance of *L. monocytogenes* in the presence of stress factors was studied by Lou and Yousef (1996). Exposing cells to sublethal environmental stresses, such as starvation, ethanol, acid and H_2O_2, greatly increased the resistance of this pathogen to heat, the heat resistance being much greater at the late than at the early stages of growth. As a consequence of the stress-induced cross-protection, these authors introduced the 'stress hardening' concept as a component of the 'hurdle' concept, remarking that stress hardening may counterbalance the benefits of the hurdle approach.

Heat resistance of *Escherichia coli* O157:H17 in a nutrient medium and in ground beef patties was influenced by storage and holding temperatures. Cultures stored frozen (−18°C) without holding at elevated temperatures had greater heat resistance than those stored under refrigeration (3°C) or at 15°C, perhaps due to physiological changes within bacterial cell as a result of freezing (Jackson *et al.*, 1996). This result indicates that food-handling conditions should be optimized for maximizing the effectiveness of cooking treatments.

Increases in D-values (up to 220% compared to the control) for *Salmonella enteritidis* following heat shock (42°C for 60 min) were reported by Xavier and Ingham (1997). This study suggested that: (i) short-term temperature abuse of food containing *S. enteritidis* may render these cells much more resistant to subsequent heat treatments; (ii) anaerobic microenvironments may enhance survival of heat-stressed cells (i.e. increases in D-values up to 28% compared to the aerobic value); and (iii) heat shock results in the overexpression of proteins that may be related to increased thermotolerance. Increased heat resistance after exposure to a supra-optimal but non-lethal temperature prior to heating has been also found for *Listeria monocytogenes, Salmonella thompson, Escherichia coli* O157:H7 and non-pathogenic *E. coli*. Again, the thermotolerance of *L. monocytogenes, S. enteritidis* and *Staphylococcus aureus* was enhanced in anaerobic conditions during and after heating.

After adaptation to mild stresses, cells behave differently from unadapted ones and can grow at values outside the traditional known ranges of temperature, water activity and pH determined under nearly optimal conditions (Hill *et al.*, 1995). Therefore, microbiological challenge testing to assess the risk of food poisoning or to establish minimally processed product stability needs careful design and stressed known or potential pathogens would be preferably selected for food inoculation.

11.4.2 Interaction between stress factors and food matrix

The intensity of the hurdles may change along product storage and/or the initial intensity of the hurdles may be less than the levels applied. Many phenomena affect the efficacy of the barriers, among others binding to food components such as proteins and fats; chemical degradation; inactivation and/or biological destabilization by other ingredients or components; pH and temperature effects on hurdle stability and activity; physical losses by mass transport from the food to the environment; poor solubility and uneven distribution in the food. Some examples are illustrated below.

It is well known that sorbates, commonly used as preservatives in many food products, can be destroyed through an autoxidation mechanism, its destruction affecting food product safety (Gerschenson and Campos, 1995). Sorbic acid stability during processing and storage is largely dependent, among others, upon pH, a_w, temperature, light, presence of oxygen, type of packaging material and other components of the system. For example, when studying the loss of sorbic acid in meat products of an a_w of 0.91, rate constants of meat systems were higher than the ones obtained for aqueous systems of similar composition, revealing that meat components played an important role in the destruction of sorbates (Campos *et al.*, 1995).

Many natural antimicrobials (e.g. bacteriocins, some phenolics and essential oils) are less effective inhibitors *in vivo* (e.g. food matrix) than *in vitro* (Schillinger *et al.*, 1996; Gould, 1996; Davidson, 1993), proteins, lipids, salts, pH and temperature affecting their antimicrobial activity. For instance, the presence of lipids in foods or test media dramatically decreases the effectiveness of BHA against *V. parahaemolyticus, C. perfringens* and *A. flavus*, and nisin activity against *L. monocytogenes* in fluid milk decreases with increasing fat concentration. These preservatives probably bind to those compounds thereby reducing their availability to inhibit microorganisms. Heating also results in some reduction of phenolic activity because they react with different substances of the food (Lindberg *et al.*, 1995).

The presence of di- and trivalent ions (such as Mg^{2+}, Ca^{2+}) has been reported to reduce the efficiency of nisin against Gram-positive bacteria due to inhibition of electrostatic interactions between positive charges on the bacteriocins and the negatively charged headgroups of the phospholipid molecules of the cytoplasmic membrane (Abee *et al.*, 1995). Also, nisin is more resistant to heat under acidic conditions and its stability and solubility decrease as the pH increases (Delves-Broughton, 1990; Hurst and Hoover, 1993).

11.5 Guidelines in the design of safe *sous vide* and cook–chill foods by combined methods

The remainder of this chapter will focus on introducing some guidelines for a rational design of safe *sous vide* and cook–chill foods using the hurdle approach. Figure 11.1 presents 11 steps suggested to select the barriers and implement the preservation process.

The first steps of the guidelines are coincident with three of the seven HACCP principles as defined by the National Advisory Committee on Microbiological Criteria for Foods (1989). Once described (step 1), the product and its intended use must be subjected to a hazard analysis (step 2). A microbiological hazard analysis requires the knowledge of the food ecosystem (i.e. microorganisms present and their responses to stresses, pH, type of acidulant, a_w, O_2, redox potential, presence of antimicrobial compounds, structure of the food, etc.), and the estimation of (a) the probability that spoilage and pathogenic microorganisms are present in the food and (b) the levels of these microorganisms that can be expected after processing. It is important to consider the types of microorganisms of concern likely to be present in raw materials or introduced during handling or processing and the effect of each processing or handling stage on their ability to grow, survive or die. Major concerns over microbiological safety of *sous vide* and

1. Define the product: description of the product, raw materials and ingredients; development of a preliminary process flow diagram; establishment of desired sensory properties and target shelf life, usual marketing conditions, intended consumer (immuno-competent or immuno-comprised), consumer practices, potential abuse of the food.

2. Assess hazards and risk associated with raw materials and ingredients, processing, manufacturing, distribution, marketing, preparation and consumption of the food.

3. Determine critical control points (CCPs) required to control the identified hazards and document the location and type of CCP in the flow diagram.

4. Analyse critically each CCP, including the type of hazard, procedures or processes to control the hazard and define the critical limits or tolerances that applied to each CCP.

5. Determine which CCPs are difficult, impossible or impractical to control in a direct way. In these points, multiple barriers must be introduced into the formulation and/or in the process of the product as a 'back-up' measure if the CCP is out of control.

6. Screen and select potential hurdles and their levels and the way of application and/or process conditions taking into account:
   (a) the sensory requirements of the product;
   (b) the literature data about microbial reaction to food related stresses and predictive microbiology in order to estimate the consequences of the processing and formulation changes on the growth, survival and inactivation of food-poisoning and food-spoilage microorganisms;
   (c) the understanding of physiological basis of microorganisms' survival and resistance to multiple stresses imposed by the use of factors in combination, including the 'stress hardening' effect;
   (d) the possible inactivation of the hurdles by interaction with food components.

7. Manufacture the *sous vide* or the cook–chill food with the additional hurdle(s) selected and evaluate in the final product and during storage:
   (a) sensory/chemical characteristics;
   (b) microbial evaluation of native flora at the CCPs of the whole process.

8. Perform challenge tests with key food-poisoning and spoilage microorganisms to determine the safety and the shelf life of the product. These tests must include fluctuations in environmental conditions (i.e. temperature during storage) and the use of the stressed relevant microorganisms.

9. Modify (if applicable) the hurdle and their levels of the process conditions; if this applies, go back to 7.

10. When 9 does not apply, scale up the process for validating at the industrial scale.

11. Implement a HACCP program for the industrial process.

Figure 11.1 Suggested guidelines for the formulation of *sous vide* and cook–chill foods with additional safety hurdles. (Adapted from Leistner, 1995a, Tapia de Daza *et al.*, 1996, Corlett, 1991 and Hanlin *et al.*, 1995.)

cook–chill foods have been detailed by various authors (Brackett, 1992; Brown, 1992; Brown and Gould, 1992; Walker, 1992; Hanlin et al., 1995; Snyder, 1995). Table 11.3 lists the pathogens that can be hazardous during cold storage (mainly psychrotrophic bacteria) as well as other pathogens that are able to grow in temperature abuse conditions. The minimum levels (all other environmental conditions are optimal) of a_w, pH and temperature permitting growth in laboratory media and their thermal resistance are also indicated.

During the cooking or pasteurization step, any heat-sensitive microorganisms (eg. spoilage bacteria, infectious pathogens and some spore-formers) are killed and growth of spores that have survived the heating treatment is inhibited by chilled storage. For short shelf life (<10–14 days) products, the main microbiological risk is the presence of infectious pathogens and the heat process should cause at least a 6 log reduction ($\approx 70°C$ for 2 minutes in the slowest heating point) in the number of these microorganisms (*Salmonella* and *Listeria*). Handling after heating and packaging should prevent recontamination in cook–chill foods. For long shelf life products, the heat process must eliminate any spores capable of growth during prolongued storage and must be at least equal to 6D for cold-growing strains of *C. botulinum* ($\approx 90°C$ for 10 minutes), or greater if spores of psychrotrophic *Bacillus* species must be also eliminated (Brown and Gould, 1992). As these types of products usually have neutral pH, high a_w and contain no preservatives; neither pH nor a_w is an inhibiting factor and the storage of the finished product at temperature below 3.3°C is the most important factor in controlling microorganisms that have either resisted the pasteurization step or recontaminated the product. As can be seen in Table 11.3, storage at chilled temperatures cannot prevent all pathogenic growth but can inhibit the growth of some pathogens and decrease the rate of growth in others.

Not only minimum values of stress factors for growth are important to know, but the time–temperature relationships for growth and for toxin production are essential to establish the shelf life. For instance, botulinum spores are of no consequence unless they are able to grow out and produce toxin. There is a general agreement than non-proteolytic strains of *C. botulinum* can grow and produce toxin within 31 to 129 days at temperatures as low as 3.3°C if all other conditions are favorable (Simunovic et al., 1985). This time for toxin production is dependent, among others, on the substrate, the previous heat treatment, the storage temperature and the salt concentration. Unfortunately, there are not many reported studies on these relationships for other microorganisms of concern in these chilled foods.

Steps 3 and 4 refer to the identification of the Critical Control Points (CCPs) in the process and to the establishment of the critical limits for preventive measures associated with each CCP.

A multitarget approach can be applied at every stage in the food chain,

Table 11.3 Minimum values of temperature, pH and a_w for growth and thermal resistance of pathogenic microorganisms of concern in *sous vide* and cook–chill foods

Critical hazard	Microorganism	T (°C)	pH	a_w	D (T°C)	z (°C)
Infectious pathogens	*Salmonella* species	5.1	3.8–4.0	0.92–0.95	1.7 min (60)	5.6
	Listeria monocytogenes	−0.1	4.4	0.90–0.92	2.85 min (60)	5.8–6.3
	Vibrio parahaemolyticus	5	4.8	0.94–0.96	0.8–48 min (47)	
	Aeromonas hydrophila	0	4.0	0.94	0.19 min (55)	
	Campylobacter jejuni and *coli*	25–30	5.3	0.985	12–21 s (58.3)	6.0–6.4
	Escherichia coli O157	4–7.1	4.4	0.935–0.95		5.1–5.8
	Yersinia enterocolitica	0	4.6	0.95	0.24–0.96 min (62.8)	
Toxinogenic pathogens	*Staphyloccus aureus*	5.0–7.7	4.0	0.86	5.2–7.8 min (60)	5.4–5.8
Toxinogenic spore-forming pathogens	*Clostridium botulinum* type E and non-proteolytic strains B & F	3.3	5.0	0.96–0.97	0.49–0.74 min (82.2)	5.6–10.7
	Bacillus cereus	<4	4.4–4.9	0.90–0.93	1 min (60)	6.9
	Clostridium perfringens	15	5.5	0.93–0.95	7.2 min (59)	3.8
	Clostridium botulinum type A and proteolytic B strains	10–12	4.6	0.94	0.3–0.23 min (121.1)	10

Adapted from Brown (1992), Brown and Gould (1992) and Snyder (1995).

from primary production of raw materials through manufacture, distribution, retail and consumer use. Of course, if CCPs are ranked in terms of overall concern in relation to severity of a hazard that needs to be controlled and the possibility of the occurrence of the hazard, the efforts must be concentrated on the storage stage as temperature and also post-process contamination cannot be effectively controlled beyond the point of sale. But changes in the intrinsic factors – produced in the formulation when additional hurdles are added to act during abuse temperature conditions – may affect the other steps increasing or decreasing their microbiological risks, specially during the pasteurization step. Thus, the whole system must be considered to avoid food safety problems.

Although specific hazards and CCPs differ depending on the product and processing techniques, there are three major control points for HACCP procedures of *sous vide* and cook–chill products where the hurdle approach may contribute to safety maintenance:

- the incoming product control (i.e. by decontamination of raw material and ingredients to reduce initial microbial load);
- the cooking/pasteurization step (i.e. by addition of heat-sensitizing additives, reduction of pH, etc., for reducing the thermal resistance);
- the storage stage (i.e. by addition of hurdles that inhibit growth of pathogens or increase the lag phase for growth).

After selecting the hurdles for integrating the combined preservation system, the effectiveness of barrier(s) must be verified for each product, not only by analyzing the native flora but by conducting appropiate challenge studies with key microorganisms. These experiences must contemplate conditions of product abuse and the possibility of induced tolerance in pathogenic microorganisms.

11.6 Conclusions

The combined factors technology has enormous potential to improve the margin of safety of low-acid cook–chill and *sous vide* foods. But many key issues need to be addressed in order for combination preservation to be useful in the food industry. The combined technology, to be successful, must be based on the HACCP concept, the application of predictive microbiology and the literature data on microbial response to multiple stresses, the knowledge of the interaction stress factor–food matrix and on a careful challenge testing program as the interactive effects of a number of factors are not always predictable. The so-called 'hardening' effect – that is, the induced tolerance of microorganisms as response to sublethal stresses – could affect the safety of the food. The formulation of the products and the process must be designed in such a way as to minimize the possibility of this occurring.

References

Abee, T., Krockel, L. and Hill, C. (1995) Bacteriocins: modes of action and potentials in food preservation and control of food poisoning. *International Journal of Food Microbiology*, **28**, 169–185.

Advisory Commitee on the Microbiological Safety Of Food (1992) *Report on Vacuum Packaging and Associated Processes*, HMSO, London, UK.

Alzamora, S.M., Cerrutti, P., Guerrero, S. and López-Malo, A. (1995) Minimally processed fruits by combined methods, in *Food Preservation by Moisture Control. Fundamentals and Applications*, 1st edn (eds G.V. Barbosa-Cánovas and J. Welti-Chanes), Technomic Publishing Co., Lancaster, USA, pp. 463–492.

Archer D.L. (1996) Preservation microbiology and safety: evidence that stress enhances virulence and triggers adaptive mutations. *Trends in Food Science and Technology*, **7**, 91–95.

Bailey, J.D. (1995) *Sous vide*: past, present and future, in *Principles of Modified Atmosphere and Sous Vide Product Packaging*, 1st edn (eds J.M. Farber and K.L. Dodds), Technomic Publishing Co., Lancaster, USA, pp. 243–262.

Baird-Parker, A.C. and Freame, B. (1967) Combined effect of water activity, pH and temperature on the growth of *Clostridium botulinum* from spore and vegetative cell inocula. *Journal of Applied Bacteriology*, **30**, 420–429.

Bal'a, M.F.A. and Marshall, D.L. (1996) Use of double-gradient plates to study combined effects of salt, pH, monolaurin, and temperature on *Listeria monocytogenes*. *Journal of Food Protection*, **59**, 601–607.

Bond, S. (1992) Marketplace product knowledge – from the consumer view point, in *Chilled Foods: A Comprehensive Guide*, 1st edn (eds C. Dennis and M. Stringer), Ellis Horwood Ltd, Chichester, England, pp. 16–37.

Brackett, R.E. (1992) Microbiological safety of chilled foods: current issues. *Trends in Food Science and Technology*, **3**, 81–85.

Brown, H.M. (1992) Shelf life determination and challenge testing, in *Chilled Foods: A Comprehensive Guide*, 1st edn (eds C. Dennis and M. Stringer), Ellis Horwood Ltd, Chichester, England, pp. 289–307.

Brown, M.H. and Gould, G.W. (1992) Processing, in *Chilled Foods: A Comprehensive Guide*, 1st edn (eds C. Dennis and M. Stringer), Ellis Horwood Ltd, Chichester, England, pp. 111–146.

Buchanam, R.L. and Klawitter, L.A. (1992) Effectiveness of *Carnobacterium piscicola* LKS for controlling the growth of *Listeria monocytogenes* Scott A in refrigerated storage. *Journal of Food Safety*, **12**, 217–234.

Campos, C.A., Alzamora, S.M. and Gerschenson, L.N. (1995) Sorbic acid stability in meat products of reduced water activity. *Meat Science*, **41**, 37–46.

Corlett, D.A. Jr (1991) Regulatory verification of Industrial HACCP systems. *Food Technology*, **45**, 144–146.

Davidson, P.M. (1993) Parabens and phenolic compounds, in *Antimicrobials in Foods*, 2nd edn (eds P.M. Davidson and A.L. Branen), Marcel Dekker , New York, pp. 263–306.

Delves-Broughton, J. (1990) Nisin and its uses as a food preservative. *Food Technology*, **44**, 100–117.

Emodi, A.S. and Lechowich, R.V. (1969) Low temperature growth of type E *Clostridium botulinum* spores. 2. Effects of solutes and incubation temperature. *Journal of Food Science*, **34**, 82–87.

Farber, J.M. (1995) Regulations and guidelines regarding the manufacture and sale of MAP and *sous vide* products, in *Principles of Modified Atmosphere and Sous Vide Product Packaging*, 1st edn (eds J.M. Farber and K.L. Dodds), Technomic Publishing Co., Lancaster, USA, pp. 425–458.

Foster, J.W. (1995) Low pH adaptation and the acid tolerance response of *Salmonella typhimurium*. *Critical Reviews in Microbiology*, **21**, 215–237.

Gerschenson, L.N. and Campos, C.A. (1995) Sorbic acid stability during processing and storage of high moisture foods, in *Food Preservation by Moisture Control*, 1st edn (eds G.V. Barbosa-Cánovas and J. Welti-Chanes), Technomic Publishing Co., Lancaster, pp. 761–790.

Gibson, A.M. and Roberts, T.A. (1986) The effect of pH, water activity, sodium nitrite and storage temperature on the growth of enteropathogenic *Escherichia coli* and Salmonellae in a laboratory medium. *International Journal of Food Microbiology*, 3, 183–194.

Goff, J.H., Bhunia, A.K. and Johnson, M.G. (1996) Complete inhibition of low levels of *Listeria monocytogenes* on refrigerated chicken meat with pediocin ACH bound to heat-killed *Pediococcus acidilactiti* cells. *Journal of Food Protection*, 59, 1187–1192.

Gould, G.W. (1977) Recent advances in the understanding of resistance and dormancy in bacterial spores. *Journal of Applied Bacteriology*, 42, 297–309.

Gould, G.W. (1989) *Mechanisms of Action of Food Preservation Procedures*, 1st edn, Elsevier, Essex.

Gould, G.W. (1995) Overview, in *New Methods of Food Preservation*, 1st edn (ed. G.W. Gould), Blackie Academic & Professional, Great Britain, pp. XV–XIX.

Gould G.W. (1996) Industry perspectives on the use of natural antimicrobials and inhibitors for food applications. *Journal of Food Protection*, **Suppl.**, pp. 82–86.

Gould, G.W., Abee, T., Granum, P.E. and Jones, M.V. (1995) Physiology of food poisoning microorganisms and the major problems in food poisoning control. *International Journal of Food Microbiology*, 28, 121–128.

Gould, G.W., Brown, M.H. and Fletcher, B.C. (1983) Mechanisms of action of food preservation procedures, in *Food Microbiology: Advances and Prospects*, 1st edn (eds T.A. Roberts and F.A. Skinner), Academic Press, London, pp. 67–84.

Gould, G.W. and Jones, M.V. (1989) Combination and synergistic effects, in *Mechanisms of Action of Food Preservation Procedures*, 1st edn (ed. G.W. Gould), Elsevier Science Publishers Ltd, Great Britain, pp. 401–421.

Graham, A.F. and Lund, B.M. (1987) The combined effect of sub-optimal temperature and sub-optimal pH on growth and toxin formation from spores of *Clostridium botulinum*. *Journal of Applied Bacteriology*, 63, 387–393.

Grant, I.R. and Patterson, M.F. (1995) The potential use of additional hurdles to increase the microbiological safety of MAP and *sous vide* products, in *Principles of Modified Atmosphere and Sous Vide Product Packaging*, 1st edn (eds J.M. Farber and K.L. Dodds), Technomic Publishing Co., Lancaster, USA, pp. 263–286.

Hanlin, J.H. and Evancho, G.M. (1992) The beneficial role of microorganisms in the safety and stability of refrigerated foods, in *Chilled Foods: A Comprehensive Guide*, 1st edn (eds C. Dennis and M. Stringer), Ellis Horwood Ltd, Chichester, England, pp. 227–259.

Hanlin, J.H., Evancho, G.M. and Slade, P.J. (1995) Microbiological concerns associated with MAP and *sous vide* products, in *Principles of Modified Atmosphere and Sous Vide Product Packaging*, 1st edn (eds J.M. Farber and K.L. Dodds), Technomic Publishing Co., Lancaster, USA, pp. 69–104.

Hill, C., O'Driscoll, B. and Booth, I. (1995) Acid adaptation and food poisoning microorganisms. *International Journal of Food Microbiology*, 28, 245–254.

Houtsma, P.C., Jacora, C. and Rombouts, F.M. (1996) Minimum inhibitory concentration (MIC) of sodium lactate and sodium chloride for spoilage organisms and pathogens at different pH values and temperatures. *Journal of Food Protection*, 59, 1300–1304.

Hurst, A. and Hoover, D.G. (1993) Nisin, in *Antimicrobials in Foods*, 2nd edn (eds P.M. Davidson and A.L. Branen), Marcel Dekker , New York, pp. 369–394.

Jackson, T.C., Hardin, M.D. and Acuff, G.R. (1996) Heat resistance of *Escherichia coli* O157:H7 in a nutrient medium and in ground beef patties as influenced by storage and holding temperatures. *Journal of Food Protection*, 59, 230–237.

Juneja, V.K. and Majka, W.M. (1995) Outgrowth of *Clostridium perfringens* spores in cook-in-bag beef products. *Journal of Food Safety*, 15, 21–34.

Knochel, S. and Gould, G.W. (1995) Preservation microbiology and safety: quo vadis? *Trends in Food Science and Technology*, 6, 127–131.

Leitsner, L. (1994) Further developments in the utilization of hurdle technology for food preservation. *Journal of Food Engineering*, 22, 411–422.

Leitsner, L. (1995a) Principles and applications of hurdle technology, in *New Methods of Food Preservation*, 1st edn (ed. G.W. Gould), Blackie Academic & Professional, UK, pp. 1–21.

Leitsner L. (1995b) Use of hurdle technology in food processing: recent advances, in *Food Preservation by Moisture Control: Fundamentals and Applications*, 1st edn (eds G.V. Barbosa-Cánovas and J. Welti-Chanes), Technomics Publishing Co., Lancaster, USA, pp. 377–396.

Leyer, G.J. and Johnson, E.A. (1992) Acid adaptation promotes survival of *Salmonella* spp. in cheese. *Applied and Environmental Microbiology*, **58**, 2075–2080.

Lindberg, Madsen, H. and Bertelsen, G. (1995) Spices as antioxidants. *Trends in Food Science and Technology*, **6**, 271–277.

Lou, Y. and Yousef, A.E. (1996) Resistance of *Listeria monocytogenes* to heat after adaptation to environmental stresses. *Journal of Food Protection*, **59**, 465–471.

Maas, M.R., Glass, K.H. and Doyle, M.P. (1989) Sodium lactate delays toxin production by *Clostridium botulinum* in cook-in-bag turkey products. *Applied and Environmental Microbiology*, **55**, 2226–2229.

Meng, J. and Genigeorgis, C.A. (1994) Delaying toxigenesis of *C. botulinum* by sodium lactate in *sous-vide* products. *Letters in Applied Microbiology*, **19**, 20–23.

NACMCF (National Advisory Commitee on Microbiological Criteria for Foods) (1990) 'Recommendations for refrigerated foods containing cooked, uncured meat or poultry products that are packed for extended, refrigerated shelf life and that are ready-to-eat or prepared with little or no aditional heat treatment'. Adopted January 31, 1990, Washington, DC.

National Advisory Committee on Microbiological Criteria for Foods (1992) Hazard Analysis and Critical Control Point System. *International Journal of Food Microbiology*, **16**, 1–23.

NYDAM (New York State Department of Agriculture and Markets) (1992) Refrigerated Foods in Reduced Oxygen Packages. Division of Food Safety and Inspection, Albany, NY.

Prabhu, G.A., Molins, R.A., Kraft, A.A., Sebranek, J.G. and Walker, H.W. (1988) Effect of heat treatment and selected antimicrobials on the shelf life and safety of cooked, vacuum-packaged, refrigerated pork chops. *Journal of Food Science*, **53**, 1270–1272, 1326.

Rhodehamel, E.J. (1992). 'FDA's concerns with sous vide processing' *Food Technol.* **46**, 73–76.

Rowley, D.B., Firstenberg-Eden, R., Powers, E.M., Shattuck, G.E., Wasserman, A.E. and Wierbicki, E. (1983) Effect of irradiation on the inhibition of *Clostridium botulinum* toxin production and the microbial flora in bacon. *Journal of Food Science*, **48**, 1016–1021, 1030.

Schellekens, M. (1996) New research issues in *sous vide* cooking. *Trends in Food Science and Technology*, **7**, 256–262.

Schillinger, U., Geisen, R. and Holzapfel, W.H. (1996) Potential of antagonistic microorganisms and bacteriocins for the biological preservation of foods. *Trends in Food Science and Technology*, **7**, 158–164.

Seward, R.A., Lin, C.F. and Melachouris, N. (1986) Heat-sensitization of *Salmonella* by polyphosphates. *Journal of Food Science*, **51**, 471–473, 476.

Shamsuzzaman, K., Chuaqui-Offermanns, N., Lucht, L., McDougall, T. and Borsa J. (1992) Microbiological and other characteristics of chicken breast meat following electron-beam and *sous-vide* treatments. *Journal of Food Protection*, **55**, 528–533.

Shelef, L.A., Jyothi, E.K. and Bulgarelli, M. (1984) Effect of sage on growth of enteropathogenic bacteria in foods and culture media. *Journal of Food Science*, **45**, 29–44.

Simpson, M.V., Smith, J.P., Dodds, K., Ramaswamy, H.S., Blanchfield, B. and Simpson, B.K. (1995) Challenge studies with *Clostridium botulinum* in a *sous-vide* spaghetti and meat-sauce product. *Journal of Food Protection*, **58**, 225–234.

Simunovic, J., Oblinger, J.L. and Adams, J.P. (1985) Potential for growth of non-proteolytic types of *Clostridium botulinum* in pasteurized restructured meat products: a review. *Journal of Food Protection*, **48**, 265–276.

Snyder, O.P. Jr (1995) The applications of HACCP for MAP and *sous vide* products, in *Principles of Modified Atmosphere and Sous Vide Products Packaging*, 1st edn (eds J.M. Farber and K.L. Dodds), Technomic Publishing Co., Lancaster, USA, pp. 325–384.

Tapia de Daza, M.S., Alzamora, S.M. and Welti-Chanes, J. (1996) Combination of preservation factors applied to minimal processing of foods. *Critical Reviews in Food Science and Nutrition*, **36**, 629–659.

Turner, A. (1992) Legislation, in *Chilled Foods: A Comprehensive Guide*, 1st edn (eds C. Dennis and M. Stringer), Ellis Horwood Ltd, Chichester, England, pp. 39–57.

Unda, J.R., Molins, R.A. and Walker, H.W. (1991) *Clostridium sporogenes* and *Listeria monocytogenes* survival and inhibition in microwave-ready beef roasts containing selected antimicrobials. *Journal of Food Science*, **56**, 198–205, 219.

Wagner, M.K. and Busta, F.F. (1984) Inhibition of *Clostridium botulinum* growth from spore inocula in media containing sodium acid pyrophosphate and potassium sorbate with or without added sodium chloride. *Journal of Food Science*, **49**, 1588–1594.

Walker S.J. (1992) Chilled foods microbiology, in *Chilled Foods: A Comprehensive Guide*, 1st edn (eds C. Dennis and M. Stringer), Ellis Horwood Ltd, Chichester, England, pp. 165–195.

Winkowski, K., Crandall, A.D. and Montville, T.J. (1993) Inhibition of *Listeria monocytogenes* by *Lactobacillus bavaricus* MN in beef systems at refrigeration temperatures. *Applied and Environmental Microbiology*, **59**, 2552–2557.

Xavier, I.J. and Ingham, S.C. (1997) Increased D-values for *Salmonella enteritidis* following heat shock. *Journal of Food Protection*, **60**, 181–184.

12 Hurdle and HACCP concepts in *sous vide* and cook–chill products

SUE GHAZALA and ROBERT TRENHOLM

12.1 Introduction

Over the last number of years, urban and suburban consumer demand has increasingly been placed on high-quality convenience foods. *Sous vide* and cook–chill processes, otherwise known as Minimally Processed Refrigerated (MPR) foods, provide two product types meeting this demand. *Sous vide*, literally meaning 'under vacuum', is a process introduced in France that utilizes vacuum packaging, pasteurization and controlled temperature storage to provide an unfrozen extended shelf life. Cook–chill processing is similar to *sous vide* but does not necessarily use a vacuum (Bertelsen and Juncher, 1996). It involves preparing a partially cooked food and placing it into a hermetically sealed container, where a thermal process is given to the food, rendering it completely cooked. The container product then is rapidly chilled and stored in refrigeration conditions until used. The products prepared by these processing methods have been proven to be more palatable and nutritious than traditionally sterilized products.

There are a number of challenges facing the chilled food industry in general, and the *sous vide* and cook–chill products in particular. There can be no doubt in the public's mind regarding the safety of chilled foods (Bangay, 1996). It is necessary to improve production turn-around times and place better controls on distribution channels to remove concerns about shelf life and 'best before' dates. The implementation of HACCP procedures at the supplier, producer, distributor and food service/retail outlet levels are necessary to ensure proper handling of *sous vide* and cook–chill products before, during and after production.

Since there is doubt whether refrigeration alone is enough to ensure safety of MPR foods, such as *sous vide* and cook–chill products, a lot of work is now focused on formulation of products with additional hurdles (pH, a_w, organic acids, protective cultures) to reduce microbiological hazards. According to Schellekens (1996), the use of additional hurdles merits further consideration, despite the complex nature of low-acid and high-moisture levels of typical *sous vide* and cook–chill products. Hurdles in foods are substances or processes inhibiting deteriorative processes. In many cases prevention of deteriorative processes and maintenance of quality are opposing actions. Therefore, to maintain optimal quality, the level of each hurdle must be kept as low as possible, i.e. by the use of combination of various hurdles. Hurdle technology as a concept has proved to be a useful tool in the optimization of traditional food products as well as in the development of novel food products. However, it should be combined, if possible, with the HACCP concept and predictive microbiology.

Three related concepts for quality assurance of foods were investigated within several FLAIR (Food Linked Agro-Industrial Research) projects of the European Union: hurdle technology (used for food design), the HACCP concept (used for process control) and predictive microbiology (used for process refinement). By considering these three different approaches, an overall strategy for securing stable, safe and high total quality foods is now in place. This strategy could be applied in an effective food design, for which a user guide is tentatively suggested (FLAIR, 1997). Based on this FLAIR project (FLAIR, 1997), a 10-step procedure is suggested which could be quite suitable for an effective food design, since it combines the three related concepts (hurdle technology, predictive microbiology and HACCP). The 10 steps still should be considered as a tentative procedure, until further practical experiences with the application of this user guide have accumulated in the food industry. The recommended ten steps for proper food design are listed below (Leistner, 1997):

1. For the modified or novel food product the desired sensory properties and desired shelf life must be defined.
2. A tentative technology for the production of food should be suggested.
3. The food is then manufactured according to this technology, and the resulting product is analyzed for pH, a_w, preservatives or other inhibitory factors, and the temperature for heating (if intended) and storage as well as the expected shelf life are defined.
4. For preliminary stability testing of the suggested food product, predictive microbiology should be employed.
5. The product should now be challenged with food poisons and/or spoilage microorganisms to study a worst case scenario. A somewhat higher level of poisons and/or spoilage microorganisms than would normally occur are inoculated into the product. These inoculated products are then exposed to abusive storage temperatures.
6. If necessary, the hurdles in the product are modified, taking the homeostasis of the microorganisms and the total quality of the food into consideration.
7. The modified product is again challenged with relevant microorganisms, and if necessary the hurdles are modified once more. Predictive microbiology for assessing the safety of the food might be helpful at this stage too.
8. Now the established hurdles of the modified or novel food are exactly

defined, including tolerances. Then the methods for monitoring the process are defined (preferably physical methods should be used).
9. Thereafter, the designed food should be produced under industrial conditions, because the possibilities for a scale-up of the proposed process must be validated.
10. Finally, for the industrial process the critical control point (CCP) and their monitoring have to be established, and thus the manufacturing process should be controlled by HACCP.

12.1.1 Sous vide *and cook–chill health concerns*

The hermetic seal placed on the *sous vide* and cook–chill products, coupled with the mild heat treatment utilized, makes these products highly susceptible to spoilage if proper spoilage controls are not used. It is possible to create a health hazard if pathogens such as *Clostridum botulinum* or *Listeria monocytogenes* are permitted to grow in abused packages. The following concerns are prevalent:

1. *Sous vide* products are generally formulated with little or no chemical preservatives.
2. *Sous vide* products receive minimal thermal processes and are not shelf-stable.
3. Vacuum packaging provides an anaerobic environment, which extends the shelf life by inhibiting normal aerobic spoilage microorganisms but promoting the growth of anaerobic or facultatively anaerobic microorganisms (such as *Clostridium botulinum*) surviving the process.
4. Adequate refrigeration must be maintained at all times to prevent the outgrowth of pathogenic or spoilage microflora.

12.1.2 *Maintaining safety in* sous vide *and cook–chill foods*

It is possible to create and maintain safe *sous vide* products using the following basic principles of *sous vide* production and marketing (Schellekens and Martens, 1993):

1. Only high-quality ingredients are to be used in the product preparation with low overall microbe counts.
2. The package integrity must be ensured through the use of appropriate sealing and packaging systems. The product must survive the physical abuse given during heating, cooling and marketing.
3. The product must be given a thermal process adequate to reduce at least the vegetative forms of pathogens by 4 to 6 log reductions (i.e. reduce numbers from 10^6 to 10^0). The most heat-resistant pathogens are examined when developing thermal processes.
4. The hot thermally processed products must be rapidly cooled to prevent

the outgrowth of surviving microorganisms such as psychrotrophic spores, mesophillic or thermophilic bacteria.
5. Package leaks must be identified quickly and the affected product isolated due to the potential for recontamination.
6. The product must be stored and marketed under a controlled temperature of less than 3°C to prevent surviving psychrotrophic and psychrophilic bacteria outgrowth. Control must be maintained over product temperature at least until the consumer purchases the product.
7. The product must be consumed within a specified time period. This time period must include storage, shipping, handling, sales and consumer consumption delay.
8. The final, and probably most important, principle is that the production process should use a properly administered HACCP system to ensure correct processing preventative measures.

12.2 *Sous vide* and hurdles

In the previous brief discussion of *sous vide* and cook–chill processing systems, it is apparent that the use of a hermetically sealed package and a thermal process coupled with a reduced storage temperature enhances the shelf life of these products. These are three examples of a technology called 'hurdle' technology. The word 'hurdle' describes the obstacles placed in front of sprinters in a track and field sport with the sole purpose of making the journey more difficult, slowing the pace and increasing the complexity of the sport. In the same way, hurdle technology places obstacles in front of microbial growth progress. Many different kinds of 'hurdles' exist and can be used in *sous vide* and cook–chill processing to reduce the potential for spoilage (Schellekens, 1996).

12.2.1 *Types of hurdle used in* sous vide *and cook–chill*

Hurdles are processes, or substances, which inhibit deterioration in foods. Usually the hurdle functions at the molecular or cellular level, disrupting the normal processes found there and inhibiting microbial growth or enzymatic function. Three main types of hurdle are used in the manufacture of foods. These are physical, physicochemical and microbially derived hurdles. Although some hurdles are used alone, more often they are used in a combination to give greater storage life, quality and safety to foods.

(a) Physical hurdles   Physical hurdles are those hurdles using a physical process or material to inhibit microbial growth. Many of the hurdles, which can be used in *sous vide* and cook–chill processing, have been developed and are used in other processes. Many hurdles are in the experimental stages and

cannot be fully utilized in these processes but hold great promise for the future. Examples of physical hurdles are as follows (Bogh-Sorensen, 1997):

1. Heat processing (pasteurizing, blanching)
2. Storage temperature (chilling, freezing)
3. Radiation (ultraviolet (UV), ionizing radiation (irradiation))
4. Electromagnetic energy, EME (microwave energy, radio frequency energy, oscillating magnetic field pulses, high electric field pulses)
5. Photodynamic inactivation
6. Ultrahigh pressure
7. Ultrasonication
8. Packaging (vacuum packaging, active packaging, edible coatings)
9. Modified atmosphere packaging (MAP)
10. Modified atmosphere storage
11. Controlled atmosphere (CA) storage (packaging and storage)
12. Hypobaric storage
13. Aseptic packaging
14. Microstructure.

Of these, the most common physical hurdles used in *sous vide* and cook–chill processes are:

• Heat processing
• Storage temperature
• Packaging
• Modified atmosphere packaging (MAP).

Heat processing. Sous vide and cook–chill products often use a pasteur-ization process to maintain control over a target microorganism that is usually a pathogenic microorganism. Two such target pathogen examples would include *Clostridium botulinum* type E or *Listeria monocytogenes.*

Pasteurization is usually performed in water heated to temperatures of from 60 to 90°C. The aim in pasteurizing the food is to reduce the numbers of target microorganisms, usually pathogenic microorganisms, by 4 to 6 decimal reduction times (D-values). A decimal reduction time (D) is the time in minutes, at a reference temperature, required to obtain a 90% reduction in the number of target microorganisms.

Heat processing may also refer to the cooking step where the food is pre-pared for packaging. The heat utilized at this stage does a lot to reduce the numbers of viable microorganisms in the sample. If the *sous vide* or cook–chill food packages are filled with hot food then there is a greater like-lihood of sanitary conditions prevailing inside the package.

Storage temperature. Reduced storage temperature is one of the most widely utilized hurdles in all product types. The activity of microorganisms subjected to lower temperatures reduces. Reduced microbial activity results in a longer shelf life for the food.

Sous vide and cook–chill foods utilize a pasteurization treatment followed by rapid cooling to control microorganisms not affected by the pasteurization. The target microorganisms, important to the pathogenic spoilage of *sous vide* and cook–chill foods, are normally psychrotrophic or psychrophyllic bacteria. Microorganisms surviving the pasteurization process are largely mesophyllic and thermophyllic bacteria, which do not grow well at reduced temperatures. Refrigeration prevents the growth of these surviving microorganisms. Temperatures less than 3.0°C are normally used.

Packaging. *Sous vide* and cook–chill products rely heavily on convenience as a selling point. Several factors are important to selecting the appropriate *sous vide* package:

- Appearance: the retail consumer requires that the package be attractive and functional.
- Size: the portions of food provided must be sufficient to feed a given quantity of individuals.
- Hermetic seal: the package must provide a hermetic seal over the product separating it from recontamination and post process spoilage.
- Wear resistance: the package must be able to withstand abuse, protecting the contents against spoilage.
- Heat resistance: the packaging must be able to withstand pasteurizing temperatures while maintaining a hermetic seal.

Packaging is considered a hurdle because it prevents microorganisms outside of the hermetic seal from touching the food. It also promotes conditions within the package inhibiting microbial growth.

Oxygen permeability is a packaging factor helping to control the growth of either anaerobic or aerobic bacteria. Low oxygen permeability, or impermeability, promotes anaerobic microorganism growth. A high oxygen permeability reduces anaerobic growth and promotes aerobic growth over time. *Sous vide* products use an impermeable film to restrict oxygen and thereby eliminate the growth of aerobic microorganisms.

Modified atmosphere packaging. Modified atmosphere packaging (MAP) is physical hurdle utilizing a process where the headspace gases of a product are replaced with a known artificial gas mixture. The gas mixture used normally provides an environment which is not favorable for the growth of spoilage microorganisms. Three gases are normally used: nitrogen, oxygen and carbon dioxide. The vacuum package is not, strictly speaking, a modified atmosphere package but has habitually been included as one due to the similarity in process. *Sous vide* products utilize a vacuum to eliminate oxygen in the package and thereby inhibit aerobic microorganism growth.

It is possible to use some modified atmosphere to provide the same result, however, the gases will expand, increasing the internal pressure, during pasteurization. The expanded gases, in a hermetically sealed flexible pouch or tray, will reduce heat transfer into the product requiring longer process times. In addition, it is possible that the increased pressure may burst the flexible containers. For these reasons, gases are not usually used in *sous vide* or cook–chill processing.

(b) Physicochemical hurdles Physicochemical hurdles are those hurdles related to the actual physical make-up and chemistry of the product. General examples of these hurdles are as follows (Bogh-Sorensen, 1997):

- Water activity (a_w)
- pH
- Redox potential (E_h)
- Salt (NaCl)
- Nitrite ($NaNO_2$)
- Nitrate ($NaNO_3$ or KNO_3)
- Carbon dioxide (CO_2)
- Oxygen (O_2)
- Ozone
- Organic acids
- Ascorbic acid
- Sulphite or SO_2
- Smoking
- Phosphates
- Glucono-δ-lactone (GDL)
- Phenols
- Chelators
- Surface treatment agents
- Ethanol
- Propylene glycol
- Maillard reaction products (MRPs)
- Spices and herbs
- Lactoperoxidase
- Lysozyme.

Physicochemical hurdles important to *sous vide* and cook–chill products are as follows:

- Water activity (a_w)
- pH
- Redox potential (E_h)
- Salt (NaCl)
- Carbon dioxide (CO_2)
- Oxygen (O_2)
- Organic acids
- Ascorbic acid
- Spices and herbs.

Water activity (a_w). Water activity is the ratio of the water vapor pressure over the food to that of pure water at the same temperature and pressure. As the water activity reduces, microbial growth becomes more challenged. A water activity below 0.86 reduces the growth of *Clostridium botulinum*. A water activity below 0.7 eliminates the growth of all pathogenic microorganisms. Water activity values below 0.6 render the food shelf-stable.

In *sous vide* and cook–chill foods the water activity can play an important role if used in conjunction with other hurdles by reducing the potential for microbial growth.

pH. Reducing pH challenges the growth of microorganisms. A value of pH below 4.6 inhibits the growth of *Clostridium botulinum* entirely. In *sous vide* and cook–chill foods, typically low-acid foods with a pH higher than 4.6, reduced pH coupled with other hurdles can be an effective means of microbial control.

Redox potential (E_h). The oxidation-reduction potential, or redox potential, denotes the tendency of the food to gain or lose electrons. It affects the way microbial cells function and the enzymatic cellular level and tends to limit microbial growth of either the aerobic or anaerobic set.

Salt (NaCl). Salt is one of the oldest means of preserving food. An increased salt content reduces microbial activity by reducing water activity and by inherent bacteriostatic effects. Salt contents in excess of 4.5% w/v are required to control *Clostridium botulinum* type E. Salt contents in excess of 27% w/v control all microorganisms.

Although salt is a good microbe inhibitor, it does not satisfy consumer demand that requires a reduced salt content in the food. Salt content must then be used in conjunction with other hurdles to increase food safety.

Carbon dioxide (CO_2). Carbon dioxide gas is mainly used in the cook–chill style of food product. *Sous vide* does not require gases in the headspace due to the utilization of a vacuum. In cook–chill foods modified atmospheres can be used to reduce the activity of any surviving microorganisms. An increase in the partial pressure of carbon dioxide in the headspace, over that normally available in the atmosphere, reduces the activity of microorganisms. Carbon dioxide concentration levels over 20% reduce or eliminate the potential for the growth of pathogens.

The use of carbon dioxide in a modified atmosphere is dependent on whether or not the food in question respires. Carbon dioxide can also have detrimental effects on the flavor of the product. The use of carbon dioxide gas is usually combined with other hurdles to reduce the activity of microorganisms in cook–chill foods.

Oxygen (O_2). As with CO_2, oxygen manipulation is used mainly in cook–chill systems utilizing a modified atmosphere. A reduction in the levels of oxygen around a food, over those normally available in the atmosphere (21%), tends to reduce microbial activity and vegetable respiration. The elimination of oxygen around the food creates an anaerobic environment promoting the growth of pathogenic anaerobes such as *Clostridium* *botulinum*. For this reason reduced oxygen levels must be used with other hurdles to assist in maintaining control over these pathogens.

Organic acids. The presence of organic acids such as propionic acid and sorbic acid inhibits the growth of microorganisms. Organic acids such as these used in *sous vide* or cook–chill foods can have the potential of reducing the viability of any surviving microflora. Organic acids are normally used in conjunction with other hurdles for microbial control.

Ascorbic acid. Ascorbic acid mainly acts as a pH-reducing mechanism in *sous vide* and cook–chill foods and has two important advantages. Firstly, ascorbic acid is an antioxidant reducing the potential for flavor and color changes in the food product. Secondly, it is an excellent nutrient.

Spices and herbs. Spices and herbs have been used in food as preservatives for thousands of years. They act in some capacity as antioxidants and as microbe inhibitors. Phenolic compounds and essential oils seem to be the most active ingredients. Microbially effective levels of these ingredients tend to seriously affect the organoleptic quality of the food. Reduced levels of herb and spice coupled with other hurdles can serve to effectively reduce microbial activity in *sous vide* and cook–chill foods.

(c) Microbially based hurdles   Microbially based hurdles are those hurdles that are created through the existence and function of microorganisms in the food product. They generally comprise the following groups (Bogh-Sorensen, 1997):

1. Competitive microflora
2. Starter cultures
3. Bacteriocins
4. Antibiotics.

Sous vide and cook–chill foods can potentially utilize only the first three microbial hurdles provided the desired microflora are either maintained alive throughout the pasteurization process or are added after the pasteurization process is complete.

An earlier chapter in this book by Betts (Chapter 6) discusses the use of competitive microflora in detail beyond the statements made here.

Competitive microflora. It is possible to inhibit pathogenic microorganisms through the proliferation of a non-pathogenic microorganism with a much greater metabolic activity in the same food. The food, as media for the growth of microorganisms, can be rendered unavailable for pathogen activity provided a more aggressive, non-pathogenic microorganism is utilizing it.

Sous vide and cook–chill foods can utilize competitive microflora provided a guaranteed type and number of non-pathogenic microorganisms can be maintained within the food such as starter cultures.

Starter cultures. Starter cultures are microorganisms commonly used for the manufacture of cheese, yogurt and other fermented products. These cultures are safe to eat by nature and are generally regarded as safe. The introduction of these cultures into *sous vide* or cook–chill foods serves both to reduce the pH, when using acid-producing starter cultures such as lactic acid bacteria (LAB), and to provide microbial competition as previously described. Some lactic acid starter cultures also produce antimicrobial compounds, referred to as bacteriocins, capable of inhibiting pathogenic microorganisms.

Bacteriocins. The only bacteriocin compound presently approved by the United States Food and Drug Administration is Nicin. This antimicrobial can be added to *sous vide* and cook–chill foods to inhibit microorganisms such as *Listeria monocytogenes*. Other bacteriocins are being investigated but have not been approved for use. Bacteriocins are not completely effective by themselves and require other hurdles to maintain control over microorganisms.

(d) Miscellaneous hurdles There are several miscellaneous hurdles not included in those previously discussed, for example:

1. Monolaurin
2. Free fatty acids
3. Chitosan
4. Chlorine.

Of these only chlorinated water would be used as a pre-wash for raw materials prior to manufacture.

12.3 Hurdle technology and HACCP

12.3.1 HACCP (What is it?)

HACCP is an acronym for Hazard Analysis Critical Control Point. It is a systematic approach to identifying and assessing potential microbiological hazards and risks associated with a food operation coupled with the definition and means for their control (Silliker *et al.*, 1988).

HACCP generally follows these steps:

• Hazard identification and assessment
• Critical control point (CCP) determination

- Specification of control criteria at each CCP
- Definition and implementation of CCP monitoring procedures
- Verification and corrective action implementation.

(a) Hazard identification and assessment All procedures concerned with the production, distribution and use of raw materials must be evaluated for microbiological hazards.

- Potentially hazardous raw materials and foods must be identified.
- Specific and potential sources of contamination must be identified in the food process.
- The potential for microbial growth during the life of the raw material and product at all stages prior to consumption must be defined.
- Risk and severity must be established for each hazard identified.

(b) Critical control point determination A critical control point (CCP) is a point in the preparation procedure of a product which, when controlled, eliminates or reduces microbial risk and/or severity of hazards identified. Usually the control points are given two levels of importance: a CCP1 is a critical control point completely eliminating one or more hazards; a CCP2 is a critical control point reducing the potential of a hazard but not completely controlling the hazard.

(c) Specification of control criteria at each CCP A means for controlling the hazard at each CCP must be defined. Some examples include:

- Water activity (a_w)
- pH
- Chlorine level in cooling water
- Distribution temperatures
- Headspace from filling operations
- Finished product preparation procedures for consumers.

(d) Definition and implementation of CCP monitoring procedures Five main types of monitoring are usually employed:

1. Visual observation
2. Sensory evaluation
3. Physical measurements
4. Chemical testing
5. Microbiological examination.

These monitoring types should be used at each of the critical control points, as required, to provide timely feedback on the performance of the implemented HACCP system. It is important that a means is provided to indicate that the process is failing in sufficient time to eliminate the possibility of unsafe product production and distribution.

(e) Verification and corrective action implementation   It is important that any information obtained through monitoring procedures and other feedback should be examined in detail regularly both to verify that the HACCP plan is working and to identify areas where the procedure is not working. When and where problem areas occur, a means of altering and improving the existing HACCP plan must be made available.

12.3.2 HACCP and hurdles

Microbial hurdles are a means for eliminating or reducing the potential for microbial growth in a food product. As such they are an important component of any HACCP plan and will be included in the statement of the critical control points of the plan. The following is a brief discussion of the general HACCP requirements of a *sous vide* and cook–chill process as it relates to these hurdles.

(a) Process flow   Table 12.1 (adapted from Bertelsten and Juncher, 1996) describes the processes normally followed and hurdles used in cook–chill and *sous vide* systems.

The differences between each of the processes in Table 12.1 essentially lie in the packaging style and timing as it relates to the heating/pasteurization step. *Sous vide* products utilize a vacuum environment which is put in place on the food immediately prior to pasteurization. The other cook–chill systems utilize a cooking/pasteurizing stage prior to packaging in an aseptic style with cooling prior to or after packaging.

The preparation stage of each process can take the form of a pre-cooking step where ingredients are combined for the product, but not necessarily so. Heating at this stage can involve any cooking procedure such as frying, baking or boiling. Preparation can also include an initial treatment of the raw materials to some physical hurdle such as UV or other radiation.

Heating, as pasteurization, requires levels of heat capable of destroying the vegetative forms of microorganisms. The heating medium normally used is hot water.

The packaging utilized in each of the processes should prevent the re-contamination by microorganisms prior to consumer utilization. The packaging should maintain hermetic conditions within the package.

Chilling and subsequently chilled storage is an essential step in the *sous vide*/cook–chill process establishing a low product temperature (<3.0°C) as quickly as possible to prevent uncontrolled microorganisms from growing (such as thermophiles). It is essential that this temperature be maintained over the life of the product to prevent spoilage.

(b) Hazard analysis   The most important microbial hazard associated with all cook–chill and *sous vide* systems is the presence and proliferation of

Table 12.1 Processes normally followed and hurdles used in cook–chill and *sous vide* systems (adapted from Bertelsten and Juncher, 1996)

Conventional cook–chill	MAP cook–chill	Sous vide	Hot filling
Preparation *Physical* Temperature, radiation, UV, packaging *Physicochemical* All *Microbial* All	**Preparation** *Physical* Temperature, radiation, UV, packaging *Physicochemical* All *Microbial* All	**Preparation** *Physical* Temperature, radiation, UV, packaging *Physicochemical* All *Microbial* All	**Preparation** *Physical* Temperature, radiation, UV, packaging *Physicochemical* All *Microbial* All
Heating/pasteurization *Physical* Temperature	**Heating/pasteurization** *Physical* Temperature	**Heating/pasteurization** *Physical* Temperature	**Heating/pasteurization** *Physical* Temperature
Chilling *Physical* Temperature	**MA Packaging** *Physical* MAP, packaging	**Heating/pasteurization** *Physical* Temperature	**Hot filling and clip sealing** *Physical* Temperature, packaging
Packaging *Physical* Packaging	**Chilling** *Physical* Temperature	**Chilling** *Physical* Temperature	**Chilling** *Physical* Temperature
Chill storage *Physical* Temperature	**Chill storage** *Physical* Temperature	**Chill storage** *Physical* Temperature	**Chill storage** *Physical* Temperature

Clostridium botulinum types A, B and E in the prepared food. Toxins of this dangerous heat-resistant microbe can be developed even at refrigeration temperatures below 3.0°C. Other microorganisms of importance are *Salmonella*, *Staphylococcus aureus* and *Listeria monocytogenes*. Most of these microorganisms will be controlled when *C. botulinum* is controlled through heat and hurdles; however, one must recognize these as hazards if improper preparation techniques are followed.

The hazards identified above comprise only a few of the possible microbial and ingredient hazards if the preparation procedures are not correctly followed. The hazards identified are important, however, to the controlling aspect of the hurdles applied.

(c) Critical control points   The critical control points and hurdles most likely to be important to these processes are as follows (Schellekens and Martens, 1993):

1. Reception of ingredients: CCP2.
2. Cool storage of ingredients (physical hurdle: temperature): CCP2.
3. Pre-treatment of raw material (physical hurdles: radiation, UV, packaging): CCP2.
4. Heating/cooking in any preparation step (physical hurdles: temperature): CCP2.
5. Recipe: addition of hurdles to food (physicochemical and microbial hurdles: all): CCP2.
6. Packaging and vacuumization (physical hurdles: packaging, MAP): CCP2.
7. Heat treatment: CCP1 or CCP2 depending upon target microorganism (physical hurdle: temperature).
8. Chilling (physical hurdle: temperature): CCP1.
9. Storage, distribution, retail holding temperatures and environment (physical hurdles: temperature, packaging): CCP1.

As can be seen from the above example, hurdles play a very large role in the development of a HACCP plan. The very nature of a hurdle is that it inhibits or eliminates the growth of pathogenic microorganisms in the food. Each critical control point identifies a point in the manufacturing process critical to the control of pathogenic microorganisms. Each stage of the manufacturing process utilizing a hurdle must therefore be a critical control point. This, of course, may not identify all of the critical control points.

The control criteria for each of the critical control points, corresponding to a hurdle, then must be the criteria establishing the efficacy of the hurdle itself. As an example, a critical control point may be storage temperature. The hurdle utilized in this case is temperature. The control criterion is that the storage temperature must not exceed 3.0°C, identical to the requirements for efficacy of the hurdle.

(d) Practical example Figure 12.1 presents a process for a shepherd's pie product in a plastic retort tray with tear-off plastic top. Heat is used throughout the process as a cooking step, doubling as a means for reducing the microbial load. This is only effective if the times between preparation and filling/sealing are short. Extended exposure to ambient temperatures will result in increased microbial load.

(e) Monitoring, verification and adaptation Each of the critical control points must be monitored using equipment appropriate to the hurdle in question. Temperature hurdles must be monitored with thermocouples, thermisters, RTDs or thermometers. Water activity should be monitored with a standard water activity meter. Initial microbial loads should be determined using standard plating techniques, and so on. The measurement requirements of each hurdle used will provide the measurement requirements for the critical control point.

Each individual plant must identify its own system of monitoring, verification and corrective action implementation to satisfy the needs of the HACCP plan.

12.4 Summary

The safety issues of *sous vide* and cook–chill foods in general have been addressed by several legislation procedures in most countries around the world. Most legislation procedures are moving towards a more general approach that includes the application of food hygiene regulation and the obligation to use HACCP throughout the process to ensure safety of the product. However in most of the circumstances, there is still lack of national and international cooperation to obtain harmonized guidelines for *sous vide* and cook–chill products.

The combination of hurdle technology and a preventive approach to food hazards (HACCP) is the modern practice towards a safe production of *sous vide* and cook–chill foods. Implementation of the HACCP approach and employee training in HACCP principles should result in a higher degree of confidence in product safety than is possible using traditional end sampling approaches to microbiological control. The use of predictive modeling to improve safety could be very useful in future to have a quantitative insight into the combined effects of different controlling factors. Additional 'hurdles' have to be used to protect the consumer.

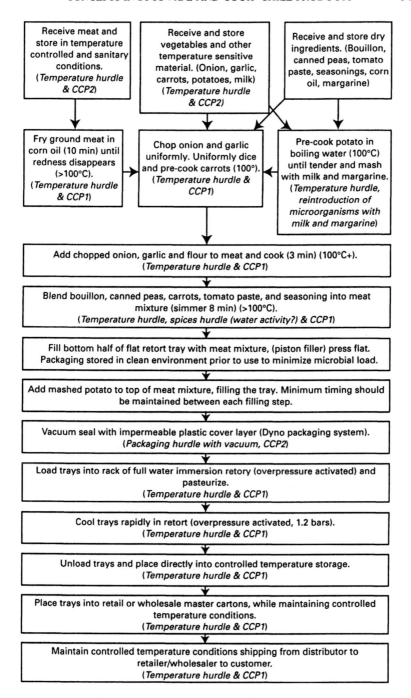

Figure 12.1 Process for a shepherd's pie product.

Acknowledgements

The authors would like to acknowledge Sheila Trenholm and Lisa Beresford for their intensive efforts in updating the literature available, and Leon Gorris for supplying FLAIR internet references.

Reviewing the manuscripts of each of the contributing authors to this book has provided us with invaluable knowledge about minimally processed foods and has provided the impetus for this chapter.

References

Bangay, L.S. (1996) The state of *sous vide* in North America, in *Proceedings of the Second European Symposium on Sous Vide,* ALMA *Sous Vide* Competence Centre, Leuven, Belgium, 10–12. April 1996.

Bertelsen, G. and Juncher, D. (1996) Oxidative stability and sensory quality of *sous vide* cooked products, in *Proceedings of the Second European Symposium on Sous Vide,* ALMA *Sous Vide* Competence Centre, Leuven, Belgium, 10–12 April 1996.

Bogh-Sorensen, L. (1997) Description of hurdles, in *Food Preservation by Combined Processes* (eds L. Leistner and L.G.M. Gorris), FLAIR Concerted Action No. 7, Subgroup B (Internet Word 6.0 Version, 1997).

FLAIR (Food Linked Agro-Industrial Research) (1997) *European Commission on Food Preservation by Combined Processes,* Directorate-General XII (Science, Research and Development).

Leistner, L. (1997) User guide to food design, in *Food Preservation by Combined Processes* (eds L. Leistner and L.G.M. Gorris), FLAIR Concerted Action No. 7, Subgroup B (Internet Word 6.0 Version, 1997).

Schellekens, M. (1996) *Sous vide* cooking: state of the art. *Second European Symposium on Sous Vide,* ALMA *Sous Vide* Competence Centre, Leuven, Belgium, 10–12 April 1996.

Schellekens, W. and Martens, T. (1993) *'Sous Vide' Cooking,* FLAIR (Food-Linked Agro-Industrial Research) 1989–1993, ALMA University Restaurants, Commission of the European Communities, Directorate-General XII (Science, Research and Development).

Silliker, J. H., Bryan, F.L. and Tompkin, R.B. (1988) Microorganisms in foods: application of the hazard analysis critical point (HACCP) system to ensure microbiological safety and quality. *ICMSF.* Blackwell Scientific Publication, London, UK.

13 Microbiological safety aspects of cook–chill foods

CATHERINE J. MOIR and ELIZABETH A. SZABO

13.1 Introduction

The increase in demand by both retail and catering industries for convenient, preservative free and apparently fresh foods has led to the development of a number of refrigerated, minimally processed products. These include cook–chill products (such as *sous vide* foods), modified atmosphere packaged (MAP) foods and prepared salads and vegetables. Most have extended shelf lives due to the application of a heat treatment, modified atmospheres (MAs), advanced packaging techniques (active packaging and selective barrier films) and/or refrigeration. There are microbiological safety concerns associated with these foods, because chilled foods are more prone to temperature abuse than frozen foods and the relatively long shelf life claimed for some products may allow ample time and conditions for a number of psychrotrophic pathogenic organisms to grow. Another risk factor associated with these types of product is that many foodborne pathogens can grow and/or produce toxins without causing any visual or sensory defects in the food.

This chapter will discuss some of the microbiological safety issues associated with cook–chill foods and approaches for controlling these hazards. For the purpose of this discussion, cook–chill is a term used to define a food processing system in which a food is cooked, chilled rapidly, and stored at controlled low temperatures (0–3°C). *Sous vide* products are examples of food manufactured using a cook–chill system: raw or partially cooked foods are vacuum packaged, usually in flexible packages, and then subjected to a pasteurisation treatment. The absence of oxygen will inhibit the growth of strictly aerobic microorganisms and prevent oxidative rancidity from occurring. For *sous vide* products, the spore-forming pathogens *Clostridium botulinum* and *Bacillus cereus* are the main food safety concerns because the heat treatments used should be sufficient to kill vegetative cells of other pathogens. However, not all foods can be manufactured using the *sous vide* method. Many other cook–chill products are packaged after receiving a pasteurisation treatment. In these products, there is the potential for microbial contamination after heat treatment and before or during packaging. In addition to *C. botulinum* and *B. cereus*, other microorganisms of concern include the psychrotrophic pathogens *Listeria monocytogenes, Yersinia enterocolitica* and *Aeromonas hydrophila.* Both *sous vide* and other cook–chill products are refrigerated after processing to control the growth of surviving microorganisms and this contributes to the increased shelf life. The microbiological hazards associated with *sous vide* and other cook–chill products are discussed.

The microbiological safety and stability of cook–chill foods may be increased by combination with other shelf life extension techniques such as modified atmosphere packaging (MAP) and this is frequently used as a non-chemical method of inhibiting microorganisms. MAP involves sealing a product in a package in an atmosphere different from air (usually one or more of CO_2, O_2 and N_2). Gas flush and vacuum packaging are the two most common techniques. CO_2 has the strongest antimicrobial effect. Though the precise mechanism of its action is not fully understood, CO_2 has a direct effect on microorganisms resulting in the extension of the lag phase and/or decrease in the growth rate during the logarithmic phase (Farber, 1991). Temperature control is critical for this antimicrobial effect. As temperature increases, the solubility of CO_2 decreases and as microorganisms inhabit the aqueous phase of biological tissue, a decrease in the antimicrobial effect of CO_2 is observed with increased temperature. Oxygen and nitrogen are usually included in packages for sensory reasons and/or as fillers. However, the inclusion of nitrogen as a filler, with the simultaneous removal of oxygen, will have an inhibitory effect on aerobic microorganisms by making the environment anaerobic.

MAP products are not static systems because the concentrations of gases in a package may change after the package is sealed. These changes may be due to metabolic activity of the food or the microorganisms in it, diffusion of gases across the package and dissolution of CO_2 in the food. Technically, *sous vide* foods are MAP products since they are vacuum packaged before receiving the pasteurisation treatment ('*sous vide*' means under vacuum). By using MAs to increase the shelf life of a cook–chill product, there can be the disadvantage of allowing certain psychrotrophic pathogens ample time and conditions to grow. Where appropriate, modified atmospheres and their effects on the microbiological safety of cook–chill and *sous vide* foods will also be discussed.

Because of the reliance on low temperature of cook–chill foods for shelf life extension, temperature control during storage is critical. As mentioned, this is also critical when applying MAs as an additional preservation technique in these products. In addition to temperature control, strict processing regimes and factory hygiene must be adhered to so that the microbiological safety of cook–chill foods can be ensured. Industry regulations, Hazard Analysis Critical Control Point (HACCP) systems, predictive modelling of microbial growth and microbiological challenge studies and their roles in assessing and ensuring the safety of cook–chill and *sous vide* foods will also be discussed in brief.

13.2 Microflora

Sous vide processing involves the cooking of foods after packaging in the absence of oxygen. Providing sufficient heat treatment is applied to inactivate vegetative cells, the background microflora present initially in the food (e.g. pseudomonads, Enterobacteriaceae, lactobacilli and other potentially pathogenic bacteria that are non-sporing) will be destroyed and this may allow for the germination and outgrowth of spore-formers such as *C. botulinum* and *B. cereus*. In cook–chill foods that are packaged after cooking, microbial recontamination by other psychrotrophic pathogens, such as *L. monocytogenes*, *Y. enterocolitica* and *Aeromonas*, may occur after cooking or during packaging. The amount of growth of contaminating microorganisms during the extended shelf life will depend on a number of factors including the storage temperature, the types of microorganisms (aerobic, anaerobic or facultative), the type of packaging system (vacuum, MAP or low barrier film) and the composition of the food.

13.2.1 Clostridium botulinum

Many of the food preservation techniques currently in use are designed specifically to either inhibit or destroy *C. botulinum*. This is largely because of the resistance of its spores to inactivation by heat and the severe nature of the illness (botulism) that it causes.

Foodborne botulism results from intoxication after the ingestion of pre-formed botulinum neurotoxin. Because the neurotoxin targets nerve cells at neuro-muscular junctions, respiratory impairment may occur in the affected host, and thus be fatal. Since 1953, *C. botulinum* has been the species designation for all organisms that produce botulinum neurotoxins (types A–G). The species, however, is heterogeneous. Strains frequently associated with human botulism can be divided into two main groups (Holdeman and Brooks, 1970; Smith and Hobbs, 1974). Group I *C. botulinum* strains are proteolytic and produce A, B or F neurotoxin. They can cause putrefaction of foods if significant growth occurs. Group II strains are non-proteolytic and produce B, E or F neurotoxin. Because group II strains do not digest complex proteins, organoleptic evidence of growth in food is less pronounced. The groups are further distinguished by their minimal growth parameters and their heat resistance (Table 13.1).

While outbreaks of botulism remain rare, changes in food processing and preparation are resulting in new *C. botulinum* hazards (Szabo and Gibson, 1997). The heat resistance of *C. botulinum* spores, the anaerobic nature of

Table 13.1 Phenotypic groups of *C. botulinum*

Properties	Group	
	I	II
Toxin types	A, B, F	B, E, F
Proteolysis	+	–
Inhibitory pH	4.6	5.0
Min. temperature for growth	10°C	3.3°C
Max. [NaCl]	10%	5%
$D_{100°C}$ value	25 min	0.1 min

the organism and the ability of non-proteolytic strains to grow at refrigeration temperatures are safety concerns in the manufacture of *sous vide* and other cook–chill products.

(a) Heat resistance The vegetative cells of *C. botulinum* should be destroyed by the heat processing used for *sous vide* and cook–chill products: a heat treatment of not less than 70°C for 2 min at the coldest spot (Australian Quarantine and Inspection Service, 1992). However, this heat treatment does not ensure the destruction of *C. botulinum* spores, if present.

D-values for *C. botulinum* spores vary with strain, method of spore production, heating menstruum, and recovery system. The spores of proteolytic *C. botulinum* strains are more heat resistant than non-proteolytic strains (Table 13.1). The heat resistance of proteolytic A and B type spores is characterised by $D_{121°C}$ values in the range of 0.10–0.20 min (Stumbo, 1973). Although the spores of proteolytic *C. botulinum* strains will survive the relatively mild heat treatment given to *sous vide* and cook–chill products, they are of limited significance to chilled foods which are properly refrigerated, because the minimum growth temperature of the vegetative cells of proteolytic *C. botulinum* strains is considered to be 10°C (Table 13.1).

The heat resistance of spores from non-proteolytic *C. botulinum* strains is much lower than that of their proteolytic counterparts (Table 13.1). The spores, however, may be able to survive mild processing temperatures (70–85°C). Because the vegetative cells of non-proteolyic strains grow at temperatures of refrigeration, survival of this organism following heat treatment and low temperature outgrowth is of particular concern to *sous vide* and other cook–chill products.

In phosphate buffer (pH 7.0), the heat resistance of non-proteolytic types E and F spores appear similar, with reported $D_{82.2°C}$ values of 0.33 and 0.25–0.84, respectively (Scott and Bernard, 1982; Lynt *et al.*, 1979; Table 13.2). While spores of some non-proteolytic *C. botulinum* type B strains have a similar heat resistance to the spores of type E and F strains in aqueous solution, Scott and Bernard (1982) reported type B spores with $D_{82.2°C}$ values as high as 32.3 min. The authors noted, however, that the apparent higher heat resistances were not consistently obtained, even with different spore suspensions of the same strain.

Differences in the reported heat resistance of *C. botulinum* spores may be attributed to the method of spore production and choice of recovery medium. Sugiyama (1951) noted that the temperature of sporulation and the composition of the sporulation medium affect heat resistance. Peck *et al.* (1995) found that heat resistance of non-proteolytic *C. botulinum* spores was not affected by spore production at temperatures between 20 and 35°C. Kralovic (1973) observed increased heat resistance of non-proteolytic type E spore crops that were treated with lysozyme and trypsin to remove cellular debris. The $D_{80°C}$ value for untreated type E spores was 1.51 min and 20.8 min for treated spores. The addition of lysozyme to the recovery medium increased the apparent heat resistance of group II spores. This was attributed to the ability of lysozyme to induce germination in heat-damaged spores (Peck *et al.*, 1992; Scott and Bernard, 1985). Peck *et al.* (1993) also found that if spores were permeable to lysozyme (naturally or after treatment with thioglycollate) there was an apparent upward shift in their heat resistance (Table 13.2). While these studies might indicate a potential protective capacity of foods containing lysozyme, the possibility of similar levels of lysozyme remaining in foods following heat processing is questionable (Lund and Notermans, 1992).

The heat resistance of spores also may be affected by the pH level and presence of other stress factors such as NaCl and nitrite concentration, and the water activity of the heating menstruum. Some studies suggest that the apparent heat resistance of non-proteolytic spores decreases with increasing NaCl concentrations or reduced pH (Juneja *et al.*, 1995; Juneja and Eblen, 1995). Juneja and Eblen (1995) postulate that heat-damaged spores may be unable to recover in the presence of salt (2–3% w/v), thus suggesting that *sous vide* and cook–chill products containing adequate levels of salt may offer protection against the germination and subsequent outgrowth of non-proteolytic spores. This may not, however, provide protection against salt-tolerant pathogens such as *Listeria monocytogenes* (see later section). Certain foods provide added instrinsic protection to spores, such as those with a high fat or protein content (Juneja *et al.*, 1995). Several studies have reported on the heat resistance of non-proteolytic *C. botulinum* spores in a range of foods. From Table 13.2, it can be seen that spore heat resistance varies greatly depending on the food type.

(b) Growth at low temperatures and in modified atmospheres   While the consumption of low numbers of spores by healthy adults poses no public health threat, the potential of botulinal spores to survive *sous vide* and cook–chill processing and subsequently grow in foods prior to consumption

Table 13.2 Reported heat resistance of non-proteolytic *C. botulinum* spores

C. botulinum type	Heating menstruum	Heating temperature (°C)	D-value (min)	Reference
B	Phosphate buffer (pH 7.0)	82.2	1.49–32.3	Scott and Bernard (1982)
B	Phosphate buffer (pH 7.0)	85.0	76.3–100[ab]	Peck et al. (1993)
B	Phosphate buffer (pH 7.0)	80.0	2.53–4.31[b]	Juneja et al (1995)
B	Cod homogenate	80.0	18.30[b]	Gaze and Brown (1990)
B	Carrot homogenate	80.0	4.24[b]	Gaze and Brown (1990)
B	Turkey slurry	80.0	15.21[b]	Juneja et al. (1995)
E	Phosphate buffer (pH 7.0)	82.2	0.33	Scott and Bernard (1982)
E	Phosphate buffer (pH 7.0)	85.0	45.6[ab]	Peck et al. (1993)
E	Phosphate buffer (pH 7.0)	80.0	1.03–4.51[b]	Juneja et al. (1995)
E	Blue crab meat	82.2	0.49–0.74	Lynt et al. (1977)
E	Tuna in oil	82.2	6.60	Bohrer et al. (1973)
E	White fish chubs	82.2	2.21	Crisley et al. (1968)
E	Cod homogenate	80.0	15.10[b]	Gaze and Brown (1990)
E	Carrot homogenate	80.0	4.33[b]	Gaze and Brown (1990)
F	Phosphate buffer	82.2	0.25–0.84	Scott and Bernard (1982)
E	Turkey slurry	80.0	13.37[b]	Juneja et al. (1995)
F	Crab meat	76.6	9.50	Lynt et al. (1979)
F	Crab meat	85.0	0.53	Lynt et al. (1979)

[a]Spores were made permeable to lysozyme.
[b]Recovery medium contained lysozyme.

is of concern. A mild heat treatment could serve potentially as a spore activation step if it were not lethal.

As mentioned previously, proper refrigeration of *sous vide* and other cook–chill products will prevent the outgrowth of the spores of proteolytic strains. However, if products are subjected to temperature abuse at any stage of their storage and distribution, there is a risk of growth of proteolytic (and non-proteolytic) *C. botulinum* with concomitant production of neurotoxin. Simpson *et al.* (1995) conducted a challenge study with proteolytic *C. botulinum* in a *sous vide* spaghetti and meat sauce product. They found that product with pH values of 5.25, 5.5, 5.75 and 6.0 stored at 15°C became toxic after 35, 21, 21 and 14 days, respectively. Of concern was the observation that product with pH values of 5.25 and 5.5 became toxic prior to visible spoilage. As also indicated by the authors, the inoculum level of *C. botulinum* used (10^3 spores/g) was much higher than 'natural' contamination levels that vary from 0.000 000 04 to 0.001 67 spores/g. Given this, the experiments represented a worst case scenario.

Non-proteolytic *C. botulinum* strains have the ability to grow at low temperatures. In 1961, Schmidt *et al.* demonstrated growth of four *C. botulinum* type E strains and toxin production within 36 days in beef stew incubated at 3.3°C. When the temperature was reduced to 2.2°C, no toxin was detected throughout the 104 days of testing. Growth of non-proteolytic type B and F strains at 3.3°C has been demonstrated with toxin detected at 129 and 52 days, respectively, in a laboratory medium (Eklund *et al.*, 1967a, b). In general, several weeks of incubation are needed for toxigenesis at temperatures <4°C (Schmidt *et al.*, 1961; Eklund *et al.*, 1967a, b; Solomon *et al.*, 1977). However, such storage periods can elapse with both *sous vide* and other cook–chill preparations.

There is concern that the mild heat treatments used for *sous vide* and other cook–chill products combined with storage in a reduced oxygen environment (vacuum or modified atmosphere packaging) may select for *C. botulinum* thus permitting toxigenesis before organoleptic spoilage, particularly if temperatures rise above 4°C. Several investigators have reported toxin production by non-proteolytic *C. botulinum* in various *sous vide* and other cook–chill products at temperatures >4°C (Table 13.3). Of those studies listed in Table 13.3, only Brown and Gaze (1990) and Brown *et al.* (1991) subjected inoculated raw food to a *sous vide* process. Other authors tended to inoculate a commercial product and then, with the exception of Notermans *et al.* (1990), deliberately vacuum seal the product. Although heat treatments were applied to spore crops prior to inoculation in these latter studies, the spores would not have received the same heat shock or damage as spores present in food that have undergone a *sous vide* treatment.

Storage of *sous vide* and other cook–chill products below 4°C is recommended so as to ensure that the safety of the product is not compromised

Table 13.3 Time to toxin production by non-proteolytic *C. botulinum* in various *sous vide* and cook–chill products at temperatures >4°C

Spore type (storage temperature)	Food	Time to toxicity (days)	Reference
B and E (8°C)	Beef	8	Meng and Genigeorgis (1994)
	Chicken	16	Meng and Genigeorgis (1994)
	Salmon	8	Meng and Genigeorgis (1994)
	Ravioli	>42	Notermans *et al.* (1990)
	Moussaka	>42	Notermans *et al.* (1990)
	Tortellini	>42	Notermans *et al.* (1990)
	Canneloni with meat	>42	Notermans *et al.* (1990)
	Canneloni with vegetables	>42	Notermans *et al.* (1990)
	Lasagna	>42	Notermans *et al.* (1990)
	Lasagna bolognese	>42	Notermans *et al.* (1990)
	Tagliatelle with pork	14–28	Notermans *et al.* (1990)
	Tagliatelle with chicken	>42	Notermans *et al.* (1990)
	Tagliatelle with mushrooms	14–28	Notermans *et al.* (1990)
	Chili con carne	>42	Notermans *et al.* (1990)
B (8°C)	Carrot	>84	Brown and Gaze (1990)
	Chicken	42	Brown *et al.* (1991)
	Cod	28	Brown *et al.* (1991)
E (8°C)	Carrot	>84	Brown and Gaze (1990)
	Chicken	28	Brown *et al.* (1991)
	Cod	21	Brown *et al.* (1991)
B and E (10°C)	Beef in gravy	6	Crandall *et al.* (1994)

as a result of the growth and subsequent neurotoxin production by non-proteolytic *C. botulinum* (Australian Quarantine and Inspection Service, 1992). However, it can be unrealistic at present to expect storage temperatures of 0–4°C for prolonged periods (weeks) in retail or domestic situations (Rose, 1986; James and Evans, 1990). Fortunately, the vegetative cells and neurotoxin are relatively sensitive to heat treatment (Szabo and Gibson, 1997). It is therefore important that the finished product is reheated sufficiently prior to consumption to ensure the destruction of both the vegetative cells of the pathogen and its neurotoxin. Heat treatments such as maintaining an internal temperature of at least 78°C for 1 minute may be used as a general guideline for convenience foods, however, the heating menstruum and its pH can markedly affect the inactivation rate of the neurotoxin (Szabo and Gibson, 1997).

13.2.2 Bacillus cereus

Bacillus cereus is a large, Gram-positive, facultatively anaerobic, spore-forming rod. It is commonly found in the environment, particularly in soil, and in foods of plant origin. *B. cereus* has been associated with foodborne illness since the 1950s and is responsible for two distinct syndromes: an emetic and a diarrhoeal illness. The emetic illness is associated with ingestion of a pre-formed, heat-stable toxin while evidence indicates that the diarrhoeal illness is caused by consumption of high numbers of *B. cereus* cells and subsequent infection and toxin production in the bowel.

(a) Heat resistance *B. cereus* spores are moderately heat resistant and this resistance will vary depending on the strain type and cultural factors such as growth and recovery medium and heating menstruum (Mazas *et al.*, 1995; Foegeding and Fulp, 1988). Heating in high fat/oil environments can provide a protective effect. Higher heat resistances may also be observed in lower water activity foods and spores have a significantly higher resistance to dry heat as opposed to moist heat. An extensive list of published heat resistance data on *B. cereus* has recently been collated by ICMSF (1996b). Some of these data are shown in Table 13.4.

Due to the heat resistance of *B. cereus* spores, it is most likely that the organism will not be completely destroyed by the heat treatment given to most *sous vide* and other cook–chill products. Therefore, the organism must be controlled in these foods by preventing its growth and/or restricting the shelf life of the product.

(b) Growth at low temperatures and in modified atmospheres *B. cereus* has long been regarded as a mesophilic organism, but psychrotrophic strains are not uncommon, particularly in raw and pasteurised milk (Griffiths, 1990). The minimum growth temperature of these strains is 4°C (van Netten *et al.*, 1990; Davis and Walker, 1994; Dufrenne *et al.*, 1994). Pre-conditioning at 8–12°C can also substantially reduce the observed lag time of *B. cereus* incubated subsequently at 8°C (Blackburn and Davies, 1994).

Many of the psychrotrophic strains have been shown to produce the diarrhoeal enterotoxin. Growth and enterotoxin production has been observed in milk at 6°C (Griffiths, 1990) and milk, minced meat, lasagne and rice meal after 11–12 days and 24 days at 7 and 4°C, respectively (van Netten *et al.*, 1990). Some strains have been shown to produce detectable toxin in aerated milk within 66.5 h at 8°C (Christiansson *et al.*, 1989) and in broth, milk and

Table 13.4 Heat resistance of *B. cereus* (adapted from ICMSF, 1996b)

Heating menstruum	Temperature (°C)	D-value (min)
Phosphate buffer (0.25 M, pH 7)	85	34.4–106
Phosphate buffer (0.067 M, pH 7)	115.6	0.13–11.3
Rice broth	100	4.2, 6.3
Pumpkin pie	100	40
Soybean oil	112	46.5

rice extract within 72 h at 10°C or 24 h at 15°C (Notermans and Tatini, 1993). However, since it is most likely that diarrhoeal food poisoning is the result of consumption of high numbers of cells and subsequent infection in the bowel (Granum *et al.*, 1993), it is the ability of psychrotrophic strains to grow to high numbers at refrigeration temperatures that is important, rather than their ability to produce toxin at low temperatures.

Most work on the effects of MAs on *B. cereus* has been performed in laboratory media. The minimum and maximum growth temperatures of *B. cereus* are increased and decreased, respectively, by 100% CO_2. Growth at optimum temperatures is inhibited by about 40% (relative to growth in nitrogen) in 100% CO_2 (Enfors and Molin, 1981). Bennik *et al.* (1995) showed that growth of *B. cereus* at 8°C occurred on an agar surface in the presence of 20% CO_2 and 1.5 or 21% O_2 (balance N_2) and near neutral pH values, but cell numbers decreased in 50% CO_2. Inhibition of growth and diarrhoeal enterotoxin production was demonstrated at 15 and 30°C under atmospheres of 20 and 40% CO_2 (approximately 0.5% O_2, balance nitrogen) and 100% CO_2 (Moir and Eyles, 1992). In this study, inhibition of enterotoxin production could be related to the reduced final population densities observed under MAs. The reduced population densities were, however, due to the limited availability of oxygen rather than the presence of CO_2. Growth rates decreased with an increase in CO_2 concentration.

The germination rate of spores is also affected by the presence of CO_2. Enfors and Molin (1978) showed that only 10–30% of *B. cereus* spores germinated in 100% CO_2 at 30°C and pH values of 5.6–6.7, with the germination rate being higher at the higher pH. This is similar to results of Moir and Eyles (1992) who showed germination rates of about 20% in 100% CO_2 and about 60% in both 20 and 40% CO_2 at 30°C. Germination rates were lower at 15°C.

As with *C. botulinum*, storage of *sous vide* and other cook–chill products below 4°C should control the potential hazards associated with *B. cereus* in these foods. Since this may be unrealistic, additional barriers such as product formulation or the use of modified atmospheres may be used to assist in the control of *B. cereus* in cook–chill foods.

13.2.3 Listeria monocytogenes

Listeria monocytogenes is a non-spore-forming, psychrotrophic pathogen that can grow and survive in diverse environments that range from soil to cool, moist areas in food-processing factories. The pathogen has been a cause for concern to both the food industry and regulatory agencies since the early 1980s. The disease manifestations in humans vary from a non-specific flu-like illness to infection involving sepsis or meningitis. It is rare for listeriosis to occur in a human host in the absence of a predisposing risk factor. Those most at risk include the elderly, pregnant women, neonates and the immunocompromised. The infectious dose for either 'healthy' or 'at risk' populations has not been determined.

The safety concerns in cook–chill products for *L. monocytogenes* relate to its widespread environmental distribution, its ability to grow at refrigeration temperatures and under vacuum or modified atmosphere packaging. Furthermore, as is discussed below, there is some evidence that the mild and prolonged cooking times for some cook–chill products might enhance the organism's thermotolerance, thus facilitating its survival during cooking. Failure to destroy *L. monocytogenes* during cooking or to prevent re-entry after cooking may lead to an unsafe product even if subsequent handling and storage are carried out properly.

(a) Distribution   *L. monocytogenes* is globally distributed. It has been isolated from silage (Gray, 1960), natural vegetation (Welshimer, 1968), soil (Botzler *et al.*, 1974), wild and domestic animals (Weis and Seeliger, 1975; Skovgaard and Morgen, 1988) and waste water (Geuenich *et al.*, 1985). As a result of this wide environmental distribution, low numbers of *L. monocytogenes* may be present on a wide variety of raw and processed food of both plant and animal origin which may be used in formulations of cook–chill products.

L. monocytogenes has been isolated from many foods. The diversity is well illustrated by those foods implicated in outbreaks of listeriosis: coleslaw (Schlech *et al.*, 1983), pasteurised milk (Fleming *et al.*, 1985), Mexican-style cheese (Linnan *et al.*, 1988), pâté (McLauchlin *et al.*, 1991), pork tongue (Jacquet *et al.*, 1995), raw milk soft cheese (Goulet *et al.*, 1995) and vacuum packaged diced cooked chicken (Hall *et al.*, 1996).

(b) Heat resistance   Table 13.5 summarises some of the published data on the heat resistance of *L. monocytogenes*. Clearly, heat resistance varies with the intrinsic properties of the heating menstruum. For example, Ben Embarek and Huss (1993) observed $D_{60°C}$ values of 1.95–1.98 min for cod fillets and 4.23–4.48 min for salmon fillets that were *sous vide* processed. The authors postulate that the increased thermal resistance observed in salmon was a protective effect due to the salmon's higher fat content compared to that of cod. Salt also has been shown to influence the heat resistance of *L. monocytogenes*. Palumbo *et al.* (1995) report $D_{64.4°C}$ values of 0.44, 8.26 and 27.3 min in egg yolk, egg yolk containing 10% salt and 5% sugar, and egg yolk containing 20% salt, respectively.

The heat resistance of *L. monocytogenes* also may be influenced by heat shock. Fedio and Jackson (1989) reported that the heat resistance of *L. monocytogenes* increased if cells were given a sublethal heat shock before heating at 60°C. An increase in D-values following heat shock has been reported by others (Linton *et al.*, 1990; Farber and Brown, 1990). There is concern that heat shock conditions may be created in cook–chill processing,

Table 13.5 Heat resistance data of *L. monocytogenes*

Heating menstruum	D-value (min)	Temperature (°C)	Reference
Minced beef	6.7	60	Hansen and Knochel (1996)
Lean (2.0% fat) beef	0.6	62.7	Fain *et al.* (1991)
Fatty (30.5% fat) beef	1.2	62.7	Fain *et al.* (1991)
Minced beef	3.8	60	Mackey *et al.* (1990)
Ground beef	3.12	60	Farber (1989)
Beef homogenate	6.27–8.32	60	Gaze *et al.* (1989)
Ground pork	1.14	60	Ollinger-Snyder *et al.* (1995)
Salmon	4.23–4.48	60	Ben Embarek and Huss (1993)
Cod	1.95–1.98	60	Ben Embarek and Huss (1993)
Crab meat	2.61	60	Harrison and Huag (1990)
Minced chicken breast meat	8.7	60	Mackey *et al.* (1990)
Minced chicken leg meat	5.6	60	Mackey *et al.* (1990)
Carrot homogenate	5.02–5.29	60	Gaze *et al.* (1989)
Chicken meat homogenate	5.02–7.76	60	Gaze *et al.* (1989)
Nutrient broth	3.6–5.4	60	Lund *et al.* (1989)
Milk	0.1–0.4	62	Donnelly *et al.* (1987)
Egg yolk	0.44	64.4	Palumbo *et al.* (1995)
Egg yolk 10% salt, 5% sugar	8.26	64.4	Palumbo *et al.* (1995)
Egg yolk 20% salt	27.3	64.4	Palumbo *et al.* (1995)

potentially facilitating an increase in the heat resistance of *L. monocytogenes*. Stephens *et al.* (1994) and Kim *et al.* (1994) have shown that heating by slowly raising the temperature enhanced the heat resistance of *L. monocytogenes* in broth and pork, respectively. Hansen and Knochel (1996) found no significant difference between slow (0.3–0.6°C/min) and rapid (>10°C/min) heating and the heat resistance of *L. monocytogenes* in low pH (<5.8) *sous vide* cooked beef prepared at a mild processing temperature. However, the latter authors did observe an increase in heat resistance of *L. monocytogenes* in high pH (6.2) *sous vide* beef.

Although the heat resistance of *L. monocytogenes* varies with heating menstruum and is influenced in some instances by a heat shock response, it is generally agreed that *L. monocytogenes* will be destroyed by proper pasteurisation. Food that is minimally heat processed at the coldest spot using a heat treatment of not less than 70°C for 2 min would ensure the destruction of *L. monocytogenes* (Ben Embarek and Huss, 1993; Mossel and Struijk, 1991; Mackey *et al.*, 1990; Gaze *et al.*, 1989). In Australia, it is recommended that the heat process applied to cook–chill products should be designed to achieve 6 decimal reductions in the count of *L. monocytogenes*, based on a $D_{70°C}$ value of 0.3 min, and a z-value of 6°C (Australian Quarantine and Inspection Service, 1992).

(c) Growth at low temperatures and in modified atmospheres   *L. monocytogenes* survives and grows at refrigeration temperatures. The reported

generation times in milk products at 4, 8 and 13°C were 29–48, 11.4 and 5.0–7.2 h, respectively (Rosenow and Marth, 1987; Donnelly and Briggs, 1986). In broth at 5°C under aerobic and anaerobic conditions, generation times of 14.4 and 4.1 h, respectively, were reported (Buchanan *et al.*, 1989). Growth is considered possible at 1°C (Varnam and Evans, 1991), and has been reported at –1.5°C on sliced roast beef (Hudson *et al.*, 1994). Johnson *et al.* (1988) report the survival of *L. monocytogenes* in meat for up to 20 days at –23°C and for a year at –20°C. Although growth is slow at refrigeration temperatures, the extended shelf life of cook–chill products and reduced competitive microflora may allow sufficient time and conditions for growth of *L. monocytogenes* to elevated levels. Failure to reheat such contaminated food properly could pose a potential health hazard.

L. monocytogenes is capable of growth in reduced oxygen environments. Buchanan *et al.* (1989) observed little difference between *L. monocytogenes* growth in air and 100% N_2 in broth systems. In food, the growth capacity of *L. monocytogenes* in vacuum-packaged products is well documented for ham, bologna, chicken and turkey products (Glass and Doyle, 1989), beef (Grau and Vanderlinde, 1990), lamb (Garcia de Fernando *et al.*, 1995), pork (van Laack *et al.*, 1993) and smoked salmon (Hudson and Mott, 1993). There is conflicting evidence regarding the effect of CO_2 on the growth and survival of *L. monocytogenes*, particularly the effectiveness of atmospheres containing <80% CO_2 in the absence of oxygen (Table 13.6).

As stated previously, *L. monocytogenes* will be destroyed by proper pasteurisation. Therefore, the pathogen is of low concern for *sous vide* products, which are vacuum sealed before cooking. However, for products that are cooked and then filled into packages, post-cooking contamination is of concern. It is recommended that products belonging to this category are stored at ≤3°C for not longer than 28 days to ensure that the product's safety is not compromised as a result of the growth of *L. monocytogenes*. Furthermore, the product should be reheated to at least 70°C prior to consumption

Table 13.6 Growth of *L. monocytogenes* in CO_2 enriched MAP food

Product	Temperature (°C)	Atmosphere	Growth	Reference
Lamb	5	100% CO_2	No	Garcia de Fernando *et al.* (1995)
	6	50% CO_2/50% N_2	Yes	Nychas (1994)
Beef	5	100% CO_2	No	Gill and Reichel (1989)
Roast beef	−1.5	100% CO_2	No	Hudson *et al.* (1994)
	3	100% CO_2	Yes	Hudson *et al.* (1994)
Chicken	1 and 6	100% CO_2	No	Hart *et al.* (1991)
	4	75% CO_2/25% N_2	No	Wimpfhimer *et al.* (1990)
	6	30% CO_2/70% N_2	No	Hart *et al.* (1991)
Pork	5	40% CO_2/60% N_2	Yes	Manu-Tawiah *et al.* (1993)

to ensure the destruction of the vegetative cells of pathogens, which may have recontaminated the cooked food.

13.2.4 Yersinia enterocolitica

Though recognised as a human pathogen since 1939 (Schliefstein and Coleman, 1939), *Yersinia enterocolitica* was identified as an agent of food-borne disease only in the 1970s. Chilled foods such as pasteurised milk (Tacket *et al.*, 1984), chocolate milk (Black *et al.*, 1978) and tofu (Tacket *et al.*, 1985) have been implicated in disease outbreaks. The resulting illness usually presents as a self-limiting enterocolitis, although some long-term complications such as arthritis have been reported.

The significance of this organism in cook–chill foods has not yet been determined. The organism is heat sensitive (Lovett *et al.*, 1982) and is destroyed by proper pasteurisation. Cook–chill products that are packaged after cooking are exposed to possible post-cooking contamination. Should *Y. enterocolitica* contaminate these types of product, it may grow given its psychrotrophic nature. Walker *et al.* (1990) reported a minimum growth temperature of $-1.3°C$ for *Y. enterocolitica*.

Y. enterocolitica has been isolated from a variety of foods including raw and packaged meats, seafood, dairy products and raw vegetables (Barton *et al.*, 1997). Only strains belonging to certain bio-serovars are associated with human disease. These are frequently isolated from pigs and sometimes pork products (Barton *et al.*, 1997). Isolates from other foods and the environment tend to belong to diverse bio-serovars. As such, the pathogenic significance of *Y. enterocolitica* food isolates needs to be ascertained before a food is deemed a potential public health threat.

The infectious dose for *Y. enterocolitica* is unknown but likely exceeds 10^4 cells in healthy adults (Robins-Browne, 1997). This implies that for food to be implicated in disease outbreaks (among 'healthy' individuals), the organism must grow in the product. We must be mindful, however, that the minimal infectious dose may not always be very high, especially for compromised individuals.

Y. enterocolitica is capable of growth under aerobic and anaerobic conditions. Kleinlein and Untermann (1990) studied the growth of *Y. enterocolitica*, serovars O:3 and O:9, in minced beef with and without protective gas and/or background flora. When packaged under air with a low level ($<10^2$ cfu/g) of competing flora, *Y. enterocolitica* grew to high levels (10^7 cfu/g) in about 2, 4 and 14 days at 15, 10 and 4°C, respectively. Under these same conditions but in the presence of a high level (10^5 cfu/g) of competing microflora, the growth of *Y. enterocolitica* serotypes was reduced substantially. At 1°C, *Y. enterocolitica* numbers only increased about 1 log (from an initial inoculum of 10^3 cfu/g) in 14 days in air. Packaging under atmospheres of 20% CO_2/80% O_2 had a significant effect on

Y. enterocolitica. At 1 and 4°C, growth was not observed during 14 days storage. However, at 10 and 15°C and under the modified atmosphere, growth of *Y. enterocolitica* was observed. When the modified atmosphere-packaged minced beef was incubated at these temperatures in the presence of a high level of competing microflora, growth of *Y. enterocolitica* could be controlled. Under these conditions, growth of the serovar O:3 strain was inhibited at both temperatures, while only a 1–2 log population increase was observed with the serovar O:9 strain. These results show the importance of the background competitive flora in inhibiting the growth of *Y. enterocolitica*: a flora which would be reduced by the heat treatments given to cook–chill foods.

Hudson *et al.* (1994) observed growth of *Y. enterocolitica* to high levels (>10^7 cfu/g) in vacuum-packaged sliced roast beef within the shelf life of the product at –1.5°C (8 weeks) and 3°C (3 weeks). In contrast, these authors found that growth of *Y. enterocolitica* could be prevented at –1.5°C if the product was packaged in an atmosphere of saturated CO_2. At 3°C, growth occurred in saturated CO_2 packs within the product's shelf life (10 weeks) but a rate slower than that observed in the vacuum-packaged product. The *Yersinia* counts on meat are usually low (<10^2 cfu/g), however to facilitate enumeration Hudson *et al.* (1994) had to use an inoculum level of 10^4 cfu/g. The authors concede that, had a smaller pathogen inoculum been used, the spoilage microflora may have retarded the growth of *Y. enterocolitica*, and the pathogen numbers measured under all their test conditions would be lower than those observed.

13.2.5 Aeromonas

The role of *Aeromonas* as a cause of foodborne disease is debatable as no fully documented outbreak has been reported. *Aeromonas* species are principally aquatic organisms and have been isolated from fresh, marine, chlorinated and polluted water. They may colonise animals used as food such as fish, sheep, pigs, cows and poultry, and have been isolated from other agricultural produce such as salads and vegetables (Kirov, 1997). Thus they may find their way into the food chain.

Some *Aeromonas* species (*A. hydrophila*, *A. caviae*, *A. veronii* biovars sobria and veronii, *A. jandaii*, *A. schubertii*, *A. trotu*) cause gastrointestinal and wound infections in humans although the infectious dose of the organism is not known (Kirov, 1997). As some pathogenic strains of *Aeromonas* grow at refrigeration temperatures, the organism is a concern for manufacturers of cook–chill foods. As with other non-spore-forming pathogens, *Aeromonas* species should be destroyed by the temperatures used in the heat treatment of cook–chill foods. Though their presence in food-processing environments has not been thoroughly investigated, it is likely that they could be present in moist areas of a factory given their aquatic nature.

Therefore, post-cooking contamination is possible in products that are cooked and then packaged.

As *Aeromonas* is a psychrotrophic organism, and there is potential for post-cooking contamination to occur, it is important to consider the significance of the presence of *Aeromonas* in extended shelf life refrigerated foods. The minimum reported growth temperature for *A. hydrophila* is –0.1 to +1.2°C (Walker and Stringer, 1987). This species is capable of growing under certain modified atmospheres that may be used for cook–chill foods. Under vacuum packaging, Hudson and Mott (1993) reported a 4.5 log increase in *A. hydrophila* from a starting population of 10^2 cfu/g on cooked beef after 12 days at 5°C. There is, however, some evidence that CO_2 may retard the growth of *Aeromonas*. Again on cooked beef, Hudson *et al.* (1994) demonstrated generation times for *A. hydrophila* that were about three times longer in saturated CO_2 packaging than in the vacuum-packaged product at 3°C.

Moderate heat treatment will inactivate *Aeromonas* species, with D-values of 1.2–2.3 min at 51°C being observed (ICMSF, 1996a). Heat injury will also affect the response of *Aeromonas* to modified atmospheres. Golden *et al.* (1989) showed an increased lag and decreased growth rate of heat-injured *A. hydrophila* growing on an agar surface under CO_2 at 30°C. The viable counts of heat-injured *A. hydrophila* under conditions that did not permit growth (97.6–94% CO_2 at 5°C) also declined more rapidly than uninjured cells.

As with the other non-spore-forming pathogens discussed, the risk of foodborne disease resulting from the growth of *Aeromonas* in cook–chill foods can be reduced by: storage of the product at < 3°C for no longer than 28 days (Australian Quarantine and Inspection Service, 1992), and adequate reheating of the product prior to consumption.

13.3 Control of microbiological hazards

Depending on the food and the source of the raw materials, a low level of contamination by *B. cereus* and, to a lesser extent, *C. botulinum* can be expected. Low numbers of these organisms in foods are not likely to cause a problem unless growth is permitted to occur. Since most non-sporing psychrotrophic pathogens can grow to levels of about 10^6 cfu/g within about 2 weeks at 3°C (Mossel and Struijk, 1991), control is achieved by ensuring pathogens are destroyed during cooking, recontamination does not occur and/or products are stored under strict temperature control. In addition, the maximum permitted shelf life must not allow time for the growth of pathogens to hazardous levels in the final product or the composition of the food or packaging system (MAs) must be designed to assist in controlling the growth of these pathogens.

Other approaches are being used by the food industry increasingly to enhance the microbiological safety of cook–chill foods. These approaches, which are discussed briefly below, include industry regulations (e.g. Codes of Practice, industry guidelines), HACCP, predictive microbiology and challenge studies.

13.3.1 Industry regulations

Betts (1992) reviewed some of the literature available on *sous vide* with respect to various regulations and industry Codes of Practice, highlighting the useful information in each. In general a minimum process of 2 min at 70°C is recommended, which is aimed at the destruction of *L. monocytogenes*. The maximum permitted shelf life is usually dependent on the heat treatment and storage temperature of the product. The greater the heat treatment and the lower the storage temperature, the greater the shelf life that is permitted.

The Australian Quarantine and Inspection Service has a Code of Hygienic Practice for heat-treated refrigerated foods (HTRF) packaged for extended shelf life (1992). As with the literature discussed by Betts (1992), the storage and distribution temperatures required depend on the thermal process applied to the product:

- Products which have not been given a 6D process for non-proteolytic *C. botulinum* and/or which have been cooked before packing and the filling operation was not aseptic, must be kept refrigerated below 3°C.
- Products which have been given a 6D process for non-proteolytic *C. botulinum* and, if cooked before packing, the filling operation was aseptic, must be kept refrigerated below 5°C.
- If the storage and distribution system cannot guarantee ≤ 3°C or ≤ 5°C (as stated above), then the shelf life of the product shall be limited to a maximum of 5 days.
- Retail product must be consumed within 10 days of purchase unless stored in a frozen state.

13.3.2 Hazard Analysis Critical Control Point (HACCP)

The application of HACCP systems during cook–chill food processing is also a useful tool to enhance the safety of these foods. The system, which identifies specific hazards and preventive measures for their control, focuses resources on hazardous points in the process and is aimed at preventing problems before they occur. Examples of critical control points for cook–chill products would include:

- the microbiological quality of raw materials – ensuring that they are of good microbiological quality and received at appropriate temperatures;

- preparation of the food – ensuring that microbial contamination and/or growth does not occur;
- filling operations (either before or after thermal process) – ensuring that contamination does not occur, appropriate fill weights are met (under-processing may occur if packages are overfilled), seals are entire;
- thermal processing – specified time and temperature regimes must be met;
- cooling – product must be cooled quickly to prevent rapid germination and growth of surviving pathogens;
- storage and distribution – temperature conditions and storage time must be controlled to minimise the growth of surviving pathogens.

13.3.3 Predictive modelling and challenge studies

Predictive modelling and challenge studies may be useful in predicting and assessing the microbiological hazards associated with cook–chill and *sous vide* foods.

Predictive modelling essentially involves the formulation of mathematical equations containing parameters for factors influencing microbial growth such as pH, salt content, water activity and storage temperature. A number of types of models have been produced to describe the growth of microorganisms (Walker and Stringer, 1990) which allow the study of the effects of several factors simultaneously and their interactive effects on the lag phase and generation times of microorganisms. Studying the growth of microorganisms in this way may more realistically reflect the conditions in foods and, hence, more accurately predict how a given microorganism may behave in a product.

Mathematical models can be useful for predicting the shelf life of cook–chill foods with respect to the growth and death of both pathogens and spoilage organisms. Models have been produced for *C. botulinum* (Genigeorgis *et al.*, 1991; Meng and Genigeorgis, 1993; Juneja *et al.*, 1995; Graham *et al.*, 1996), *B. cereus* (Baker and Griffiths, 1993; Davis and Walker, 1994), *Listeria* (Buchanan, 1991; Cole *et al.*, 1993; Fernandez *et al.*, 1997), *Pseudomonas* (Neumeyer *et al.*, 1997a, b), *Yersinia* (Jones *et al.*, 1994) and *Aeromonas* (Buchanan, 1991; McClure *et al.*, 1994). Predictive models are also useful for determining safe shelf lives when developing new products, reformulating existing products and developing HACCP systems (Elliott, 1996).

Microbiological challenge studies are used to observe what happens to the microflora of foods at any stage during preparation, processing, packaging, storage and/or distribution and can be used to estimate both the microbiological safety and stability of foods. They involve the inoculation of foods with the microorganisms of interest and subjecting the food to conditions that may be expected from preparation to consumer use. Changes in the microbiological flora are then monitored throughout the study. The predominant disadvantage with challenge studies is that they are only valid for the food type and conditions tested and so are often more appropriate as a confirmation of results of predictive modelling.

It is important that challenge tests be designed specifically for each product so that the study is appropriate to the food, preparation, distribution and storage conditions. Potential abuse conditions should also be included in studies as should different product formulations and preparation or processing conditions. The choice and level of inoculation of the test organism(s) must be appropriate to the product. Strains that are tolerant or resistant to the conditions in the product should be included in the study.

The advantage of microbiological challenge tests is that they are performed under controlled conditions. By inoculating a food with known numbers and strains of organisms and controlling the subsequent treatment of the product, it is easier to predict the potential safety hazard with the food than if relying on monitoring a food that is naturally contaminated. Guidelines for microbiological challenge testing of foods have been published (CFDRA, 1987).

13.4 Concluding remarks

With the rapidly increasing market in chilled prepared meals it is essential that manufacturers are aware of the microbiological safety hazards with these products and have methods in place for their control. The hazards can be addressed by good manufacturing practices, good quality raw materials, adherence to minimum cooking times (at least 70°C for 2 min) and shelf lives appropriate for the heat treatments, prevention of post process contamination and strict temperature control during storage and distribution. The packaging system, such as vacuum or modified atmosphere packaging, can enhance the shelf life, but may also increase microbiological safety hazards. An effective food safety plan and monitoring system (e.g. HACCP system) will assist in ensuring the above practices are carried out. Identification of the potential hazards is extremely important during product development to ensure that all of the risks can be addressed.

The Australian food industry has been slow to take up cook–chill processing compared to its overseas counterparts. Although the technique has been used for some time, *sous vide* and other cook–chill products have really only made in-roads into the Australian retail market in the last couple of years. However, many agree that the technology should be confined to skilled personnel in professional kitchens and it is expected that these products will be most popular in the food service industry such as meals on wheels, mining camps, oil rigs, large single catering events, health care outlets, schools, ships, armed forces and airlines (Pickard, 1992). In these situations, the cost savings and quality advantages associated with *sous vide* and cook–chill foods may be better accepted. Concerns still exist in the retail market due mainly to doubts about maintaining foods below 3°C throughout the distribution chain and the cost involved in the packaging and marketing of these new products.

References

Australian Quarantine and Inspection Service (1992) *Code of Hygienic Practice for Heat-treated Refrigerated Foods Packaged for Extended Shelf Life*, Commonwealth Government Publishing Service, Canberra.

Baker, J.M. and Griffiths, M.W. (1993) Predictive modelling of psychrotrophic *Bacillus cereus*. *Journal of Food Protection*, **56**, 684–688.

Barton, M.D., Kolega, V. and Fenwich, S.G. (1997) *Yersinia enterocolitica*, in *Foodborne Microorganisms of Public Health Significance*, 5th edn (eds A.D. Hocking *et al.*), AIFST (NSW Branch) Food Microbiology Group, pp. 493–519.

Ben Embarek, P.K. and Huss, H.H. (1993) Heat resistance of *Listeria monocytogenes* in vacuum-packaged pasteurized fish fillets. *International Journal of Food Microbiology*, **20**, 85–95.

Bennik, M.H.J., Smid, E.J., Rombouts, F.M. and Gorris, L.G.M. (1995) Growth of psychrotrophic foodborne pathogens in a solid surface model system under the influence of carbon dioxide and oxygen. *Food Microbiology*, **12**, 509–519.

Betts, G.D. (1992) The microbiological safety of sous-vide processing, *Technical Memorandum No. 39*, Campden Food and Drink Research Association.

Black, R.E., Jackson, R.J., Tsai, T., Medvesdy, M., Shayegani, M., Feeley, J.C., Macleod, K.I.E. and Wakelee, A.W. (1978) Epidemic of *Yersinia enterocolitica* infection due to contaminated chocolate milk. *New England Journal of Medicine*, **298**, 76–79.

Blackburn, C. de W. and Davies, A.R. (1994) Effects of pre-conditioning on pathogenic bacteria in foods. *Food Technology International Europe*, pp. 37–40.

Bohrer, C.W., Denny, C.B. and Yao, M.G. (1973) *Thermal Destruction of Type E* Clostridium botulinum, National Canners Association Research Foundation, Washington, DC.

Botzler, R.G., Cowan, A.B. and Wetzler, T.F. (1974) Survival of *Listeria monocytogenes* in soil and water. *Journal of Wildlife Diseases*, **10**, 204–231.

Brown, G.D. and Gaze, J.E. (1990) Determination of the growth potential of *Clostridium botulinum* types E and non-proteolytic B in *sous vide* products at low temperatures, *Technical Memorandum No. 593*, Campden Food and Drink Research Association.

Brown, G.D., Gaze, J.E. and Gaskell, D.E. (1991) Growth of *Clostridium botulinum* non-proteolytic type B and type E in sous vide products stored at 2–15°C. *Technical Memorandum No. 635*, Campden Food and Drink Research Association.

Buchanan, R.L. (1991) Using spreadsheet software for predictive microbiology applications. *Journal of Food Safety*, **11** (2), 123–134.

Buchanan, R.L., Stahl, H.G. and Whiting, R.C. (1989) Effects and interactions of temperature, pH, atmosphere, sodium chloride, and sodium nitrite on the growth of *Listeria monocytogenes*. *Journal of Food Protection*, **52**, 844–851.

CFDRA (1987) Microbiological challenge testing. *Technical Memorandum No. 20*, Campden Food and Drink Research Association.

Christiansson, A., Naidu, A.S., Nilsson, I. Wadström, T. and Pettersson, H.-E. (1989) Toxin production by *Bacillus cereus* dairy isolates in milk at low temperatures. *Applied and Environmental Microbiology*, **55**, 2595–2600.

Cole, M.B., Davies, K.W., Munro, G., Holyoak, C.D. and Kilsby, D.C. (1993) A vitalistic model to describe the thermal inactivation of *Listeria monocytogenes*. *Journal of Industrial Microbiology*, **12** (3/5), 232–239.

Crandall, A.D., Winkowski, K. and Montville, T.J. (1994) Inability of *Pediococcus pentasaceus* to inhibit *Clostridium botulinum* in sous vide beef with gravy at 4 and 10 degrees C. *Journal of Food Protection*, **57**, 104–107.

Crisley, F.D., Peeler, J.T., Angelotti, R. and Hall, H.E. (1968) Thermal resistance of spores of five strains of *Clostridium botulinum* type E in ground whitefish chubs. *Journal of Food Science*, **33**, 411–416.

Davis, S.C. and Walker, S.J. (1994) Factors affecting the growth of psychrotrophic *Bacillus cereus*. *Technical Memorandum No. 701*, Campden Food and Drink Research Association.

Donnelly, C.W. and Briggs, E.H. (1986) Psychrotropic growth and thermal inactivation of *Listeria monocytogenes* as a function of milk composition. *Journal of Food Protection*, **49**, 994–998.

Donnelly, C.W., Briggs, E.H. and Donnelly, L.S. (1987) Comparison of heat resistance of *Listeria monocytogenes* in milk as determined by two methods. *Journal of Food Protection*, **50**, 14–17.

Dufrenne, J., Soentoro, P., Tatini, S., Day, T. and Notermans, S. (1994) Characteristics of *Bacillus cereus* related to safe food production. *International Journal Food Microbiology*, **23**, 99–109.

Eklund, M.W., Poysky, F.T. and Wieler, D.I. (1967a) Characteristics of *Clostridium botulinum* type F isolated from the Pacific coast of the United States. *Applied Microbiology*, **15**, 1316–1323.

Eklund, M.W., Wieler, D.I. and Poysky, F.T. (1967b) Growth and toxin production of nonproteolytic type B *Clostridium botulinum* at 3.3 to 5.6°C. *Journal of Bacteriology*, **93**, 1461–1462.

Elliott, P.H. (1996) Predictive microbiology and HACCP. *Journal of Food Protection*, **Suppl.**, 48–53.

Enfors, S.-O. and Molin, A. (1978) The influence of high concentrations of carbon dioxide on the germination of bacterial spores. *Journal of Applied Bacteriology*, **45**, 279–285.

Enfors, S.-O. and Molin, A. (1981) The influence of temperature on the growth inhibitory effect of carbon dioxide on *Pseudomonas fragi* and *Bacillus cereus*. *Canadian Journal of Microbiology*, **27**, 15–19.

Fain, A.R. Jr, Line, J.E., Moran, A.B., Martin, L.M., Lechowich, R.V., Carosella, J.M. and Brown, W.L. (1991) Lethality of heat to *Listeria monocytogenes* Scott A: D-value determinations in ground beef and turkey. *Journal of Food Protection*, **54**, 756–761.

Fang, T.J. and Lo Wei Lin (1994) Inactivation of *Listeria monocytogenes* on raw pork treated with modified atmosphere packaging and nisin. *Journal of Food and Drug Analysis*, **2**, 189–200.

Farber, J.M. (1989) Thermal resistance of *Listeria monocytogenes* in foods. *International Journal of Food Microbiology*, **8**, 285–291.

Farber, J.M. (1991) Microbiological aspects of modified atmosphere packaging technology – a review. *Journal of Food Protection*, **54**, 58–70.

Farber, J.M. and Brown, B.E. (1990) Effect of prior heat shock on heat resistance of *Listeria monocytogenes* in meat. *Applied and Environmental Microbiology*, **56**, 1584–1587.

Fedio, W.M. and Jackson, H. (1989) Effect of tempering on the heat resistance of *Listeria monocytogenes*. *Letters in Applied Microbiology*, **9**, 157–160.

Fernandez, P.S., George, S.M., Sills, C.C. and Peck, M.W. (1997) Predictive model of the effect of CO_2, pH, temperature and NaCl on the growth of *Listeria monocytogenes*. *International Journal of Food Microbiology*, **37**, 37–45.

Fleming, D.W., Cochi, S.L., MacDonald, K.L., Brondum, J., Hayes, P.S., Plikaytis, B.D., Holmes, M.B., Audurier, A., Broome, C.V. and Reingold, A.L. (1985) Pasteurized milk as a vehicle of infection in an outbreak of listeriosis. *North England Journal of Medicine*, **312**, 404–407.

Foegeding, P.M. and Fulp, M.L. (1988) Comparison of coats and surface-dependent properties of *Bacillus cereus* T prepared in two sporulation environments. *Journal of Applied Bacteriology*, **65**, 249–259.

Garcia de Fernando, G.D., Nychas, G.J.E., Peck, M.W. and Ordonez, J.A. (1995) Growth/survival of psychrotrophic pathogens on meat packaged under modified atmospheres. *International Journal of Food Microbiology*, **28**, 221–231.

Gaze, J.E. and Brown, G.D. (1990) Determination of the heat resistance of a strain of non-proteolytic *Clostridium botulinum* type B and a strain of type E, heated in cod and carrot homogenate over the temperature range 70 to 92 degrees. *Technical Memorandum No. 592*, Campden Food and Drink Research Association.

Gaze, J.E., Brown, G.D., Gaskell, D.E. and Banks, J.G. (1989) Heat resistance of *Listeria monocytogenes* in homogenates of chicken, beef steak and carrot. *Food Microbiology*, **6**, 251–259.

Genigeorgis, C.A., Meng, J.H. and Baker, D.A. (1991) Behavior of nonproteolytic *Clostridium botulinum* type B and E spores in cooked turkey and modelling lag phase and probability of toxigenesis. *Journal of Food Science*, **56** (2), 373–379.

Geuenich, H., Muller, H.E., Schretten Brunner, A. and Seeliger, H.P.R. (1985) The occurrence of different *Listeria* species in municipal waste water. *Zentralblatt fur Mikrobiologie und Hygene*, **181**, 563–565.

Gill, C.O. and Reichel, M.P. (1989) Growth of cold-tolerant pathogens *Yersinia enterocolitica*, *Aeromonas hydrophila*, and *Listeria monocytogenes* on high-pH beef packaged under vacuum or carbon dioxide. *Food Microbiology*, **6**, 223–230.

Glass, K.A. and Doyle, M.P. (1989) Fate of *Listeria monocytogenes* in processed meat products during refrigerated storage. *Applied and Environmental Microbiology*, **55**, 1565–1569.

Golden, D.A., Eyles, M.J. and Beuchat, L.R. (1989) Influence of modified-atmosphere storage on the growth of uninjured and heat-injured *Aeromonas hydrophila*. *Applied and Environmental Microbiology*, **55**, 3012–3015.

Goulet, V., Jacquet, C., Vaillant, V., Rebiere, I., Mouret, E., Lorente, E., Steiner, F. and Rocourt, J. (1995) Listeriosis from consumption of raw milk cheese. *Lancet*, **345**, 1581–1582.

Graham, A.F., Mason, D.R. and Peck, M.W. (1996) Predictive model of the effect of temperature, pH and sodium chloride on growth from spores of non-proteolytic *Clostridium botulinum*. *International Journal of Food Microbiology*, **31** (1/3), 69–85.

Granum, P.E., Brynestad, S., O'Sullivan, K. and Nissen, H. (1993) Enterotoxin from *Bacillus cereus*: production and biochemical characterization. *Netherlands Milk and Dairy Journal*, **47**, 63–70.

Grau, F.H. and Vanderlinde, P.B. (1990) Growth of *Listeria monocytogenes* on vacuum-packaged beef. *Journal of Food Protection*, **53**, 739–741.

Gray, M.L. (1960) Silage feeding and listeriosis. *Journal of American Veterinary Medical Association*, **136**, 205–208.

Griffiths, M. W. (1990) Toxin production by psychrotrophic *Bacillus* spp. present in milk. *Journal of Food Protection*, **53**, 790–792.

Hall, R., Shaw, D., Lim, I., Murphy, R., Davos, D., Lanser, J., Delroy, B., Tribe, I., Holland, R. and Carman, J. (1996) A cluster of listeriosis cases in South Australia. *Communicable Disease Information*, **20**, 465.

Hansen, T.B. and Knochel, S. (1996) Thermal inactivation of *Listeria monocytogenes* during rapid and slow heating in sous vide cooked beef. *Letters in Applied Microbiology*, **22**, 425–428.

Harrison, M.A. and Huang, Y-W. (1990) Thermal death times for *Listeria monocytogenes* (Scott A) in crabmeat. *Journal of Food Protection*, **53**, 878–880.

Hart, C.D., Mead, G.C. and Norris, A.P. (1991) Effects of gaseous environment and temperature on the storage behaviour of *Listeria monocytogenes* on chicken breast meat. *Journal of Applied Bacteriology*, **70**, 40–46.

Holdeman, L.V. and Brooks, J.B. (1970) Variation among strains of *Clostridium botulinum* and related clostridia, in *Protocols of the First U.S–Japan Conference on Toxic Microorganisms*, pp. 278–286.

Hudson, J.A. and Mott, S.J. (1993) Growth of *Listeria monocytogenes*, *Aeromonas hydrophila* and *Yersinia enterocolitica* on cold-smoked salmon under refrigeration and mild temperature abuse. *Food Microbiology*, **10**, 61–68.

Hudson, J.H., Mott, S.J. and Penney, N. (1994) Growth of *Listeria monocytogenes*, *Aeromonas hydrophila* and *Yersinia enterocolitica* on vacuum and saturated carbon dioxide controlled atmosphere-packaged sliced roast beef. *Journal of Food Protection*, **57**, 204–208.

ICMSF (International Commission on Microbiological Specifications for Foods) (1996a) *Aeromonas*, in *Microorganisms in Foods 5 – Microbiological Specifications of Food Pathogens*, Blackie Academic & Professional, London, pp. 5–19.

ICMSF (International Commission on Microbiological Specifications for Foods) (1996b)

Bacillus cereus, in *Microorganisms in Foods 5 – Microbiological Specifications of Food Pathogens*, Blackie Academic & Professional, London, pp. 20–35.

Jacquet, C., Catimel, B., Brosch, R., Buchrieser, C., Dehaumont, P., Goulet, V., Lepoutre, V., Veit, P. and Rocourt, J. (1995) Investigations related to the epidemic strain involved in the French listeriosis outbreak in 1992. *Applied and Environmental Microbiology*, **61**, 2242–2246.

James, S. and Evans, J. (1990) Temperatures in the retail and domestic chilled chain, in *Processing and Quality of Foods*, Vol. 3 (eds P. Zeuthen *et al.*), Elsevier Applied Science, London, pp. 3.273–3.278.

Johnson, J.L., Doyle, M.P., Cassens, R.G. and Schoeni, J.L. (1988) Fate of *Listeria monocytogenes* in tissues of experimentally infected cattle and in hard salami. *Applied and Environmental Microbiology*, **54**, 497–501.

Jones, J.E., Walker, S.J., Sutherland, J.P., Peck, M.W. and Little, C.L. (1994) Mathematical modelling of the growth, survival and death of *Yersinia enterocolitica*. *International Journal of Food Microbiology*, **23** (3/4), 433–447.

Juneja, V.K. and Eblen, B.S. (1995) Influence of sodium chloride on thermal inactivation and recovery of nonproteolytic *Clostridium botulinum* type B strain KAP B5 spores. *Journal of Food Protection*, **58**, 813–816.

Juneja, V.K., Marmer, B.S., Phillips, J.G. and Miller, A.J. (1995) Influence of the intrinsic properties of food on thermal inactivation of spores of nonproteolytic *Clostridium botulinum*: development of a predictive model. *Journal of Food Safety*, **15**, 349–364.

Kim, K., Murano, E.A. and Olson, D.G. (1994) Heating and storage conditions affect survival and recovery of *Listeria monocytogenes* in ground pork. *Journal of Food Science*, **59**, 30–32, 59.

Kirov, S.M. (1997) *Aeromonas enterocolitica*, in *Foodborne Microorganisms of Public Health Significance*, 5th edn (eds A.D. Hocking *et al.*), AIFST (NSW Branch) Food Microbiology Group, pp. 473–492.

Kleinlein, N and Untermann, F. (1990) Growth of pathogenic *Yersinia enterocolitica* strains in minced meat with and without protective gas and consideration of competitive background flora. *International Journal of Food Microbiology*, **10**, 65–72.

Kralovic, R.C. (1973) *Clostridium botulinum* type E spores: increased heat resistance through lysozyme treatment. *Abstracts of the Annual Meeting of the American Society for Microbiology*, **73**, 41.

Larsen, M.D. and Jørgensen, K. (1997) The occurrence of *Bacillus cereus* in Danish pasteurized milk. *International Journal of Food Microbiology*, **34**, 179–186.

Linnan, M.J., Mascola, L., Lou, X.D., Goulet, V., May, S., Salminen, C., Hird, D.W., Yonekura, M.L., Hayes, P., Weaver, R., Audurier, A., Plikaytis, B.D., Fannin, S.L., Klels, A. and Broome, C.V. (1988) Epidemic listeriosis associated with Mexican-style cheese. *North England Journal of Medicine*, **319**, 823–828.

Linton, R.H., Pierson, M.D. and Bishop, J.P. (1990) Increase in heat resistance of *Listeria monocytogenes* Scott A by sublethal heat shock. *Journal of Food Protection*, **53**, 924–927.

Lovett, J., Bradshaw, J.G. and Peeler, J.T. (1982) Thermal inactivation of *Yersinia enterocolitica* in milk. *Applied and Environmental Microbiology*, **44**, 517–519.

Lund, B.M., Knox, M.R. and Cole, M.B. (1989) Destruction of *Listeria monocytogenes* during microwave cooking. *Lancet*, **1**, 218.

Lund, B.M. and Notermans, S.H.W. (1992) Potential hazards associated with REPFEDS, in *Clostridium botulinum, Ecology and Control in Foods* (eds A.H.W. Hauschild and K.L. Dodds), Marcel Dekker, New York, pp. 279–303.

Lynt, R.K., Kautter, D.A. and Solomon, H.M. (1979) Heat resistance of nonproteolytic *Clostridium botulinum* type F in phosphate buffer and crabmeat. *Journal of Food Science*, **44**, 108–111.

Lynt, R.K., Solomon, H.M., Lilly, T. Jr and Kautter, D.A (1977) Thermal death time of *Clostridium botulinum* type E in meat of the blue crab. *Journal of Food Science*, **42**, 1022–1025, 1037.

Mackey, B.M, Pritchet, C., Norris, A. and Mead, G.C. (1990) Heat resistance of *Listeria*: strain differences and effects of meat type and curing salts. *Letters in Applied Microbiology*, **10**, 251–255.

Manu-Tawiah, W., Myers, D.J., Olson, D.G. and Molins, R.A. (1993) Survival and growth of *Listeria monocytogenes* and *Yersinia enterocolitica* in pork chops packaged under modified gas atmospheres. *Journal of Food Science*, **58**, 475–479.

Mazas, M., González, I., López, M., González, J. and Sarmiento, R.M. (1995) Effects of sporulation media and strain on thermal resistance of *Bacillus cereus* spores. *International Journal of Food Science and Technology*, **30**, 71–78.

McClure, P.J., Cole, M.B. and Davies, K.W. (1994) An example of the stages in the development of a predictive mathematical model for microbial growth: the effects of NaCl, pH and temperature on the growth of *Aeromonas hydrophila*. *International Journal of Food Microbiology*, **23** (3/4) 359–375.

McLauchlin, J., Hall, S.M., Velani, S.K. and Gilbert, R.J. (1991) Human listeriosis and paté – a possible association. *British Medical Journal*, **303**, 773–775.

Meng, J.H. and Genigeorgis, C.A. (1993) Modelling lag phase of nonproteolytic *Clostridium botulinum* toxigenesis in cooked turkey and chicken breast as affected by temperature, sodium lactate, sodium chloride and spore inoculum. *International Journal of Food Microbiology*, **19** (2), 109–122.

Meng, J. and Genigeorgis, C.A. (1994) Delaying toxigenisis of *Clostridium botulinum* by sodium lactate in *sous vide* products. *Letters in Applied Microbiology*, **19**, 20–23.

Moir, C.J. and Eyles, M.J. (1992) Effect of modified atmospheres on growth and enterotoxin production of *Bacillus cereus*. *Australian Microbiologist*, **13** (3), A213.

Mossel, D.A.A. and Struijk, C.B. (1991) Public health implication of refrigerated pasteurized ('sous-vide') foods. *International Journal of Food Microbiology*, **13**, 187–206.

Neumeyer, K., Ross, T. and McMeekin, T.A (1997a) Development of a predictive model to describe the effects of temperature and water activity on the growth of spoilage pseudomonads. *International Journal of Food Microbiology*, **38**, 45–54.

Neumeyer, K., Ross, T., Thomson, G. and McMeekin, T.A. (1997b) Validation of a model describing the effects of temperature and water activity on the growth of psychrotrophic pseudomonads. *International Journal of Food Microbiology*, **38**, 55–63.

Notermans, S. and Tatini, S. (1993) Characterization of *Bacillus cereus* in relation to toxin production. *Netherlands Milk and Dairy Journal*, **47**, 71–77.

Notermans, S., Dufrenne, J. and Lund, B.M. (1990) Botulism risk of refrigerated, processed foods of extended durability. *Journal of Food Protection*, **53**, 1020–1024.

Nychas, G.J.E. (1994) Modified Atmosphere packaging of meats, in *Minimal Processing of Foods and Process Optimization, An Interface* (eds R.P. Singh and F.A.R. Oliveiria), CRC Press, London, pp. 417–436.

Ollinger-Snyder, P., El-Gazzar, F., Matthews, M.I., Harth, E.H. and Unklesbay, N. (1995) Thermal destruction of *Listeria monocytogenes* in ground pork prepared with and without soy hulls. *Journal of Food Protection*, **58**, 573–576.

Palumbo, M.S., Beers, S.M., Bhaduri, S. and Palumbo, S.A. (1995) Thermal resistance of *Salmonella* ssp. and *Listeria monocytogenes* in liquid egg yolk and egg yolk products. *Journal of Food Protection*, **58**, 960–966.

Peck, M.W., Evans, R.I., Fairbairn, D.A., Hartley, M.G. and Russell, N.J. (1995) Effect of sporulation temperature on some properties of spores of non-proteolytic *Clostridium botulinum*. *International Journal of Food Microbiology*, **28**, 289–297.

Peck, M.W., Fairbairn, D.A. and Lund, B.M. (1992) The effect of recovery medium on the estimated heat-inactivation of spores of non-proteolytic *Clostridium botulinum*. *Letters in Applied Microbiology*, **15**, 146–151.

Peck, M.W., Fairbairn, D.A. and Lund, B.M. (1993) Heat resistance of spores of non-proteolytic *Clostridium botulinum* estimated on medium containing lysozyme. *Letters in Applied Microbiology*, **16**, 126–131.

Pickard, D. (1992) Sous-vide technique taking rapid hold. *Food Industry*, January, 17–18, 20.

Robins-Browne, R.M. (1997) *Yersinia enterocolitica*, in *Food Microbiology: Fundamentals and Frontiers* (eds M.P. Doyle, L.R. Beuchat and T.J. Montville) ASM Press, Washington DC. pp. 192–215.

Rose, S. (1986) Temperature observations on chilled foods from refrigerated retail displays. *Technical Memorandum No. 423*, Campden Food Preservation Research Association.

Rosenow, E.M. and Marth, E.H. (1987) Growth of *Listeria monocytogenes* in skim, whole and chocolate milk and in whipping cream during incubation at 4,48, 13, 21 and 35°C. *Journal of Food Protection*, **50**, 452–459.

Schlech, W.F. III, Lavigne, P.M., Bortolussi, R.A., Allen, A.C., Haldane, E.V., Wort, A.J., Hightower, A.W., Johnson, A.E., King, S.H., Nicholls, E.S. and Broome, C.V. (1983) Epidemic Listeriosis – evidence for transmission by food. *North England Journal of Medicine*, **308**, 203–206.

Schliefstein, J.I. and Coleman, M.B. (1939) An unidentified microorganism resembling *B. lingieri* and *Past. pseudotuberculosis*, and pathogenic for man. *New York State Journal of Medicine*, **39**, 1749–1753.

Schmidt, C.F., Lechowich, R.V. and Folinazzo, J.F. (1961) Growth and toxin production by type E *Clostridium botulinum* below 40°F. *Journal of Food Science*, **26**, 626–630.

Scott, V.N. and Bernard, D.T. (1982) Heat resistance of non-proteolytic type B *Clostridium botulinum*. *Journal of Food Protection*, **45**, 909–912.

Scott, V.N. and Bernard, D.T. (1985) The effect of lysozyme on the apparent heat resistance of nonproteolytic type B *Clostridium botulinum*. *Journal of Food Safety*, **7**, 145–154.

Simpson, M.V., Smith, J.P., Dodds, K., Ramaswamy, H.S., Blanchfield, B. and Simpson, B.K. (1995) Challenge studies with *Clostridium botulinum* in *sous vide* spaghetti and meat-sauce product. *Journal of Food Protection*, **58**, 229–234.

Skovgaard, N. and Morgen, C. (1988) Detection of *Listeria* spp. in faeces from animals, in feeds and in raw foods of animal origin. *International Journal of Food Microbiology*, **6**, 229–242.

Smith, L.D.S. and Hobbs, G. (1974) Genus III *Clostridium* Prazmowski 1880, 23, in, *Bergey's Manual of Determinative Bacteriology*, 8th edn (eds R.E. Buchanan and N.E. Gibbons), Williams & Wilkinson, Baltimore, pp. 551–572.

Solomon, H.M., Lynt, R.K., Lilly, T. Jr and Kautter, D.A. (1977) Effect of low temperature on growth of *Clostridium botulinum* spores B in meat of the blue crab. *Journal of Food Protection*, **40**, 5–7.

Stephens, P.J., Cole, M.B. and Jones, M.V. (1994) Effect of heating rate on thermal inactivation of *Listeria monocytogenes*. *Journal of Applied Bacteriology*, **77**, 702–708.

Stumbo, C.R. (1973) *Thermobacteriology in Food Processing*, 2nd edn, Academic Press, New York.

Sugiyama, H. (1951) Studies on factors affecting the heat resistance of spores of *Clostridium botulinum*. *Journal of Bacteriology*, **62**, 81–96.

Szabo, E.A. and Gibson, A.M. (1997) *Clostridium botulinum*, in *Foodborne Microorganisms of Public Health Significance*, 5th edn (eds A.D. Hocking *et al.*), AIFST (NSW Branch) Food Microbiology Group, pp. 429–64.

Tacket, C.O., Navain, J.P., Sattin, R., Lofgren, J.P., Kongsberg, C., Redtorff, R.C., Rausa, A., Davis, B.R. and Cohen, M.L. (1984) A multistate outbreak of infections caused by *Yersinia entercolitica* transmitted by pasteurised milk. *Journal of the American Medical Association*, **51**, 483–486.

Tacket, C.O., Ballard, J., Harris, N., Allard, J., Nolan, C., Quan, T. and Cohen, M.L. (1985) An outbreak of *Yersinia enterocolitica* infections caused by contaminated tofu. *American Journal of Epidemiology*, **121**, 705–711.

van Laack, R.L.J.M., van Johnson, J.L., Van der Palen, C.J.N.M., Smulders, F.J.M. and Snijders, J.M.A. (1993) Survival of pathogenic bacteria on pork loins as influenced by hot processing and packaging. *Journal of Food Protection*, **56**, 847–851, 873.

van Netten, P., van de Moosdijk, A., van Hoensel, P., Mossel, D.A.A. and Perales, I. (1990) Psychrotrophic strains of *Bacillus cereus* producing enterotoxin. *Journal of Applied Bacteriology*, **69**, 73–79.

Varnam, A.H. and Evans, M.G. (1991) *Foodborne pathogens. An Illustrated Text*, Wolfe Publishing Ltd, London.

Walker, S.J. and Stringer, M.F. (1987) Growth of *Listeria monocytogenes* and *Aeromonas hydrophila* at chill temperatures. *Technical Memorandum No. 462*, Campden Food and Drink Research Association.

Walker, S.J. and Stringer, M.F. (1990) Microbiology of chilled foods, in *Chilled Foods – The State of the Art* (ed. T.R. Gormley), Elsevier Applied Science, London, pp. 269–304.

Walker, S.J., Archer, P. and Banks, J.G. (1990) Growth of *Yersinia enterocolitica* at chill temperatures in milk and other media. *Milchwissenschaft*, **45**, 503–506.

Weis, J. and Seeliger, H.P.R. (1975) Incidence of *Listeria monocytogenes* in nature. *Applied Microbiology*, **30**, 29–32.
Welshimer, H.J. (1968) Isolation of *Listeria monocytogenes* from vegetation. *Journal of Bacteriology*, **95**, 300–303.
Wimpfhimer, L., Altman, N.S. and Hotchkiss, J.H. (1990) Growth of *Listeria monocytogenes* Scott A, serotype 4 and competitive spoilage organisms in raw chicken packaged under modified atmospheres and in air. *International Journal of Food Microbiology*, **11**, 205–214.

Index

Printed in the United States
86152LV00002BA/142/A

9 780751 404333